Computer-Aided Analysis and Design of Switch-Mode Power Supplies

ELECTRICAL ENGINEERING-ELECTRONICS SOFTWARE

1. Transformer and Inductor Design Software for the IBM PC, *Colonel Wm. T. McLyman*
2. Transformer and Inductor Design Software for the Macintosh, *Colonel Wm. T. McLyman*
3. Digital Filter Design Software for the IBM PC, *Fred J. Taylor and Thanos Stouraitis*

Computer-Aided Analysis and Design of Switch-Mode Power Supplies

Yim-Shu Lee

Department of Electronic Engineering
Hong Kong Polytechnic
Kowloon, Hong Kong

CRC Press
Taylor & Francis Group
Boca Raton London New York

CRC Press is an imprint of the
Taylor & Francis Group, an **informa** business

Library of Congress Cataloging-in-Publication Data

Lee, Yim-Shu,
    Computer-aided analysis and design of switch-mode power supplies/
Yim-Shu Lee.
        p.   cm. -- (Electrical engineering and electronics : v. 81)
    Includes bibliographical references and index.
    ISBN 0-8247-8803-6 (acid-free paper)
    1. Switching power supplies—Design—Data processing.   2. Computer-
aided design.   I. Title.   II. Series.
TK7868.P6L43   1993
621.3'17—dc20   92-40316
                                                                                      CIP

Marcel Dekker, Inc.
270 Madison Avenue, New York, New York 10016

FOR MY WAI-CHU

Preface

This book is concerned with the computer-aided analysis and design of the switch-mode power supply (SMPS). The objective is to equip electronics engineers with the necessary background about the SMPS and computer-aided design (CAD) techniques to help them perform analytical and design work efficiently. It is also a suitable text for training courses and undergraduate and graduate programs to prepare the new generation of designers and researchers who can use CAD tools freely to analyze and design power electronic circuits.

The SMPS is actually a class of power supply that makes use of electronic switches to process the electric power. Since ideal switches do not dissipate power, the SMPS can be designed to have, in theory, 100 percent conversion efficiency and infinitely small size. How much an SMPS designer can achieve depends on the quality of the available circuit components and the designer's ability to make the best use of these components. Because of the highly nonlinear nature of the switching circuitry, it is difficult, even for expert designers, to optimize a design in terms of conversion efficiency, size, steady-state/transient regulation characteristics, reliability, and cost. Therefore, most designers rely heavily on trial-and-error work to finalize a design.

CAD tools can be used to enhance a designer's ability to analyze the operation, predict the performance, and optimize the design of an SMPS. However, there are necessary conditions for any successful CAD exercise. They are:

1. The designer must have a good and intuitive understanding, at least qualitatively, of the circuit to be designed.

2. The designer must be able to develop related models of circuits and components for simulation purposes.
3. The designer must be familiar with the available CAD tool and be able to apply it effectively.
4. The CAD tool must be reasonably bug-free and powerful.
5. The simulation speed must be reasonably fast.

With the wide spread of high-speed personal computers such as those using an 80486 processor (or 80386 CPU and 80387 mathematics coprocessor) and powerful simulation programs such as SPICE (or its variants), the conditions specified in points 4 and 5 are actually not difficult to satisfy. It is the intention of this book to help readers create the conditions mentioned in points 1, 2, and 3.

The first five chapters introduce the development, principles of operation, modeling, and manual analysis of the SMPS. The objective of these chapters is to equip the reader with an intuitive and thorough understanding of the fundamental principles. This background is absolutely necessary in order to enable use of CAD tools freely in analyzing and designing the SMPS.

Chapter 6 is concerned with the computer simulation of the switching behaviors of dc-to-dc converters and the modeling of linear and nonlinear circuit components. Such simulations are extremely useful for the design of converters.

Chapter 7 deals with the modeling and simulation of the low-frequency behaviors of converters, including current-controlled converters and converters with multiple outputs. These simulations are essential for the computer-aided design of the feedback and compensation circuits of regulators.

Chapter 8 describes the philosophy, techniques, and precautions of using CAD tools to design converters and regulators. Design examples are given to illustrate the detailed steps involved.

Chapter 9 introduces the principles and design techniques of quasi-resonant and resonant converters.

The practical techniques of design of the SMPS are covered in Chapter 10 as a separate but important topic.

In addition to these ten chapters, this book contains four appendixes designed to help the reader learn the simulation program SPICE.

I wish to thank the Hong Kong Polytechnic for support and the following staff for their contributions: Kelvin So and C. W. Li for their efforts in developing the CAD techniques, preparing the figures, and proofreading; Michael Tse for his advice in reviewing the material; David Cheng, S. C. Wong, Martin Chow, and K. W. Ma for their assistance in developing the simulation techniques; Y. L. Cheng for his experimental work to verify computer simulations; K. H. Zhang for his advice on SPICE; and Cora Au, Stella Lai, and Rhoda Lam for typing the manuscript.

I also wish to thank Robyn Flemming of Wordswork Ltd. for her editorial advice, and MicroSim Corporation for making complimentary evaluation versions of PSpice available to class instructors. Above all, I loyally give my heartfelt thanks to my truly helpful colleagues at the Polytechnic for their previously unsung support in making the publication of this book possible.

Yim-Shu Lee

Contents

Contents

1
Switch-Mode Power Supply Fundamentals

The so-called switch-mode power supply (SMPS) is a class of power supply that makes use of electronic switches to process electric power. Since ideal switches do not dissipate power, the SMPS can be designed to have a high efficiency. When a high switching frequency is used, the size of the transformers and filtering circuits in the SMPS can also be minimized. Because of these overwhelming advantages, the SMPS has today become so common that its linear counterpart is now used only in low-power circuits or in circuits where interference must be kept to an absolute minimum.

As an introduction to the subject of computer-aided analysis and design of the SMPS, the specific aims of this first chapter are:

1. To provide a brief outline of the development of the SMPS
2. To describe the switching behaviors of switches, inductors, and capacitors in the SMPS
3. To identify the role of computer-aided design (CAD) tools

The chapter explains the fundamental principles of dc-to-dc converters, which are the basic power conversion engines used in the SMPS. An understanding of such principles is essential for the modeling, analysis, and CAD work that follows in subsequent chapters.

To help introduce the essential concepts directly, effectively, and without unnecessary complications, the explanation in this chapter is based on idealized circuit elements, such as lossless inductors and capacitors and perfect switches with zero turn-on resistance and zero switching time.

The power converter circuits to be studied in this chapter include the buck converter, buck-boost converter, boost converter, Ćuk converter, and electronic transformers. Converters with transformer isolation will be

1

covered in Chapter 4, after the modeling and analysis of simple converters in Chapters 2 and 3.

1.1 THE DEVELOPMENT OF SWITCH-MODE POWER SUPPLIES

This section contains a brief history of the development of converter circuits, as well as modeling, analysis, and design techniques. The objective is to provide readers with a better perspective of the subject matter and to enable them to trace the important literature related to this field of study. However, readers who are not interested in the history may proceed to Section 1.2.

1.1.1 The Development of Power Converters

The heart of a SMPS is a dc-to-dc converter, which accepts a dc input and produces a controlled dc output. Semiconductor dc-to-dc converters have appeared in practical use since the 1960s. [10,41,43,44]. The three basic types are the buck converter, the boost converter, and the buck-boost converter, the circuits of which are shown in Fig. 1.1. (Note that a buck converter with an isolation transformer is called a forward converter, and a buck-boost converter with an isolation transformer is known as a flyback converter. The transformer-coupled push-pull converter, half-bridge converter, and full-bridge converter, which will be studied in Chapter 4, are variants of the forward converter.) In each of the converter circuits shown in Fig. 1.1, there is an electronic switch SW that is driven on and off at a high switching frequency (e.g., 5–500 kHz). It is the duty cycle of the electronic switch that controls the dc output voltage V_o. The output filtering capacitor C_L in each circuit is used to smooth out the ripple component of the output voltage due to high-frequency switching. By adding a feedback circuit in a converter, the output voltage of the converter can be regulated.

Because of the square/trapezoidal/triangular shape of the current pulses in the switching device, these converters will generally be referred to as square-wave converters or simply converters (as opposed to resonant converters or quasi-resonant converters, which will be studied in Chapter 9). In each of the converter circuits shown in Fig. 1.1, the energy-storage inductance L can be chosen to be so large that the current in it is substantially smoothed. The buck converter, shown in Fig. 1.1(a), is then characterized by a smoothed output current i_o but a pulsating input current i_i. The boost converter, shown in Fig. 1.1(b), is characterized by a smoothed input current but a pulsating output current. The buck-boost converter

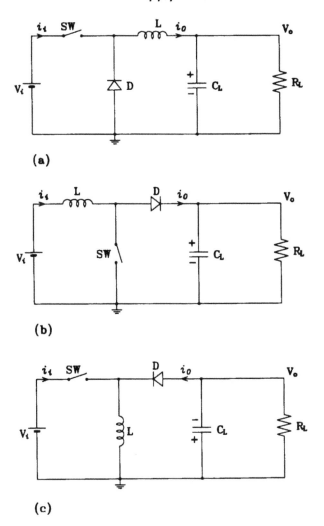

FIGURE 1.1 Basic square-wave converters: (a) buck, (b) boost, (c) buck-boost.

shown in Fig. 1.1(c) has, on the other hand, both pulsating input and output currents. In 1977, Ćuk and Middlebrook introduced a new "optimum topology switching dc-to-dc converter" [22], which is now more commonly referred to as the Ćuk converter. A special feature of the Ćuk converter is that it can be designed to have both smoothed input and output currents, although the current in the switching device remains square/

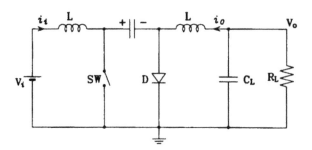

FIGURE 1.2 Ćuk converter.

trapezoidal/triangular in shape. Figure 1.2 shows the basic circuit of the Ćuk converter, which functionally is a cascaded connection of a boost converter followed by a buck converter. Various topologies of square-wave converters were also studied by Ćuk [24], Landsman [33], and Severns and Bloom [53,124]. Recently, much attention has been focused on the design of high-frequency converters with reduced weight and size [13,28,52,54,101], and the use of fast MOS (metal oxide semiconductor) power transistors as switching elements [55].

Resonant-type dc-to-dc converters using frequency modulation (FM) control were studied and implemented by Schwarz in 1970 [48]. The resonant dc-to-dc converter is characterized by the sinusoidal shape of the resonant currents in the circuit. Figure 1.3 shows the circuit of a typical resonant converter. The potential advantages of resonant converters are:

1. The resonant nature of resonant converters reduces the voltage/current spikes.
2. The leakage inductance of transformers may be absorbed as an integral part of the resonant circuit.
3. For semiconductor switches of a given speed, a higher switching frequency (and therefore a smaller converter size) is possible when used in a resonant converter because the switching time requirements are less stringent.

Some resonant converters, such as those developed by Biess, Inouye, and Shank [12], Schwarz and Klaassens [50], Ranganathan, Ziogas, and Stefanovic [45], Ebbinge [25], and Tilgenkamp, de Haan, and Huisman [69], used the SCR (silicon controlled rectifier) as the switching device and operated at a relatively low frequency, typically 5–50 kHz. Transistor resonant converters, which can operate at higher frequencies, were studied by Buchanan and Miller [15,39], Schwarz and Klaassens [51], King and

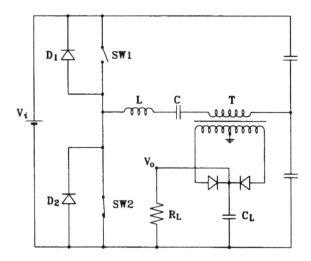

FIGURE 1.3 Resonant dc-to-dc converter.

Stuart [31,32], Vorpérian and Ćuk [59], Myers and Peck [42], Steigerwald [58], Robson [47], Witulski and Erickson [62], Redl, Molnár, and Sokal [46,70], and Divan [71]. Many of the designs described in the references were highly successful.

More recently, quasi-resonant converters have been developed and studied by Lee, Liu, Oruganti and Tabisz [73–76], Ngo [77], Vorpérian [78], Weinberg and Ghislanzoni [79], Erickson et al. [80], Barbi et al. [81], and Higashi et al. [82].

1.1.2 The Development of Modeling, Analysis, and Design Techniques

In 1970, Middlebrook and his power electronics group at Caltech (California Institute of Technology) began systematic research into the modeling and analysis of switch-mode power supplies [38]. In 1972, Wester and Middlebrook developed, using the averaging technique, a set of low-frequency behavior models for the buck, buck-boost, and boost converters [61]. In 1976, Ćuk and Middlebrook introduced a unified approach to the modeling of dc-to-dc converters using the state-space averaging technique [21,23,35,36], which combines the advantages of both the state-space and the averaging methods. In 1981, a program known as SCAP (Switching Converter Analysis Program) was developed by the Caltech power electronics group to enable small computers to carry out analysis using the state-space averaging technique [9]. This program is capable of performing dc analysis and small-signal analysis at a given dc operating point. The

SCAP was later developed into the SCAMP (Switching Converter Analysis and Measurement Program), which includes additional features to help the designer measure the frequency characteristics of the converter. Large-signal analysis of dc-to-dc converters/regulators was carried out by Edwards and Caughey [26], Lee and Yu [34], Harada and Nabeshima [29], and Erickson, Ćuk, and Middlebrook [27]. Also based on the averaging technique, Chetty proposed in 1981 the use of the CIECA (current injected equivalent circuit approach) to simplify the modeling and analysis of dc-to-dc converters [17–19]. On the other hand, accurate but much more complex approaches to analysis using z-transform and sampled-data methods were studied independently by Capel, Ferrante, and Prajoux [16] and Brown and Middlebrook [14]. Analyses using SPICE [57] were carried out by Bello [11,67]. Monteith Jr., and Salcedo [40], Hageman [96], and Kimhi and Ben-Yaakov [102]. Current-controlled converters were investigated by Deisch [92], Redl, Novak, and Sokal [93,94], Holland [95], and Ridley [103]. The use of alternor for the analysis of converters was proposed by Ioinovici [109].

The techniques of modeling and analysis of resonant and quasi-resonant converters have been studied by Schwarz [48,49], King and Stuart [31,32], Vorpérian and Ćuk [59,60,78], Witulski, Erickson, and Hernandez [62,100], Liu, Oruganti, and Lee [64,65,73], Lee and Siri [66], Redl, Molnár, and Sokal [68], Lee and Cheng [3,4,6], Kim and Youn [98], Ninomiya et al. [99], and Kang, Upadhyay, and Stephens [97].

Since power converters and the SMPS are nonlinear circuits, computer-aided analysis and design techniques for such circuits are highly desirable. So far, very little literature in this area has been published. Thus, a major objective of this book is to fill this gap of knowledge. It should be understood, however, that in order to use CAD tools efficiently, the designer must have a good and intuitive understanding of the circuit to be designed. In the following five sections, the principle of operation of the basic square-wave power converters used in the SMPS will be studied. High-frequency resonant and quasi-resonant converters will be discussed in Chapter 9.

1.2 THE BUCK CONVERTER

This section describes the operation of the buck converter, which is also known as the step-down converter. The two modes of operation, namely, the continuous mode and the discontinuous mode, will be analyzed in Subsections 1.2.1 and 1.2.2, respectively. The boundary condition between the two modes of operation and the differences in their characteristics will be discussed in Subsections 1.2.3 and 1.2.4.

1.2.1 The Buck Converter in Continuous-Mode Operation

Let us consider the operation of the switching regulator shown in Fig. 1.4, which employs a buck converter as the power-conversion engine.

Under the control of the pulse-width modulator, the buck converter shown in Fig. 1.4 receives a dc input voltage V_i and produces a regulated dc output voltage V_o. During normal operation, the electronic switch SW of the buck converter is driven on and off by the pulse-width modulator at a high switching frequency (e.g., 5–500 kHz). The state of the electronic switch SW (either fully on or fully off) and the waveforms of i_L, i_i, i_D, v_D, and V_o are shown in Fig. 1.5. In the waveform diagram, the following assumptions have been made:

1. The switching action of the converter has reached a steady state.
2. The percentage of time in which the electronic switch SW is turned to the on state within a switching period (T) is defined as the duty cycle D:

$$D = \frac{t_{ON}}{t_{ON} + t_{OFF}} = \frac{t_{ON}}{T} \tag{1.1}$$

$$t_{ON} = DT \tag{1.2}$$

3. The inductance L is so large that the inductor current i_L will not decay to zero during the time the electronic switch SW is turned off. This is known as the continuous-mode operation.
4. The output filtering capacitor C_L is so large that within a switching cycle the change in V_o is very small.

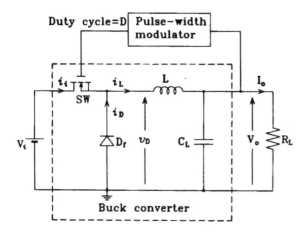

FIGURE 1.4 Buck converter in switching regulator; D = duty cycle.

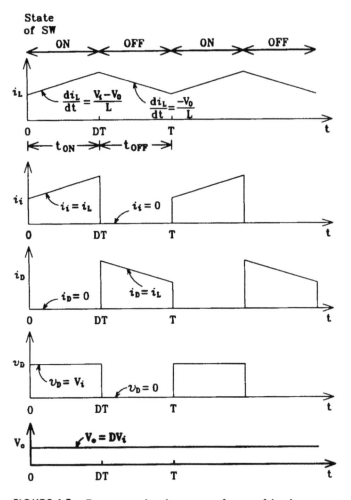

FIGURE 1.5 Current and voltage waveforms of buck converter in continuous-mode operation.

Having made the above assumptions, we can now proceed to examine the operation of the converter step by step. Refer to Fig. 1.4 for the circuit and to Fig. 1.5 for the waveforms.

1. At $t = 0$, the transistor switch SW is turned fully on. V_i assists the current i_L to build up at the rate of

$$\frac{di_L}{dt} = \frac{V_i - V_o}{L} \qquad \text{(during normal operation, } V_o < V_i) \qquad (1.3)$$

The inductor current i_L increases linearly, as shown in Fig. 1.5. The current i_D at this time is zero, because the diode D_f is reversely biased by V_i.

2. For $0 < t < DT$, SW is maintained in the on state.

$$\frac{di_L}{dt} = \frac{V_i - V_o}{L} \tag{1.4}$$

$$i_i = i_L \qquad (i_i \text{ flows through } L \text{ to charge up } C_L) \tag{1.5}$$

$$i_D = 0 \qquad (D_f \text{ is reversely biased}) \tag{1.6}$$

$$v_D = V_i \tag{1.7}$$

3. At $t = DT$, SW is turned off. Since the inductive current i_L is now unable to pass through SW, it forces its way through the diode D_f, forwardly biasing the diode. D_f is called a "flywheel" diode because it allows the inductive "flywheel" current to flow through it when SW is turned off.

4. For $DT < t < T$, SW remains off. The inductor current i_L continues to flow through D_f, but it is opposed by the output voltage V_o. Therefore, the amplitude of i_L decreases at the rate of

$$\frac{di_L}{dt} = \frac{-V_o}{L} \tag{1.8}$$

At the same time, we have

$$i_D = i_L \qquad (i_L \text{ continues to flow through } D_f \text{ to charge up } C_L) \tag{1.9}$$

$$v_D = 0 \qquad (\text{the forward voltage drop of } D_f \text{ is assumed to be}$$
$$\text{zero}) \tag{1.10}$$

5. At $t = T$, SW is turned on once more. The actions described in steps 1–4 repeat once again.

The waveform of v_D in Fig. 1.5 shows that within each cycle we have

$$v_D = V_i \qquad \text{for } 0 < t < DT \tag{1.11}$$

$$v_D = 0 \qquad \text{for } DT < t < T \tag{1.12}$$

The averaged value of v_D is therefore equal to DV_i. If we assume that the filtering circuit (L and C_L) allows only the dc component of v_D to be delivered to the output, we have

$$V_o = \text{Averaged value of } v_D = DV_i \tag{1.13}$$

The output voltage V_o can therefore be controlled by varying the duty cycle D of the driving pulses applied to the switching transistor SW. The "pulse-width modulator" shown in Fig. 1.4 is, in fact, the feedback control circuit used to control this duty cycle so as to regulate the output voltage V_o. Since, in the waveform given in Fig. 1.5, the inductor current i_L is continuous and never falls to zero, it is known as the continuous-mode operation, as opposed to the discontinuous-mode operation, which will be described in the next subsection.

1.2.2 The Buck Converter in Discontinuous-Mode Operation

The waveforms given in Fig. 1.5 for the buck converter shown in Fig. 1.4 are based on the assumption that the inductor current i_L is continuous and never falls to zero. This assumption becomes invalid if the load resistance R_L is sufficiently large or if the inductance L is too small. Under such conditions, i_L will fall to zero during the time the switch SW is turned off. The voltage and current waveforms will also change to the new shapes shown in Fig. 1.6.

In the new waveforms shown in Fig. 1.6, it is assumed that the inductance L is much smaller than the value assumed in Fig. 1.5. As a result, the following changes in the waveforms can be observed:

1. Because of the smaller inductance, the rate of change of the inductor current i_L is larger.
2. The inductor current i_L can now fall to zero at

    $$t = DT + D_2T$$

    where D_2T is the period of time during which a current flows through the flywheel diode D_f, and D_2 is effectively the duty cycle of D_f. (Since i_L is now discontinuous, it is known as the discontinuous-mode operation.)
3. Within the period of

    $$(D + D_2)T < t < T$$

    the inductor current i_L is equal to zero, as shown in Fig. 1.6. Since the average volt-second product across the inductor L should be equal to zero, we have

    $$DT(V_i - V_o) = D_2TV_o \tag{1.14}$$

    $$D_2 = D\frac{V_i - V_o}{V_o} \tag{1.15}$$

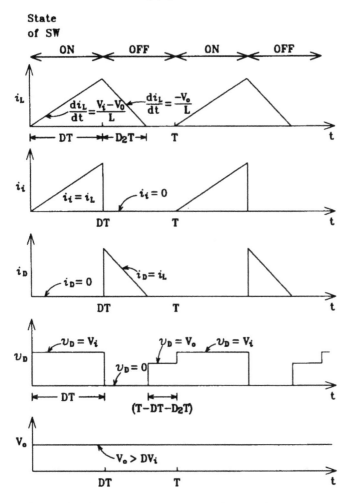

FIGURE 1.6 Current and voltage waveforms of buck converter in discontinuous-mode operation.

4. The new output voltage V_o, which is actually the averaged value of the diode voltage v_D, is now larger than DV_i. This can be deduced from the waveform for v_D, as shown in Fig. 1.6. Averaging v_D over a complete cycle, we have

$$V_o = \frac{(DT)V_i + (T - DT - D_2T)V_o}{T}$$

$$= DV_i + (1 - D - D_2)V_o \qquad (1.16)$$

Knowing that $(1 - D - D_2)$ is positive, we may therefore conclude that

$$V_o > DV_i \qquad (1.17)$$

It will be shown in Chapter 3 that the output voltage V_o for the discontinuous-mode operation is equal to

$$V_o = V_i \frac{D^2 G R_L}{2} \left\{ \left[1 + \frac{4}{D^2 G R_L} \right]^{1/2} - 1 \right\} \qquad (1.18)$$

where $G = T/(2L)$.

1.2.3 Boundary Condition Between Continuous and Discontinuous Modes of Operation

Referring to the waveform of i_L at the boundary condition between continuous and discontinuous modes of operation, as shown in Fig. 1.7, it can be found that, under this critical condition, we have

$$I_o = \frac{V_o}{R_L} = \frac{1}{2} \frac{V_o}{L} (1 - D)T \qquad (1.19)$$

The corresponding critical loading resistance is therefore given by

$$R_{Lcrit} = \frac{2L}{(1 - D)T} \qquad (1.20)$$

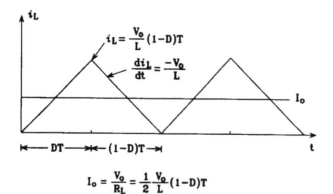

$$I_o = \frac{V_o}{R_L} = \frac{1}{2} \frac{V_o}{L} (1-D)T$$

FIGURE 1.7 Waveform of i_L for boundary condition between continuous and discontinuous modes of operation.

When R_L is larger than the critical value given in Eq. (1.20), the converter will enter into the discontinuous-mode operation. From Eq. (1.19) we can also determine the critical inductance L_{crit} for a given value of R_L:

$$L_{crit} = (1/2)R_L(1 - D)T \tag{1.21}$$

When L is smaller than the critical value given in Eq. (1.21), the converter will also enter into the discontinuous-mode operation.

1.2.4 Comparison Between Continuous-Mode and Discontinuous-Mode Characteristics

We shall now study the characteristics of, and differences between, the two modes of operation.

A comparison of the waveforms for the two modes of operation, as shown in Figs. 1.5 and 1.6, reveals the following differences:

1. If the converter is operated in the continuous-mode operation and the inductance L is sufficiently large, the current i_L can be made very smooth. The continuous-mode operation therefore tends to produce smaller ripple in the output voltage V_o and less interference.
2. The amplitudes of the ripple components of the output voltage V_o for the two modes of operation can be found and compared by examining the waveforms of the ripple currents in the filtering capacitor C_L, as shown in Fig. 1.8 (ripple current $= i_L - I_o$).

    In Fig. 1.8(a), which is for the continuous-mode operation, the area of the triangle ABC represents the charge pumped into the output capacitor C_L during the time the inductor current i_L is larger than the dc output current I_o in the load R_L. The same amount of charge will be discharged from C_L during the time i_L is smaller than I_o. The peak-to-peak ripple voltage ΔV_o is therefore given by

$$\Delta V_o = \frac{\text{Area of triangle } ABC}{C_L} \tag{1.22}$$

$$= \frac{\frac{1}{2}\frac{T}{2}\frac{1}{2}\frac{V_o}{L}(1 - D)T}{C_L} \tag{1.23}$$

$$= \frac{T^2 V_o}{8C_L L}(1 - D) \tag{1.24}$$

Alternatively, ΔV_o may also be expressed as

$$\Delta V_o = \frac{T^2}{8C_L L} D(V_i - V_o) \tag{1.25}$$

If a large inductance L or capacitance C_L is chosen, the ripple in V_o can be made very small.

For the discontinuous-mode operation, the area of the triangle ABC shown in Fig. 1.8(b) represents the charge pumped into C_L when i_L is larger than I_o. We therefore have

$$\Delta V_o = \frac{\Delta ABC}{C_L} \tag{1.26}$$

$$= \frac{1}{C_L} \Delta BDE \left[\frac{BG}{BF} \right]^2 \tag{1.27}$$

$$= \frac{1}{C_L} \Delta BDE \left[\frac{BF - FG}{BF} \right]^2 \tag{1.28}$$

$$= \frac{1}{C_L} \Delta BDE \left[1 - \frac{FG}{BF} \right]^2 \tag{1.29}$$

where

$$\Delta BDE = \frac{1}{2}(D + D_2)T \frac{V_o}{L} D_2 T \tag{1.30}$$

$$BF = \frac{V_o}{L} D_2 T \tag{1.31}$$

$$FG = I_o = \frac{1}{2}(D + D_2)T \frac{V_o}{L} D_2 \tag{1.32}$$

$$D_2 = D \frac{V_i - V_o}{V_o} \quad \text{[From Eq. (1.15)]} \tag{1.33}$$

Substitution of Eq. (1.33) into Eqs. (1.30–1.32) and the resultant equations into Eq. (1.29) gives

$$\Delta V_o = \frac{T^2 D^2}{8C_L L} (V_i - V_o) \frac{V_i}{V_o} \left[2 - D \frac{V_i}{V_o} \right]^2 \tag{1.34}$$

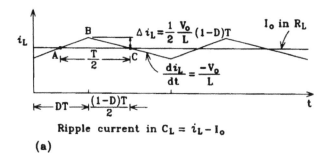

Ripple current in $C_L = i_L - I_o$

(a)

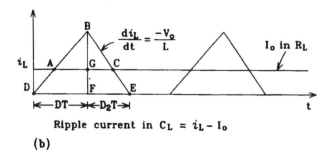

Ripple current in $C_L = i_L - I_o$

(b)

FIGURE 1.8 Waveforms of ripple current in C_L for buck converter: (a) ripple current in C_L for continuous-mode operation; (b) ripple current in C_L for discontinuous-mode operation.

When actual circuit parameters are substituted into Eqs. (1.25) and (1.34), it can be found that the continuous-mode operation will give a lower value of ripple voltage.

3. Given the same output current I_o, the peak current in the inductor and switching devices is larger for the discontinuous-mode operation than for the continuous-mode operation.

It is also interesting to note that the two modes of operation result in two very different low-frequency equivalent circuits. As will be shown in Section 2.3, the low-frequency output equivalent circuit of a buck converter in the continuous-mode operation is an LCR circuit, as shown in Fig. 1.9(a), which has two pole frequencies. But that of the discontinuous-mode operation is an RC circuit, as shown in Fig. 1.9(b), which has only a single pole frequency. From these equivalent circuits, it is quite obvious that the continuous-mode operation should give a better steady-state load regulation because the dc voltage drop across the inductance L should be zero.

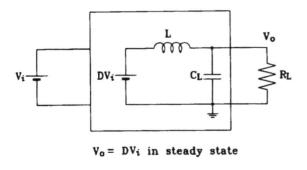

$$V_o = DV_i \text{ in steady state}$$

(a)

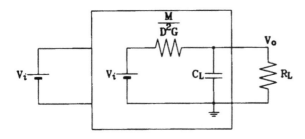

$$V_o = M V_i \text{ in steady state}$$

$$M = \frac{D^2 G R_L}{2}\left[\sqrt{1+\frac{4}{D^2 G R_L}} - 1\right] \qquad G = \frac{T}{2L}$$

(b)

FIGURE 1.9 Low-frequency output equivalent circuits of buck converter: (a) output equivalent circuit for continuous-mode operation; (b) output equivalent circuit for discontinuous-mode operation.

However, when feedback is added to the converter to form a regulator, the discontinuous-mode operation, having a single-pole characteristic, tends to be more stable and can provide a better transient regulation.

1.3 THE BUCK-BOOST CONVERTER

In this section, the operation and characteristics of the buck-boost converter will be studied. Figure 1.10 shows the circuit of a buck-boost converter. Figure 1.11 shows the voltage and current waveforms of the converter for the continuous-mode operation.

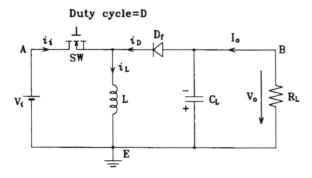

FIGURE 1.10   Buck-boost converter; D = duty cycle.

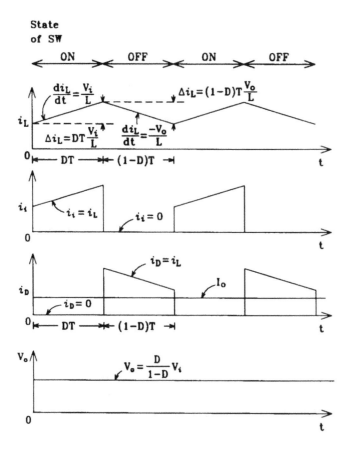

FIGURE 1.11   Waveforms of buck-boost converter in continuous-mode operation.

1.3.1 The Buck-Boost Converter in Continuous-Mode Operation

In the waveform diagram shown in Fig. 1.11, assumptions similar to those given for the buck converter in Subsection 1.2.1 have been made. With reference to Figs. 1.10 and 1.11, the continuous-mode operation of the buck-boost converter can be explained as follows:

1. At $t = 0$, SW is turned on. V_i assists the inductor current i_L to build up linearly:

$$\frac{di_L}{dt} = \frac{V_i}{L} \tag{1.35}$$

2. For $0 < t < DT$, SW is maintained in the on state.

$$\frac{di_L}{dt} = \frac{V_i}{L} \tag{1.36}$$

$$i_i = i_L \tag{1.37}$$

$$i_D = 0 \qquad \text{(because diode } D_f \text{ is reversely biased)} \tag{1.38}$$

3. At $t = DT$, SW is turned off. Since the inductive current i_L cannot flow through SW, it forces its way through the flywheel diode D_f to charge up the output filtering capacitor C_L.

4. For $DT < t < T$, SW remains off. The inductor current i_L continues to flow through D_f, charging up the output capacitor C_L. Since the flow of i_L is opposed by the output voltage V_o, i_L decreases linearly:

$$\frac{di_L}{dt} = \frac{-V_o}{L} \tag{1.39}$$

At the same time, we have

$$i_D = i_L \qquad (i_L \text{ flows through } D_f \text{ to charge up } C_L) \tag{1.40}$$

5. At $t = T$, SW is turned on again. The actions described in steps 1–4 repeat once more.

During the time when $i_D = 0$, the charge stored in C_L maintains the output voltage V_o. If C_L is large, the ripple in V_o can be maintained low. Note that the output voltage polarity at node B is opposite to that of the input at node A (all with respect to the common node E).

The steady-state output voltage V_o can be determined by examining the waveform of i_L, as shown in Fig. 1.11. It is found that the increase in i_L between $t = 0$ and $t = DT$ is given by

$$\text{Increase of } i_L = DT \frac{V_i}{L} \qquad\qquad (1.41)$$

It can also be found that the decrease of i_L between $t = DT$ and $t = T$ is equal to

$$\text{Decrease of } i_L = (1 - D)T \frac{V_o}{L} \quad \text{(in amplitude)} \qquad (1.42)$$

Since, in the steady state, the increase of i_L in Eq. (1.41) and the decrease of i_L in Eq. (1.42) are equal, we have

$$DT \frac{V_i}{L} = (1 - D)T \frac{V_o}{L}$$

$$\therefore V_o = \frac{D}{1 - D} V_i \qquad\qquad (1.43)$$

By varying the duty cycle D of the driving pulses applied to the electronic switch SW, we can achieve control of the output voltage V_o. Either a step-up or a step-down of the voltage can be obtained.

1.3.2 The Buck-Boost Converter in Discontinuous-Mode Operation

When the load resistance R_L is large or the inductance L is small, the buck-boost converter may enter into the discontinuous-mode operation. This is similar to the case of a buck converter. The critical inductance L_{crit} separating the two modes of operation can be found from Fig. 1.12, which shows the waveform of the inductor current i_L at the boundary (critical) condition:

$$I_o = \frac{\Delta ABC}{T} = \frac{1}{2} \frac{V_o}{L_{crit}} (1 - D)^2 T$$

$$\frac{V_o}{R_L} = \frac{1}{2} \frac{V_o}{L_{crit}} (1 - D)^2 T$$

$$\therefore L_{crit} = (1/2) R_L (1 - D)^2 T \qquad\qquad (1.44)$$

Figure 1.13 shows the typical waveforms for the discontinuous-mode operation, in which it is assumed that the inductance L is sufficiently small

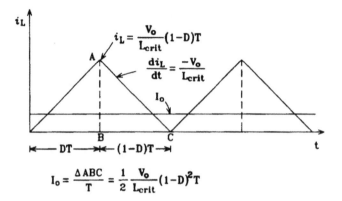

FIGURE 1.12 Waveform of i_L at boundary condition between continuous and discontinuous modes of operation.

to cause the converter to enter into the discontinuous-mode operation. The waveform diagram should be easy to understand. Note that in the discontinuous-mode operation we have

$$DTV_i = D_2TV_o \qquad \text{(zero volt-second product across } L)$$

$$D_2 = D\frac{V_i}{V_o} \tag{1.45}$$

It is a straightforward exercise to find the output voltage of a buck-boost converter operating in the discontinuous-mode operation. All that is required is to equate the input energy with the output energy for a complete switching cycle. Referring to the waveform of i_L, shown in Fig. 1.13, we have

$$\text{Input energy} = (1/2)L(\Delta i_L)^2 \tag{1.46}$$

$$\text{Output energy} = \frac{V_o^2}{R_L}T \tag{1.47}$$

Equating Eqs. (1.46) and (1.47), we have

$$\frac{1}{2}L(\Delta i_L)^2 = \frac{V_o^2}{R_L}T$$

$$\frac{1}{2}L\left[DT\frac{V_i}{L}\right]^2 = \frac{V_o^2}{R_L}T$$

$$\therefore V_o = DV_i\left[\frac{R_LT}{2L}\right]^{1/2} \tag{1.48}$$

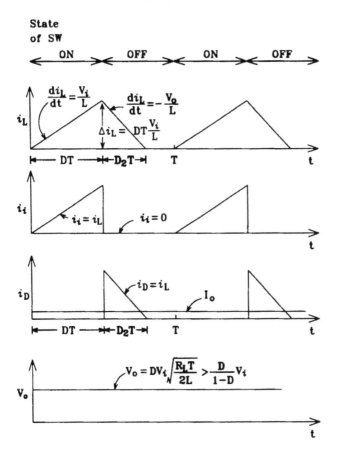

FIGURE 1.13 Waveforms of buck-boost converter in discontinuous-mode operation.

Equation (1.48) indicates that, for a buck-boost converter operating in the discontinuous-mode operation, the output voltage is proportional to the duty cycle D.

1.3.3 Comparison Between Continuous-Mode and Discontinuous-Mode Characteristics

A comparison between the two modes of operation of the buck-boost converter reveals the following:

1. Unlike the buck converter, the continuous-mode operation in a buck-boost converter does not help to smooth out the output current.

2. In the continuous-mode operation, the peak-to-peak ripple voltage in V_o, denoted as ΔV_o, can be found as (refer to Figs. 1.10 and 1.11):

$$\Delta V_o = \frac{\text{Charge taken from } C_L \text{ for } 0 < t < DT}{C_L}$$

$$= \frac{I_o T}{C_L} D \tag{1.49}$$

assuming that the minimum value of i_L is larger than I_o.
Since, from Eq. (1.43), we have

$$V_o = \frac{D}{1 - D} V_i \tag{1.50}$$

$$(1 - D)V_o = DV_i \tag{1.51}$$

$$D = \frac{V_o}{V_i + V_o} \tag{1.52}$$

substitution of Eq. (1.52) into Eq. (1.49) gives the following alternative expression for the peak-to-peak ripple voltage for continuous-mode operation:

$$\Delta V_o = \frac{V_o}{V_i + V_o} \frac{I_o T}{C_L} \tag{1.53}$$

For the discontinuous-mode operation, an exact expression for the ripple voltage can be derived by applying an analysis similar to the one given in Subsection 1.2.4 for a buck converter in discontinuous-mode operation. However, an approximate, but practically more useful, expression can be found by assuming that C_L has to supply all the output current during the period $(T - D_2 T)$, as shown in Fig. 1.13. We then have

$$\Delta V_o = \frac{I_o(T - D_2 T)}{C_L} = \frac{I_o T}{C_L}(1 - D_2) \tag{1.54}$$

where $D_2 = D(V_i/V_o)$. It should be noted that the continuous-mode operation of a buck-boost converter would not help much to reduce the ripple in the output voltage.

3. For a given output current I_o, the peak current in the switching devices is larger for the discontinuous-mode operation than for the continuous-mode operation. (This characteristic is similar to those of the buck converter and others.)
4. The low-frequency output equivalent circuits for the two modes of operation, as will be derived from Section 2.4, are shown in Fig. 1.14. It is interesting to note that, in the continuous-mode equivalent circuit, there is a nonlinear inductance $L/(1 - D)^2$.

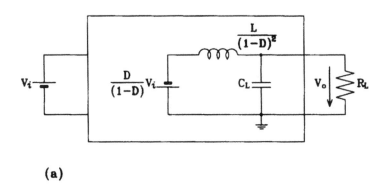

(a)

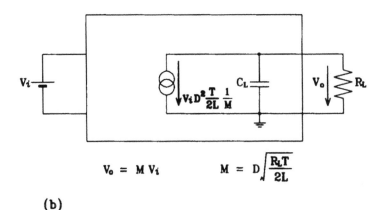

$$V_o = M\,V_i \qquad M = D\sqrt{\frac{R_L T}{2L}}$$

(b)

FIGURE 1.14 Low-frequency output equivalent circuits of buck-boost converter: (a) low-frequency equivalent circuit for continuous-mode operation; (b) low-frequency equivalent circuit for discontinuous-mode operation.

1.4 THE BOOST CONVERTER

In this section, the operation and characteristics of the boost converter will be studied. Figure 1.15 shows the circuit of a boost converter. Figure 1.16 shows the waveforms. The solid lines are for the continuous-mode operation and the dashed lines for the discontinuous-mode operation.

Having examined the buck and buck-boost circuits, it should be rather straightforward to understand the operation of the boost converter. The following is a summary of the key points:

1. During the time the electronic switch SW is turned on (refer to Fig. 1.15), the inductor current i_L increases linearly at the rate of

$$\frac{di_L}{dt} = \frac{V_i}{L} \tag{1.55}$$

   Meanwhile, magnetic energy is stored up in inductance L. This is similar to what happens in a buck-boost converter.

2. During the time the electronic switch SW is turned off, the inductor current i_L decreases at the rate of

$$\frac{di_L}{dt} = \frac{V_i - V_o}{L} \tag{1.56}$$

   At the same time, the energy stored in the inductance L, together with the energy from the input voltage source V_i, is delivered to the output circuit. Since the supply voltage V_i is added in series with the inductor voltage to charge up the output capacitor C_L, the output voltage V_o is larger than V_i. It is for this reason that the converter is known as a boost converter.

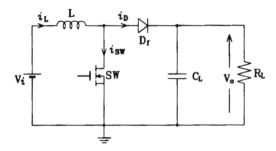

FIGURE 1.15 Boost converter.

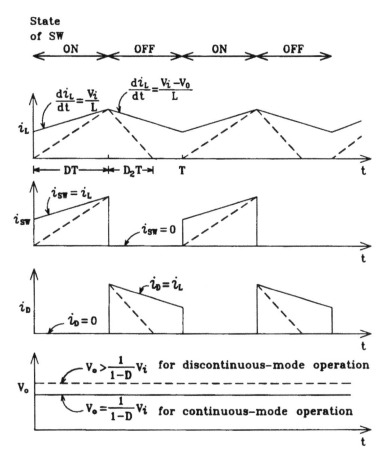

FIGURE 1.16   Waveforms of boost converter. Solid lines indicate continuous-mode operation; dashed lines indicate discontinuous-mode operation.

3. The dc output voltage V_o of the boost converter in the continuous-mode operation can be found from the solid-line waveform of i_L, shown in Fig. 1.16.

$$\frac{V_i}{L} DT = \frac{-(V_i - V_o)}{L} (1 - D)T \qquad (1.57)$$

$$= \frac{-V_i}{L} (1 - D)T + \frac{V_o}{L} (1 - D)T \qquad (1.58)$$

$$V_i = (1 - D)V_o \qquad\qquad (1.59)$$

$$V_o = \frac{V_i}{1 - D} \qquad\qquad (1.60)$$

Note that the output voltage for the boost converter in discontinuous-mode operation, as found in Section 3.7, is

$$V_o = V_i \left[\frac{1 + (1 + 4D^2 G R_L)^{1/2}}{2} \right] \qquad\qquad (1.61)$$

where $G = T/(2L)$.

4. The critical inductance separating the continuous-mode operation from the discontinuous-mode operation can be obtained by referring to Fig. 1.17, which shows the waveform of i_L at the boundary condition. From the waveform, we find that

$$I_o = \frac{\Delta ABC}{T} = \frac{1}{2} \frac{V_i}{L_{crit}} DT(1 - D) \qquad\qquad (1.62)$$

Since Eq. (1.59) is still valid under the critical condition, substitution of Eq. (1.59) and $I_o = V_o/R_L$ into Eq. (1.62) gives

$$\frac{V_o}{R_L} = \frac{1}{2} \frac{(1 - D)^2 V_o}{L_{crit}} DT \qquad\qquad (1.63)$$

$$\therefore L_{crit} = (1/2)R_L D(1 - D)^2 T \qquad\qquad (1.64)$$

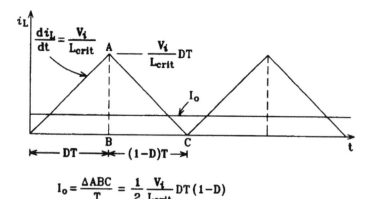

$$I_o = \frac{\Delta ABC}{T} = \frac{1}{2} \frac{V_i}{L_{crit}} DT(1-D)$$

FIGURE 1.17 Waveform of i_L at the boundary condition between continuous and discontinuous modes of operation.

5. If the converter is operated in continuous mode and a large inductance L is used, the input current i_L of the boost converter can be made very smooth. However, the output current i_D continues to pulsate. This is similar to what happens in a buck-boost converter.

6. Because of the similar output current waveforms, Eqs. (1.49) and (1.54), found in Subsection 1.3.3 for the output ripple voltage of a buck-boost converter, also apply to a boost converter, that is,

$$\Delta V_o = \frac{I_o T}{C_L} D \qquad (1.65)$$

for the continuous-mode operation, and

$$\Delta V_o = \frac{I_o T}{C_L} (1 - D_2) \qquad (1.66)$$

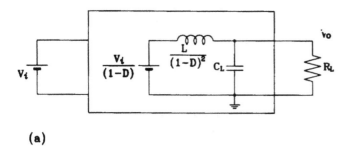

(a)

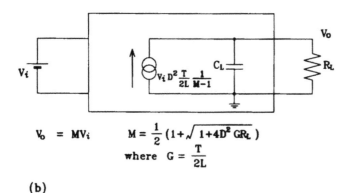

$$V_o = M V_i \qquad M = \frac{1}{2} \left(1 + \sqrt{1 + 4 D^2 G R_L} \right)$$

$$\text{where} \quad G = \frac{T}{2L}$$

(b)

FIGURE 1.18 Low-frequency equivalent circuits of boost converter: (a) in continuous-mode operation; (b) in discontinuous-mode operation.

for the discontinuous-mode operation. Note, however, that while the D_2 for the buck-boost converter is equal to DV_i/V_o, that of the boost converter is equal to $DV_i/(V_o - V_i)$.

7. The low-frequency equivalent circuits for the two modes of operation of the boost converter, as found in Chapter 2, are shown in Fig. 1.18. Note that a nonlinear inductance $L/(1 - D)^2$ exists in the continuous-mode equivalent circuit. This is similar to what happens in a buck-boost converter.

1.5 THE ĆUK CONVERTER

In this section, the operation and characteristics of the Ćuk converter will be examined.

The Ćuk converter is effectively a cascaded connection of a boost converter followed by a buck converter. Figure 1.19 shows the circuit of a Ćuk converter. Figure 1.20 shows how a Ćuk converter may be decomposed into a boost converter and a buck converter. The two converters shown share the same electronic switch SW and flywheel diode D_f. While the energy storage capacitor C is used as the output filtering capacitor for the boost converter, it is also effectively used as a battery providing power to the buck converter. The load resistance R_L' shown in the boost converter is used to represent the loading effect of the buck converter.

The boost and buck parts of the Ćuk converter should be compared to the original buck and boost converters shown in Figs. 1.15 and 1.4 to see how the circuits correspond to each other.

The main advantage of the Ćuk converter is the possibility of designing it to give a very smooth input current i_{LA} and output current i_{LB}. In other

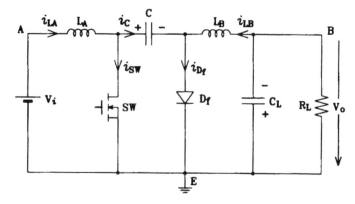

FIGURE 1.19 Ćuk converter.

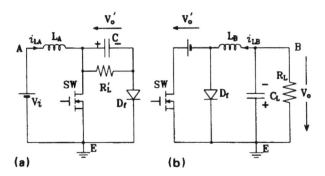

FIGURE 1.20 Decomposition of a Ćuk converter into (a) a boost converter and (b) a buck converter.

words, it combines the merits of a boost converter, which has a smoothed input current, and a buck converter, which has a smoothed output current. However, it should be noted that, in order to achieve this advantage, both the boost and buck parts of the Ćuk converter need to operate in the continuous-mode operation (i.e., a current is maintained in the flywheel diode D_f during the entire period of $DT < t < T$, when the switch SW is turned off). That being the assumption, we shall have (refer to Fig. 1.20)

$$V_o' = \frac{1}{1 - D} V_i \qquad \text{for the boost part} \tag{1.67}$$

$$V_o = DV_o' \qquad \text{for the buck part} \tag{1.68}$$

$$V_o = DV_o' = \frac{D}{(1 - D)} V_i \qquad \text{for the Ćuk converter} \tag{1.69}$$

Note that the output voltage at node B, as shown in Fig. 1.19, is opposite in polarity to the input voltage at node A (all with respect to the common node E).

An interesting feature of the Ćuk converter is that the ripples in i_{LA} and i_{LB} can, in fact, be further reduced by coupling L_A and L_B together, as shown in Fig. 1.21. The reasons for the reduced ripples are:

1. When the switch SW is turned on, both i_{LA} and i_{LB} tend to increase at the same time. However, since they oppose each other in the coupled circuit, they can now increase only at a slower rate.
2. When the switch SW is turned off, both i_{LA} and i_{LB} tend to drop at the same time. However, as a result again of the coupling, they can now drop only at a slower rate.

A slower rate of rise and fall implies a reduced ripple.

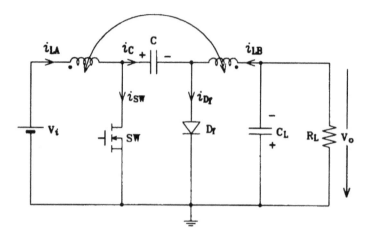

FIGURE 1.21 Ćuk converter with coupled inductors.

1.6 ELECTRONIC TRANSFORMERS

The term electronic transformer is used here to mean a class of power conversion circuits that allows bidirectional flow of energy. All the power converters we have considered so far allow the energy to flow in only one direction, that is, from input to output but not vice versa. If energy can flow in both directions in a power-conversion circuit, the circuit will effectively function as a transformer. The concept of electronic transformers, and the practical methods of implementation, will be explained in the following three subsections.

1.6.1 The Concept of Electronic Transformers

Figure 1.22 shows the equivalent circuit of an electronic transformer, in which SWA and SWB are two bidirectional electronic switches controlled by a pulse-width modulator. SWA and SWB are always in opposite states. Thus, if we assume that the duty cycle of SWA is D, the duty cycle of SWB will be $(1 - D)$.

It is interesting to note that if an input voltage V_i is connected between nodes A and E, and a loading circuit (consisting of C_L and R_L in parallel) is connected between nodes B and E, as shown in Fig. 1.23, the electronic transformer will provide the same function as a Ćuk converter in continuous-mode operation (shown in Fig. 1.19). The reasons are simple:

1. Within each switching cycle when $0 < t < DT$, the switch SW in the Ćuk converter (Fig. 1.19) and the switch SWA in the electronic

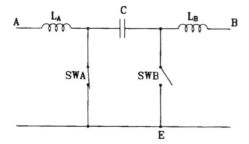

FIGURE 1.22 Equivalent circuit of an electronic transformer.

transformer (Fig. 1.23) are both in the on state. At the same time, the diode D_f in the Ćuk converter and the switch SWB in the electronic transformer are both in the off state.

2. During the time $DT < t < T$, both SW and SWA are in the off state. At the same time, both D_f and SWB are in the on state.

Since the equivalent circuits are the same for all stages of the switching cycle, as described above, both circuits would provide the same function. We have, therefore, for both circuits,

$$V_o = \frac{D}{(1 - D)} V_i \qquad\qquad (1.70)$$

where D is the duty cycle of SW or SWA, and $(1 - D)$ is the duty cycle of SWB or D_f.

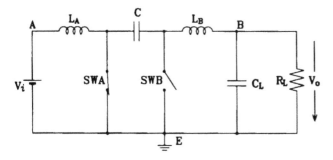

FIGURE 1.23 Electronic transformer performing the function of a Ćuk converter.

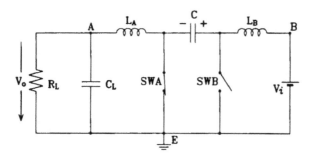

FIGURE 1.24   Electronic transformer with input and output swapped.

If the input voltage V_i and the output loading circuit (C_L and R_L) of the electronic transformer shown in Fig. 1.23 are swapped, as shown in Fig. 1.24, the circuit will also function equally well as a converter because the circuit is symmetrical. The only difference is that we now have

$$V_o = \frac{\text{Duty cycle of } SWB}{1 - \text{duty cycle of } SWB} V_i \qquad (1.71)$$

The flow of energy in both directions is therefore possible in an electronic transformer.

Since SWA and SWB are assumed to be bidirectional switches, even the polarity of the input V_i to the electronic transformer, shown in Figs. 1.23 and 1.24, can be reversed. This will result in only a reversed polarity of the output V_o. From this, we can further postulate that this electronic transformer is not only a dc-to-dc transformer but can also function as an ac-to-ac transformer (because it works in both polarities). There is, however, a limitation on the ac operation frequency of the electronic transformer; that is, it must be much lower than the internal switching frequency of the electronic transformer.

Unlike electronic transformers, if the input and output of a Ćuk converter are swapped or if the polarity of the V_i input is reversed, the Ćuk converter will not operate properly. A Ćuk converter is effectively a unipolar and unidirectional electronic transformer.

1.6.2 Implementation of the Electronic Transformer

A problem encountered in the implementation of the electronic transformer, shown in Fig. 1.22, is the difficulty of finding suitable bidirectional

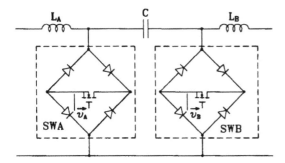

FIGURE 1.25 Implementation of electronic transformer.

switches to function as *SWA* and *SWB*. Most of the existing electronic switches, such as the bipolar transistor, the MOS, and the SCR, are uni-directional devices that normally allow current to flow in only one direction. Although the Triac allows current to flow in both directions, it is too slow and is difficult to turn off. A practical method of implementing a bidirectional switch is to use a combined transistor and full-wave rectifier circuit, as shown in Fig. 1.25. The circuit within each dashed-line box is functionally a bidirectional switch, the on–off of which is determined by the control voltage applied to the gate of the MOS transistor. Similar circuits using bipolar transistors may also be used. (Note that the driving voltages v_A and v_B, shown in Fig. 1.25, are pulse-width modulated pulses in complementary states.)

1.6.3 Electronic dc-to-dc Transformer

The electronic transformer circuit shown in Fig. 1.25 can operate either as a dc-to-dc transformer or as an ac-to-ac transformer because both *SWA* and *SWB* are bidirectional switches. If the circuit is to operate only as a unipolar dc-to-dc transformer, a simpler circuit such as that shown in Fig. 1.26 can be used.

Ignoring the diode D_{fa} and the transistor *SWB*, the circuit shown in Fig. 1.26 is actually a Ćuk converter, similar to the circuit in Fig. 1.19. Note that, under the operating condition shown in Fig. 1.26, D_{fa} does not affect the operation of the circuit because v_D is either positive or zero and D_{fa} never conducts. The switch *SWB* will also not affect the circuit operation because it is turned off whenever the switch *SWA* is turned on.

When the input and output circuits are swapped, as shown in Fig. 1.27, the switches *SWB* and D_{fa} will become active to form a Ćuk converter in

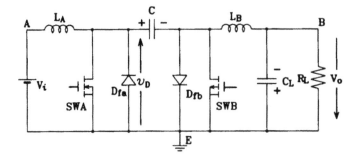

FIGURE 1.26 Electronic dc-to-dc transformer. *SWA* = *N*-channel enhancement-mode MOS transistor; *SWB* = *P*-channel enhancement-mode MOS transistor; *SWA* and *SWB* are in complementary states.

the reverse direction. At the same time, the switches *SWA* and D_{fb} will become inactive.

Referring back to Fig. 1.26, if we interpret the voltage at node *A* (with respect to node *E*) as the primary voltage, and the voltage at node *B* (also with respect to node *E*) as the secondary voltage, the circuit is effectively a dc-to-dc transformer with a secondary to primary turns ratio equal to

$$N = \frac{-D}{(1 - D)} \tag{1.72}$$

where *D* is the duty cycle of switch *SWA*, and $(1 - D)$ is the duty cycle of switch *SWB*. The negative sign is to indicate a polarity inversion between nodes *A* and *B*.

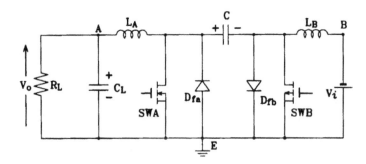

FIGURE 1.27 Electronic dc-to-dc transformer with input and output swapped.

1.7 THE ROLE OF CAD TOOLS IN ANALYSIS AND DESIGN OF THE SMPS

The simplified analyses given in Sections 1.2–1.6 of this chapter are helpful for understanding the basic principles of the SMPS. However, when an accurate analysis is to be carried out on a practical circuit that contains imperfect components (and possibly also a transformer together with its leakage inductance and primary inductance), the analysis can become much more complicated. This is why computer simulations are often employed in the accurate analysis of such circuits. Also, in the design of feedback and compensation circuits for regulators, the nonlinear nature of the low-frequency equivalent circuits (such as those shown in Figs. 1.9, 1.14, and 1.18) can make manual analysis very difficult. Computer-aided design (CAD) tools are therefore also desirable for the analysis and design of such feedback circuits.

The CAD tools currently available for analog circuits, such as SPICE and its variants, do not actually have the intelligence to perform design work. Creative design ideas must still come from circuit designers, while CAD tools are more appropriately used for simulation, analysis, verification, and documentation purposes.

To create a satisfactory environment for computer-aided design, the following are the basic requirements:

1. Designers know reasonably well, at least qualitatively, the circuit they are designing.
2. Designers are able to develop appropriate computer models for the circuit to be designed.
3. The computer hardware is reasonably fast.
4. The CAD software is reasonably bug-free and powerful.

With the widespread use of high-performance personal computers and the easy availability of software tools such as SPICE, PSpice, and HSPICE, the requirements stated above should not be difficult to meet.

1.8 SUMMARY AND FURTHER REMARKS

In this chapter, the development of switch-mode power supplies has been briefly outlined. Based on the idealized voltage and current waveforms, the switching operations of various converters (including buck, buck-boost, boost, and Ćuk) and electronic transformers have been studied. The differences between continuous and discontinuous modes of operation of converters have also been compared. The fundamental switching principles covered in this chapter are essential for the analytical and design work to be discussed in later chapters.

Two important aspects of the converter characteristics that have not been discussed are the small-signal transfer function from the duty cycle to the output voltage, $\delta V_o(s)/\delta D(s)$, and the small-signal transfer function from the input voltage to the output voltage, $\delta V_o(s)/\delta V_i(s)$. These frequency-domain transfer functions are essential to the design of the feedback circuit of a regulator, but they cannot be obtained directly from the switching waveforms considered in this chapter. We therefore need also to develop the so-called low-frequency behavior models of converters in order to find these $\delta V_o(s)/\delta D(s)$ and $\delta V_o(s)/\delta V_i(s)$ characteristics.

The next chapter examines the modeling of the low-frequency behaviors of square-wave power converters.

EXERCISES

1. Not all switch-mode power supplies can achieve a high conversion efficiency. List the necessary conditions to ensure a high conversion efficiency.
2. Compare the characteristics of the following types of converters:
   a. buck
   b. buck-boost
   c. boost
   d. Ćuk
3. Compare the advantages and disadvantages of the continuous-mode operation with those of the discontinuous-mode operation for
   a. a buck converter
   b. a buck-boost converter
   c. a boost converter
   Consider aspects of
   a. voltage and current stress/rating of components
   b. ripple components of output voltage
   c. feedback control when used in a regulator
4. Sketch the typical waveforms of i_{LA}, i_{SW}, i_C, i_{LB}, and i_{Df} for the Ćuk converter circuit shown in Fig. 1.19. Assume that a nonzero i_{Df} flows in diode D_f during the entire period when the switch SW is turned off (the condition for continuous-mode operation).
5. Compare the characteristics of the following types of power-conversion circuits:
   a. Ćuk converter
   b. electronic transformer
   c. dc-to-dc electronic transformer

2
Low-Frequency Behavior Models of Square-Wave Power Converters

This chapter introduces a simple and elegant approach to the modeling of the low-frequency behavior of power switches in square-wave power converters, which is known as the MISSCO (minimum separable switching configuration) method [2]. The MISSCO method will be used to derive the low-frequency behavior models of buck, buck–boost, boost, and Ćuk converters, as well as electronic transformers. These models will later be used for analysis and simulation of converter characteristics such as the relationship between the duty cycle and the output voltage, $\delta V_o(s)/\delta D(s)$, and the relationship between the dc input voltage and the output voltage, $\delta V_o(s)/\delta V_i(s)$.

2.1 MODELING THE LOW-FREQUENCY BEHAVIOR OF SWITCHES

In the modeling of the low-frequency behavior of a converter, one school of thought makes use of discrete methods using z-transforms and difference equations to describe the switching operation, but this approach usually produces complex equations that are difficult to manipulate. Another school of thought makes use of the averaging technique to obtain the low-frequency behavior model of the converter. A well-known method using this approach is the state-space averaging method introduced by S. Ćuk and R. D. Middlebrook [21,23,35,36]. Based on a similar method, P. R. K. Chetty introduced a current injected equivalent circuit approach (CIECA) to simplify the process in order to obtain the equivalent circuit [17–19]. Efforts were also made to use Y-parameters to represent the converter [1] in an attempt to simplify the analysis. However, the lack of a procedure to reduce the switching circuitry to a minimum configuration before the modeling process often results in a needlessly complex analysis.

The method introduced in this chapter is also based on the averaging technique but includes the following unique features in order to simplify the modeling process [2]:

1. Before embarking on the modeling of the converter, the switching part of the converter is first separated into a minimum separable switching configuration (MISSCO). A MISSCO is here defined as a minimum circuit configuration that satisfies both of the following requirements:
   a. It can be separated from the converter circuit and be represented by an identifiable low-frequency equivalent circuit.
   b. It contains all the power switches (including the diode switches) of the converter circuit but a minimum number of other circuit components (as explained in Section 2.2).
2. Once the MISSCO of the converter is identified, constant voltage or current sources are connected to the external nodes of the MISSCO in order to evaluate the responses of the switching circuit. Based on the averaged-response equations, the low-frequency equivalent circuit of the MISSCO is then synthesized.

The above unique features combine to make the modeling and analysis of the switching circuit much easier for practical circuit designers.

2.2 IDENTIFYING MISSCO AND ITS EQUIVALENT CIRCUIT

Based on the requirements stated in Section 2.1, the following procedures can be performed to identify a MISSCO:

1. Draw the equivalent circuits of the converter for all the stages of a switching cycle. Assume that inductors and capacitors are ideal cur-

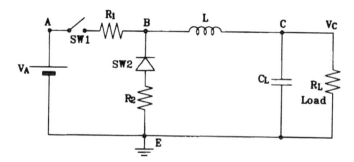

FIGURE 2.1 Buck converter.

rent and voltage sources, respectively. As an example, the equivalent circuits for the buck converter in Fig. 2.1 are shown in Fig. 2.2. (In this example, it is assumed that the converter is in the continuous-mode operation.)

2. Assuming that each of the voltage and current sources in the equivalent circuits is a given and independent excitation, examine the voltage across, and the current flowing through, each circuit component except the switches and their series resistances. Those components that have finite and determinable voltage and current during all stages of the switching cycle are not qualified to be components of a MISSCO. Only the components that tend to have infinite or indeterminable voltage or current during any stage of the switching cycle, together with the switches (and their series resistances), are qualified to be the components of a MISSCO. In the circuit example

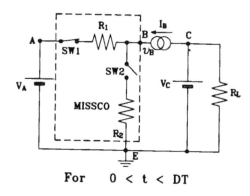

For   0 < t < DT

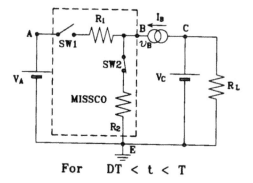

For   DT < t < T

FIGURE 2.2   Equivalent circuits of buck converter in continuous-mode operation.

shown in Fig. 2.2, since all node voltages and branch currents are finite and determinable (as functions of I_B, V_A, and V_C), for all stages of the switching cycle, only the switches (together with their series resistances) are qualified to be components of a MISSCO.

3. The MISSCO is a minimum switching circuitry that includes only the following components:

    a. all switches (together with their series resistances)

    b. any component qualified to be a component of the MISSCO

    In Fig. 2.2, the circuit within the dashed line is the MISSCO of the converter.

For any MISSCO, it is possible for a set of equations to be formulated that describes its external behaviors. From these equations, the equivalent circuit for the MISSCO can be established. The detailed procedures required to find an equivalent circuit of a MISSCO are as follows:

1. Assume that all voltage and current sources (including capacitors that are assumed to be ideal voltage sources, and inductors that are assumed to be ideal current sources) are separate and independent excitations in order to evaluate the responses of the MISSCO.

2. Determine and average the responses of the MISSCO at different stages of a switching cycle to produce a set of response equations. This set of equations defines the low-frequency characteristics of the MISSCO.

3. Based on the equations obtained in step 2, synthesize a low-frequency equivalent circuit of the MISSCO.

Finally, a low-frequency behavior model of the converter can be obtained by replacing the MISSCO with its low-frequency equivalent circuit in the complete converter circuit.

2.3   MODELING THE BUCK CONVERTER

This section describes how the low-frequency behavior model of a buck converter can be obtained.

2.3.1   The Buck Converter in Continuous-Mode Operation

Reconsider the buck converter shown in Fig. 2.1 and the MISSCO identified in Fig. 2.2. The two different-stage equivalent circuits of the MISSCO shown in Fig. 2.3 may be used to determine its low-frequency parameters.

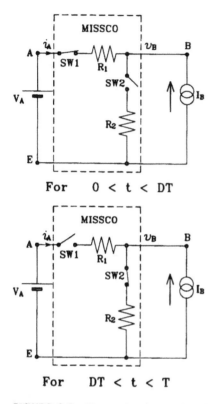

For   0 < t < DT

For   DT < t < T

FIGURE 2.3 Determination of low-frequency parameters of MISSCO for buck converter in continuous-mode operation.

The dc sources V_A and I_B are applied as excitations for evaluating the responses in i_A and v_B.

The responses in i_A and v_B can be found as follows (refer to Fig. 2.3):

1. For $0 < t < DT$, D being the duty cycle of SW1; SW1 is closed and SW2 is opened:

$$i_A = -I_B \tag{2.1}$$
$$v_B = V_A + I_B R_1 \tag{2.2}$$

2. For $DT < t < T$; SW1 is opened and SW2 is closed:

$$i_A = 0 \tag{2.3}$$
$$v_B = I_B R_2 \tag{2.4}$$

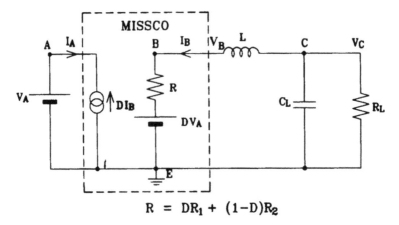

$$R = DR_1 + (1-D)R_2$$

FIGURE 2.4 Low-frequency equivalent circuit of MISSCO and complete behavior model of buck converter in continuous-mode operation.

Since the events described above repeat once for each switching cycle, the averaged responses of i_A and v_B, denoted as I_A and V_B, may be found from Eqs. (2.1–2.4) as

$$I_A = -DI_B \tag{2.5}$$
$$V_B = D(V_A + I_B R_1) + (1 - D)R_2 I_B \tag{2.6}$$
$$= DV_A + RI_B \tag{2.7}$$

where $R = DR_1 + (1 - D)R_2$.

The low-frequency equivalent circuit of the MISSCO, as derived from Eqs. (2.5) and (2.7), is shown within the dashed-line box in Fig. 2.4. This equivalent circuit may be combined with other components of the buck converter to form a complete low-frequency behavior model of the converter, as shown in Fig. 2.4.

2.3.2 The Buck Converter in Discontinuous-Mode Operation

The converter circuit is the same as that shown in Fig. 2.1, but the equivalent circuits at different stages of a switching cycle and the MISSCO are different. Figure 2.5 shows the discontinuous-mode equivalent circuits, in which:

1. For $0 < t < DT$, SW1 is on and SW2 is off.
2. For $DT < t < (D + D_2)T$, SW1 is off and SW2 is on.
3. For $(D + D_2)T < t < T$, both SW1 and SW2 are off.

Note that D and D_2 are the duty cycles of SW1 and SW2, respectively.

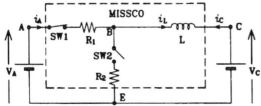

(a) Equivalent circuit for $0 < t < DT$.

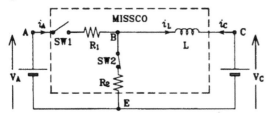

(b) Equivalent circuit for $DT < t < (D+D_2)T$.

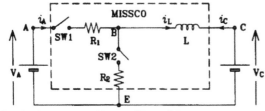

(c) Equivalent circuit for $(D+D_2)T < t < T$.

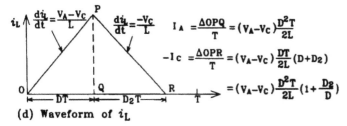

(d) Waveform of i_L

FIGURE 2.5 Equivalent circuits of buck converter in discontinuous-mode operation.

From the equivalent circuits shown in Fig. 2.5, it can be found that if the inductor L is assumed to be an ideal current source (i_L), it will tend to produce an infinite voltage at node B during the time $(D + D_2)T < t < T$. Therefore, L is qualified to be a component of the MISSCO. The circuit within the dashed lines is the MISSCO. The dc voltages V_A and V_C are used as excitations for evaluating the responses in i_A and i_C.

From Fig. 2.5, we can find the following (assuming $R_1 = R_2 = 0$):

1. For $0 < t < DT$; $SW1$ is closed and $SW2$ is opened:

$$i_A = (V_A - V_C)\frac{t}{L} \qquad (2.8)$$

$$i_C = -(V_A - V_C)\frac{t}{L} \qquad (2.9)$$

2. For $DT < t < (D + D_2)T$; $SW1$ is opened and $SW2$ is closed:

$$i_A = 0 \qquad (2.10)$$

$$i_C = V_A\frac{-DT}{L} + V_C\frac{t}{L} \qquad (2.11)$$

3. For $(D + D_2)T < t < T$; both $SW1$ and $SW2$ are opened:

$$i_A = 0 \qquad (2.12)$$
$$i_C = 0 \qquad (2.13)$$

The current waveform of i_L is shown in Fig. 2.5(d).

It can be found from Eqs. (2.8–2.13) that the averaged i_A and i_C over each complete switching cycle, denoted as I_A and I_C, are given by

$$I_A = V_A\frac{D^2T}{2L} - V_C\frac{D^2T}{2L} \qquad (2.14)$$

$$I_C = -V_A\left[\frac{D^2T}{2L}\left(1 + \frac{D_2}{D}\right)\right] + V_C\left[\frac{D^2T}{2L}\left(1 + \frac{D_2}{D}\right)\right] \qquad (2.15)$$

where D_2 can be found by inspecting the inductor current waveform i_L, shown in Fig. 2.5(d):

$$\frac{V_A - V_C}{L}DT = \frac{V_C}{L}D_2T$$

$$\therefore D_2 = \frac{V_A - V_C}{V_C}D \qquad (2.16)$$

It should be noted that Eqs. (2.14) and (2.15) can also be found graphically from Fig. 2.5(d):

$$I_A = \frac{\text{Area of triangle } OPQ}{T} = (V_A - V_C)\frac{D^2 T}{2L} \tag{2.17}$$

[which is actually the same as Eq. (2.14)]

$$-I_C = \frac{\text{Area of triangle } OPR}{T} = (V_A - V_C)\frac{DT}{2L}(D + D_2)$$

$$I_C = (V_C - V_A)\frac{D^2 T}{2L}\left(1 + \frac{D_2}{D}\right) \tag{2.18}$$

[which is actually the same as Eq. (2.15)].

The low-frequency equivalent circuit of the MISSCO, as derived from Eqs. (2.14) and (2.15), is shown in Fig. 2.6(a). To visualize better the

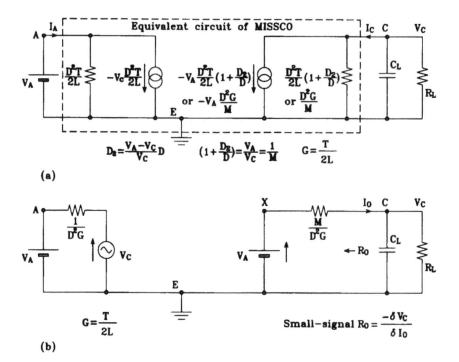

(a)

(b)

FIGURE 2.6 Low-frequency behavior models of buck converter in discontinuous-mode operation: (a) model derived from Eqs. (2.14) and (2.15); (b) alternative model.

operation of the circuit, an alternative form of the equivalent circuit is also given, as shown in Fig. 2.6(b). When compared with the original equivalent circuit in Fig. 2.6(a), the following modifications in Fig. 2.6(b) should be noted:

1. The original Norton's equivalent circuits of current sources in parallel with conductances are now changed into Thévenin's equivalent circuits of voltage sources in series with resistances.
2. The term $T/(2L)$ is denoted as G, which is a conductance.
3. The voltage ratio V_C/V_A is denoted as M.
4. Since $D_2 = [(V_A - V_C)/V_C]D$, we have

$$\left[1 + \frac{D_2}{D}\right] = \frac{V_A}{V_C} = \frac{1}{M} \tag{2.19}$$

$$\frac{D^2 T}{2L}\left[1 + \frac{D_2}{D}\right] = \frac{D^2 G}{M} \tag{2.20}$$

2.3.3 Interpretation of the Behavior Model

Once the low-frequency behavior model of the buck converter is obtained, it is important to interpret it properly. The model shown in Fig. 2.4 for continuous-mode operation is easy to understand, but the discontinuous-mode models shown in Fig. 2.6 need to be interpreted carefully.

Consider, for example, the output part of the equivalent circuit given in Fig. 2.6(b). What it actually means is that, for any given steady values of V_A, V_C, and D, the effective dc resistance between node X and node C is equal to

$$\frac{M}{D^2 G} = \frac{V_C}{V_A}\frac{1}{D^2 G} \tag{2.21}$$

where $G = T/(2L)$. Figure 2.7 illustrates graphically the output V_C versus I_o characteristic of the circuit. The operating points Q_1, Q_2, and Q_3 in Fig. 2.7 can be found through the following steps:

1. For a given $M(M_1, M_2, \text{ or } M_3)$, draw the $V_C = MV_A$ horizontal line.
2. Calculate $M/(D^2 G)$.
3. Draw a straight line with a slope equal to $-M/(D^2 G)$ through the $V_C = V_A$ point of the Y axis.
4. The interception of the two straight lines drawn in steps 1 and 3 is then the operating point (Q point) of the converter.

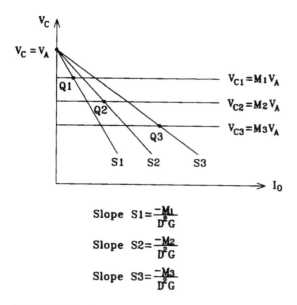

FIGURE 2.7 Output voltage V_C versus output current I_o characteristic of buck converter in discontinuous-mode operation.

When all the Q points in Fig. 2.7 are joined together, they form the V_C versus I_o characteristic of the converter.

Figure 2.8 shows a sample of the V_C versus I_o output characteristic of a buck converter for both the discontinuous and continuous modes of operation. The curved characteristic between points a and b is for the discontinuous-mode operation, and the portion between points b and c is for

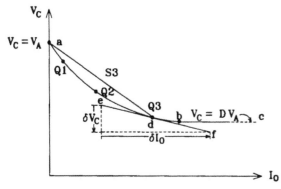

FIGURE 2.8 Complete V_C versus I_o characteristic of buck converter.

the continuous-mode operation. The following points related to Fig. 2.8 should be noted:

1. In the continuous-mode operation, it is assumed that the inductor L has zero resistance and that the effective resistance of the switch, shown as R in Fig. 2.4, is also equal to zero. V_C is therefore constantly equal to DV_A for the continuous-mode operation.

2. The small-signal output resistance R_o for discontinuous-mode operation is actually not equal to the dc resistance $M/(D^2G)$ shown in Fig. 2.6(b). An example to illustrate this phenomenon is shown in Fig. 2.8 for the operating point Q_3, where we can see that

$$\text{Small-signal } R_o = \frac{-\delta V_C}{\delta I_o} \text{ (slope of straight line } ef) \qquad (2.22)$$

$$\text{Slope } S_3 = \frac{-M}{D^2G} \text{ (slope of straight line } ad) \qquad (2.23)$$

$$\therefore R_o \neq S_3 \qquad (2.24)$$

The reason for the difference is that, while $M/(D^2G)$ is meant to be a dc resistance, R_o is the small-signal ac resistance.

A method for finding the small-signal output resistance R_o from the equivalent circuit will be studied in the next chapter.

2.4 MODELING THE BUCK-BOOST CONVERTER

In this section, the low-frequency behavior model of the buck-boost converter will be developed for both the continuous and discontinuous modes of operation.

2.4.1 The Buck-Boost Converter In Continuous-Mode Operation

The circuit of a buck-boost converter is shown in Fig. 2.9. The equivalent circuits at different stages of a switching cycle for continuous-mode operation are shown in Fig. 2.10. Since all node voltages and branch currents are finite and determinable for both stages of the switching cycle, only the switches (together with their series resistances) are qualified to be the components of the MISSCO. The circuit within the dashed lines in Fig. 2.10 is identified as the MISSCO for continuous-mode operation.

If dc excitation sources V_A, I_B, and V_C are assumed and techniques similar to those mentioned in Subsection 2.3.1 are applied, the averaged

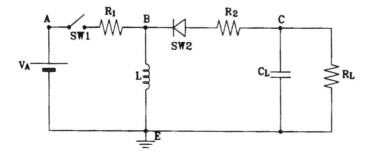

FIGURE 2.9 Buck-boost converter.

responses of i_A, i_C, and v_B, denoted as I_A, I_C, and V_B, can be found as (refer to Fig. 2.10):

Averaged value of $i_A = I_A = -DI_B$ (2.25)

Averaged value of $i_C = I_C = -(1 - D)I_B$ (2.26)

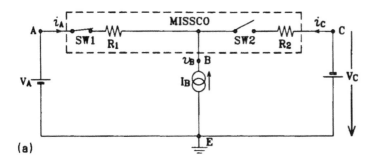

(a)

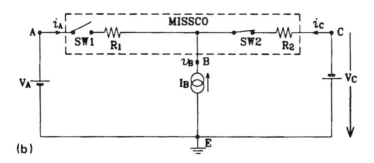

(b)

FIGURE 2.10 Determination of MISSCO parameters for buck-boost converter in continuous-mode operation: (a) equivalent circuit for $0 < t < DT$; (b) equivalent circuit for $DT < t < T$.

Averaged value of $v_B = V_B$

$$= D(V_A + R_1 I_B) + (1 - D)(-V_C + R_2 I_B)$$
$$= DV_A - (1 - D)V_C + I_B[DR_1 + (1 - D)R_2]$$
$$= DV_A - (1 - D)V_C + I_B R \qquad (2.27)$$

where $R = DR_1 + (1 - D)R_2$.

Figure 2.11(a) shows the equivalent circuit of the MISSCO, as derived from Eqs. (2.25–2.27), and the complete behavior model of the converter. From Eq. (2.26) and Fig. 2.11(a), we have

$$I_A(s) = -DI_B(s)$$
$$= D\frac{1}{(1 - D)} I_C(s) \qquad (2.28)$$

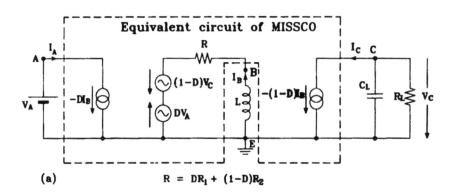

(a) $R = DR_1 + (1-D)R_2$

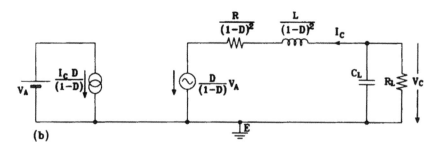

(b)

FIGURE 2.11 Low-frequency behavior models of buck-boost converter in continuous-mode operation: (a) equivalent circuit of MISSCO and complete behavior model; (b) alternative behavior model.

$$I_B(s) = [-DV_A(s) + (1 - D)V_C(s)]\frac{1}{sL + R} \qquad (2.29)$$

$$I_C(s) = -(1 - D)I_B(s)$$

$$= [(1 - D)DV_A(s) - (1 - D)^2V_C(s)]\frac{1}{sL + R}$$

$$= \left[\frac{D}{(1 - D)}V_A(s) - V_C(s)\right]\frac{1}{\dfrac{sL + R}{(1 - D)^2}} \qquad (2.30)$$

Based on Eqs. (2.28) and (2.30), an alternative behavior model of the buck-boost converter in continuous-mode operation can be found, as shown in Fig. 2.11(b).

2.4.2 The Buck-Boost Converter in Discontinuous-Mode Operation

Assume that the buck-boost converter circuit shown in Fig. 2.9 is now operating in the discontinuous-mode operation. The equivalent circuits of the converter at different stages of a switching cycle are found as shown in Fig. 2.12. From these equivalent circuits, it is seen that if the inductor L is assumed to be an ideal current souce (i_L), it will tend to produce an infinite voltage at node B during the time $(D + D_2)T < t < T$. Therefore, L is qualified to be a component of the MISSCO. The circuit within the dashed lines is the MISSCO. The dc voltage sources V_A and V_C in Fig. 2.12 are assumed to be excitations for evaluating the currents i_A and i_C.

From Fig. 2.12, we have (assuming that $R_1 = R_2 = 0$)

1. For $0 < t < DT$; $SW1$ is closed and $SW2$ is opened:

$$i_A = V_A\frac{t}{L} \qquad (2.31)$$

$$i_C = 0 \qquad (2.32)$$

2. For $DT < t < (D + D_2)T$; $SW1$ is opened and $SW2$ is closed:

$$i_A = 0 \qquad (2.33)$$

$$i_C = V_A\frac{DT}{L} - V_C\frac{1}{L}(t - DT) \qquad (2.34)$$

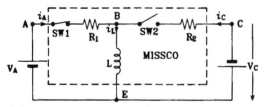

(a) Equivalent circuit for $0 < t < DT$.

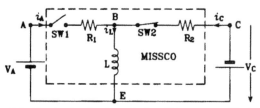

(b) Equivalent circuit for $DT < t < (D+D_2)T$.

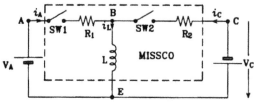

(c) Equivalent circuit for $(D+D_2)T < t < T$.

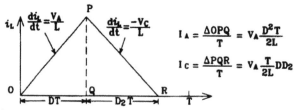

(d) Current waveform of i_L

FIGURE 2.12 Equivalent circuits of buck-boost converter in discontinuous-mode operation.

3. For $(D + D_2)T < t < T$; both $SW1$ and $SW2$ are opened:

$$i_A = 0 \tag{2.35}$$
$$i_C = 0 \tag{2.36}$$

It can be found from Eqs. (2.31–2.36) or from Fig. 2.12(d) that the averaged i_A and i_C over a complete switching cycle, denoted as I_A and I_C, are given by

$$\text{Averaged value of } i_A = I_A = V_A \frac{D^2T}{2L} \tag{2.37}$$

$$\text{Averaged value of } i_C = I_C = V_A \frac{T}{2L} DD_2 \tag{2.38}$$

where D_2 can be found by inspecting the inductor current waveform shown in Fig. 2.12(d):

$$\frac{V_A}{L} DT = \frac{V_C}{L} D_2T \tag{2.39}$$

$$D_2 = \frac{V_A}{V_C} D \tag{2.40}$$

Substitution of $D_2 = (V_A/V_C)D$ and $(V_C/V_A) = M$ into Eq. (2.38) gives an alternative expression for I_C:

$$I_C = V_A D^2 \frac{T}{2L} \frac{1}{M} \tag{2.41}$$

The low-frequency behavior model, as derived from Eqs. (2.37), (2.38), and (2.41), is shown in Fig. 2.13. It should be noted that $D^2T/(2L)$ is a nonlinear dc conductance and that $V_A[T/(2L)]DD_2$ is a nonlinear dc current source.

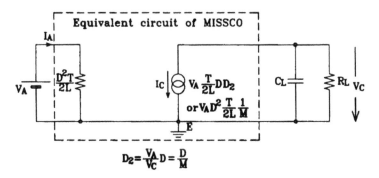

FIGURE 2.13 Low-frequency behavior model of buck-boost converter for discontinuous-mode operation.

2.5 MODELING THE BOOST CONVERTER

This section develops the behavior model of the boost converter for both the continuous and discontinuous modes of operation.

2.5.1 The Boost Converter in Continuous-Mode Operation

Figure 2.14 shows the circuit of a boost converter. The equivalent circuits at different stages of a switching cycle for continuous-mode operation are shown in Fig. 2.15. The circuit within the dashed lines is identified as the MISSCO because all node voltages and branch currents are finite and determinable for both stages of the switching cycle.

If the dc excitation sources I_A and V_C are assumed, the averaged responses in i_C and v_B, denoted as I_C and V_B, can be found as

$$\text{Averaged value of } i_C = I_C = -(1 - D)I_A \qquad (2.42)$$
$$\text{Averaged value of } v_B = V_B = DI_AR_1 + (1 - D)(I_AR_2 + V_C) \qquad (2.43)$$
$$= I_A[DR_1 + (1 - D)R_2] + (1 - D)V_C \qquad (2.44)$$
$$= I_AR + (1 - D)V_C \qquad (2.45)$$

where $R = DR_1 + (1 - D)R_2$.

Figure 2.16 shows the low-frequency behavior model of the complete converter, as derived from Eqs. (2.42) and (2.45). From this model, it can be found that

$$I_A(s) = \frac{V_A(s) - (1 - D)V_C(s)}{sL + R} \qquad (2.46)$$

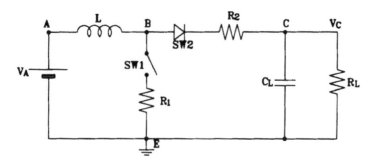

FIGURE 2.14 Boost converter.

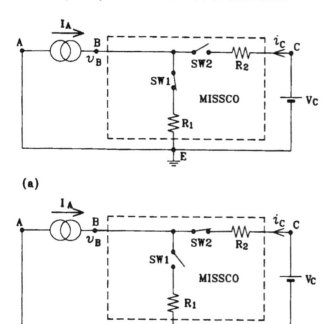

(a)

(b)

FIGURE 2.15 Determination of MISSCO parameters for boost converter in continuous-mode operation: (a) equivalent circuit for $0 < t < DT$; (b) equivalent circuit for $DT < t < T$.

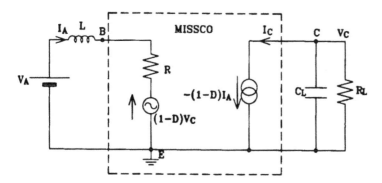

FIGURE 2.16 Low-frequency behavior model of boost converter in continuous-mode operation.

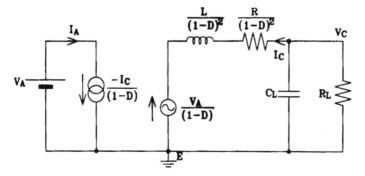

FIGURE 2.17 Alternative form of low-frequency behavior model of boost converter in continuous-mode operation.

$$I_C(s) = -(1 - D)I_A(s) \tag{2.47}$$

$$= -(1 - D)\frac{V_A(s) - (1 - D)V_C(s)}{sL + R} \tag{2.48}$$

$$= (1 - D)^2 \frac{V_C(s) - \dfrac{V_A(s)}{(1 - D)}}{sL + R} \tag{2.49}$$

$$I_C(s) = \frac{V_C(s) - \dfrac{V_A(s)}{(1 - D)}}{s\dfrac{L}{(1 - D)^2} + \dfrac{R}{(1 - D)^2}} \tag{2.50}$$

Starting from Eqs. (2.42) and (2.50), an alternative form of the low-frequency behavior model of the boost converter in continuous-mode operation can also be found, as shown in Fig. 2.17.

2.5.2 The Boost Converter in Discontinuous-Mode Operation

Assume that the boost converter circuit shown in Fig. 2.14 is now operating in the discontinuous-mode operation. The equivalent circuits of the converter at different stages of a switching cycle are shown in Fig. 2.18. From these equivalent circuits, it is found that if the inductor L is assumed to be an ideal current source (i_L), it will tend to produce an infinite voltage at node B during the time $(D + D_2)T < t < T$. Therefore, L is qualified to be a component of the MISSCO. The circuit within the dashed lines is the MISSCO. The dc voltages V_A and V_C in Fig. 2.18 are assumed to be excitations for evaluating the current responses i_A and i_C.

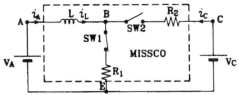

(a) Equivalent circuit for $0 < t < DT$.

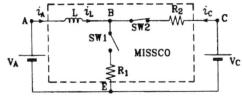

(b) Equivalent circuit for $DT < t < (D+D_2)T$.

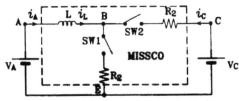

(c) Equivalent circuit for $(D+D_2)T < t < T$.

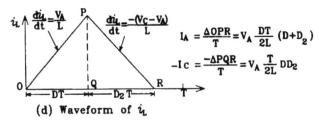

(d) Waveform of i_L

FIGURE 2.18 Equivalent circuits of boost converter in discontinuous-mode operation.

From Fig. 2.18, we have (assuming $R_1 = R_2 = 0$):

1. For $0 < t < DT$; $SW1$ is closed and $SW2$ is opened:

$$i_A = V_A \frac{t}{L} \tag{2.51}$$

$$i_C = 0 \tag{2.52}$$

2. For $DT < t < (D + D_2)T$; SW1 is opened and SW2 is closed:

$$i_A = V_A \frac{DT}{L} - (V_C - V_A)\frac{1}{L}(t - DT) \tag{2.53}$$

$$i_C = -i_A \tag{2.54}$$

$$= -V_A \frac{DT}{L} + (V_C - V_A)\frac{1}{L}(t - DT) \tag{2.55}$$

3. For $(D + D_2)T < t < T$; both SW1 and SW2 are opened:

$$i_A = 0 \tag{2.56}$$
$$i_C = 0 \tag{2.57}$$

From Eqs. (2.51–2.57) or from Fig. 2.18(d), the averaged values of i_A and i_C over a complete switching cycle, denoted as I_A and I_C, are found as

Averaged value of $i_A = I_A = V_A \frac{T}{2L} D(D + D_2) \tag{2.58}$

Averaged value of $i_C = I_C = -V_A \frac{T}{2L} DD_2 \tag{2.59}$

where D_2 can be found by referring to the inductor current waveform shown in Fig. 2.18(d):

$$\frac{V_A}{L} DT = \frac{V_C - V_A}{L} D_2 T \tag{2.60}$$

$$D_2 = \frac{V_A}{V_C - V_A} D \tag{2.61}$$

Substitution of $D_2 = DV_A/(V_C - V_A)$ and $V_C = MV_A$ into Eq. (2.59) gives an alternative expression for I_C:

$$I_C = -V_A D^2 \frac{T}{2L} \frac{1}{M - 1} \tag{2.62}$$

The low-frequency behavior model as derived from Eqs. (2.58), (2.59), and (2.62) is shown in Fig. 2.19. Note the similarities and differences between the behavior model of a boost converter in discontinuous-mode operation (Fig. 2.19) and that of a buck-boost converter also in discontinuous-mode operation (Fig. 2.13).

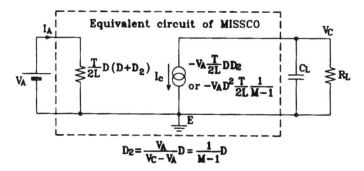

$$D_2 = \frac{V_A}{V_C - V_A} D = \frac{1}{M-1} D$$

FIGURE 2.19 Low-frequency behavior model of boost converter in discontinuous-mode operation.

2.6 MODELING ELECTRONIC TRANSFORMERS AND THE ĆUK CONVERTER

The basic circuit of an electronic transformer is shown in Fig. 2.20. The same circuit may also be interpreted as an equivalent circuit of an electronic dc-to-dc transformer, provided that the polarities of the input and output are properly observed. Furthermore, the very same circuit may be interpreted as an equivalent circuit of the Ćuk converter, too, provided that the constraint on the input polarity is also observed and that the converter operates in the continuous-mode operation.

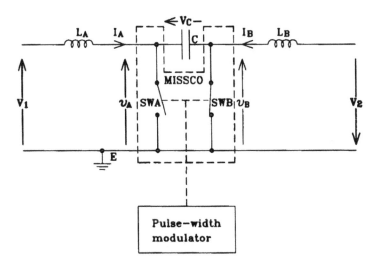

FIGURE 2.20 Basic circuit of electronic transformer.

By applying the method introduced in Section 2.2, the MISSCO of the electronic transformer can be identified as the circuit shown within the dashed lines in Fig. 2.20. The equivalent circuits of the MISSCO at different stages of a switching cycle, given in Fig. 2.21, are used to determine the MISSCO parameters.

The current i_C and voltages v_A and v_B in Fig. 2.21 can be found through the following steps:

1. For $0 < T < DT$; SWA is closed and SWB is opened:

$$i_C = -I_B \tag{2.63}$$
$$v_A = (I_A + I_B)R_A \tag{2.64}$$
$$v_B = (I_A + I_B)R_A - V_C \tag{2.65}$$

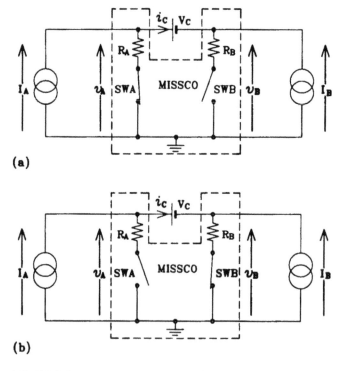

(a)

(b)

FIGURE 2.21 Determination of MISSCO parameters for electronic transformer: (a) equivalent circuit for $0 < t < DT$; (b) equivalent circuit for $DT < t < T$. R_A, R_B = series resistances of SWA and SWB.

2. For $DT < t < T$; SWA is opened and SWB is closed:

$$i_C = I_A \tag{2.66}$$
$$v_A = (I_A + I_B)R_B + V_C \tag{2.67}$$
$$v_B = (I_A + I_B)R_B \tag{2.68}$$

It is found from Eqs. (2.63–2.68) that the averaged values of i_C, v_A, and v_B, denoted as I_C, V_A, and V_B, are given by

$$I_C = (1 - D)I_A - DI_B \tag{2.69}$$
$$\begin{aligned} V_A &= D(I_A + I_B)R_A + (1 - D)(I_A + I_B)R_B + (1 - D)V_C \\ &= (I_A + I_B)[DR_A + (1 - D)R_B] + (1 - D)V_C \\ &= (I_A + I_B)R + (1 - D)V_C \end{aligned} \tag{2.70}$$
$$\begin{aligned} V_B &= D(I_A + I_B)R_A + (1 - D)(I_A + I_B)R_B - DV_C \\ &= (I_A + I_B)R - DV_C \end{aligned} \tag{2.71}$$

where $R = DR_A + (1 - D)R_B$.

An equivalent circuit that satisfies Eqs. (2.69–2.71) is shown in Fig. 2.22. When the equivalent circuit of the MISSCO shown in Fig. 2.22 is

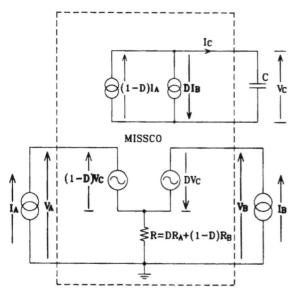

FIGURE 2.22 Low-frequency equivalent circuit of MISSCO.

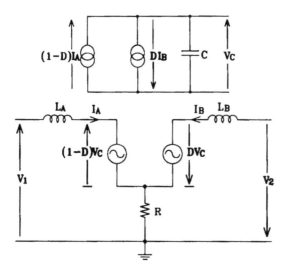

FIGURE 2.23 Low-frequency behavior model of electronic transformer (and Ćuk converter in continuous-mode operation).

substituted into the original dc-to-dc transformer in Fig. 2.20, the complete low-frequency behavior model of the electronic transformer (or that of the Ćuk converter in continuous-mode operation) is obtained, as shown in Fig. 2.23.

2.7 SUMMARY AND FURTHER REMARKS

In this chapter, a simple and effective method for modeling the low-frequency behaviors of switches in square-wave dc-to-dc converters has been introduced. Simplification of the switching circuit to a minimum configuration (MISSCO) before the modeling process has proved to be an efficient way of simplifying the modeling process. The low-frequency behavior models for various converter circuits and electronic transformers have also been developed.

Maximum care should be taken, however, in interpreting the behavior models derived. Since the MISSCO parameters are based on the averaged dc quantities, all the nonlinear parameters in the model should also be understood as large-signal dc quantities only. That is,

1. Voltages are dc voltages.
2. Currents are dc currents.

3. Conductances are ratios of

$$\frac{\text{dc current}}{\text{dc voltage}}$$

4. Resistances are ratios of

$$\frac{\text{dc voltage}}{\text{dc current}}$$

The linear circuit elements in the model are, of course, valid for both small-signal and large-signal conditions.

It should also be noted that the low-frequency behavior models of power converters are valid only for signal frequencies up to less than half of the switching frequency.

The behavior models derived in this chapter will be used for analysis and computer simulation of the low-frequency behaviors of converters in subsequent chapters.

EXERCISES

1. For what purposes are low-frequency behavior models of power converters developed?
2. What is the range of frequency in which the low-frequency behavior model of a converter is valid?
3. What is a MISSCO? Why is the MISSCO for discontinuous-mode operation different from that for continuous-mode operation?
4. Explain why the small-signal output resistance R_o shown in Fig. 2.6 is not equal to $M/(D^2G)$.
5. Describe how the MISSCO of the electronic transformer shown in Fig. 2.20 can be found.
6. Assuming that L_A and L_B are now coupled together with a coupling coefficient of 0.9 (to reduce the current ripples), develop a low-frequency behavior model for the new electronic transformer.

3
Analysis of Square-Wave Power Converters

This chapter is concerned with the analysis of square-wave power converters based on the low-frequency behavior models derived in Chapter 2. The objective is to develop a good understanding of the converter characteristics for various operating conditions. Knowledge of such characteristics is essential for the design of closed-loop regulators. The main aspects to be considered include:

1. The steady-state output voltage
2. The small-signal equivalent circuit
3. The transfer function between a small change in the dc input voltage and the resultant change in the dc output voltage
4. The transfer function between a small change in the duty cycle and the resultant change in the dc output voltage

Readers familiar with the above-mentioned characteristics of buck, buck-boost, boost, and Ćuk converters, and electronic transformers, may proceed to Chapter 4.

3.1 APPROXIMATIONS, ASSUMPTIONS, AND NOTATIONS

It will be found during the following analysis that some approximations and assumptions have been made in order to simplify the analytical work. This is necessary because of the limited capability of manual analysis. However, when we carry out computer-aided analysis and design in later chapters, we shall use models that are as complete as possible in order to improve their accuracy. Most of the converter models used in this chapter are derived from Chapter 2. They have, however, been modified to make the analysis more appropriate and convenient. Such modifications include the following:

1. Instead of V_A, the input supply voltage is now denoted as V_i.
2. Instead of V_C, the output voltage is denoted as V_o.
3. The voltage ratio V_o/V_i is denoted as M.
4. The averaged current in the inductor L is denoted as I_L.
5. The term $T/(2L)$ is denoted as G, which is a conductance.
6. The resistance R, which is defined as $R = D$ [series resistance of switch $SW1$ (or SWA)] $+ (1 - D)$ [series resistance of switch $SW2$ (or SWB)], is assumed to be constant.
7. The polarities of voltages and currents may be differently defined.

3.2 THE BUCK CONVERTER IN CONTINUOUS-MODE OPERATION

The circuit of a buck converter, together with its low-frequency behavior model for continuous-mode operation, is shown in Fig. 3.1. The equivalent circuit in Fig. 3.1(b) is obtained from Fig. 2.4, with the modifications as

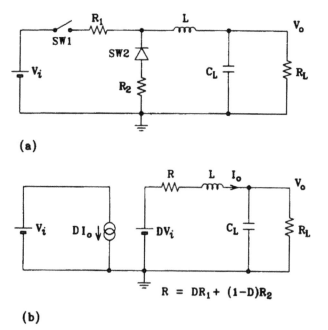

(a)

(b)

FIGURE 3.1 (a) Buck converter and (b) continuous-mode behavior model.

stated in Section 3.1. From the equivalent circuit, it is found that in the steady state we have the output voltage V_o given by

$$V_o = DV_i \frac{R_L}{R_L + R} \qquad (3.1)$$

If $R_L \gg R$, we have

$$V_o = DV_i \qquad (3.2)$$

To find the $\delta V_o(s)/\delta V_i(s)$ transfer function for a constant duty cycle D, we now assume a perturbation δV_i in V_i, as shown in Fig. 3.2(a). The corresponding changes in I_o and V_o are denoted as δI_o and δV_o, respectively. If only the small-signal components are considered, we can have a small-signal equivalent circuit, as shown in Fig. 3.2(b). From this equivalent circuit, we have

$$\delta V_o(s) = D\delta V_i(s) \frac{\dfrac{1}{(1/R_L) + sC_L}}{sL + R + \dfrac{1}{(1/R_L) + sC_L}} \qquad (3.3)$$

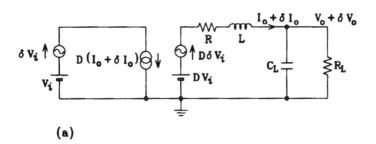

(a)

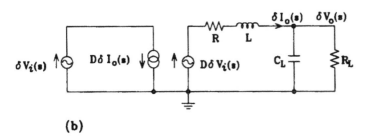

(b)

FIGURE 3.2 (a) Buck converter with perturbation δV_i and (b) continuous-mode, small-signal equivalent circuit.

$$= D\delta V_i(s)\,\frac{R_L/(1 + sC_LR_L)}{sL + R + [R_L/(1 + sC_LR_L)]} \tag{3.4}$$

$$= D\delta V_i(s)\,\frac{1}{s^2LC_L + s\left(\dfrac{L}{R_L} + C_LR\right) + 1 + \dfrac{R}{R_L}} \tag{3.5}$$

Therefore, the $\delta V_o(s)/\delta V_i(s)$ transfer function is given by

$$\frac{\delta V_o(s)}{\delta V_i(s)} = D\,\frac{1}{s^2LC_L + s\left(\dfrac{L}{R_L} + C_LR\right) + \dfrac{R}{R_L} + 1} \tag{3.6}$$

To continue the analysis, let us now determine the $\delta V_o(s)/\delta D(s)$ transfer function. We first assume a perturbation δD as shown in Fig. 3.3(a). Correspondingly, I_o, DI_o, and V_o are changed to $(I_o + \delta I_o)$, $(D + \delta D)(I_o + \delta I_o)$, and $(V_o + \delta V_o)$. Considering only small-signal changes, we have the small-signal equivalent circuit as shown in Fig. 3.3(b). From this equivalent circuit, $\delta V_o(s)$ can be determined as

$$\delta V_o(s) = \delta D(s)V_i\,\frac{\dfrac{1}{(1/R_L) + sC_L}}{sL + R + \dfrac{1}{(1/R_L) + sC_L}} \tag{3.7}$$

$$= \delta D(s)V_i\,\frac{1}{s^2LC_L + s[(L/R_L) + C_LR] + 1 + (R/R_L)} \tag{3.8}$$

Therefore, the transfer function $\delta V_o(s)/\delta D(s)$ is given by

$$\frac{\delta V_o(s)}{\delta D(s)} = V_i\,\frac{1}{s^2LC_L + s[(L/R_L) + C_LR] + (R/R_L) + 1} \tag{3.9}$$

Equations (3.6) and (3.9) show that both the $\delta V_o(s)/\delta V_i(s)$ and $\delta V_o(s)/\delta D(s)$ transfer functions have two pole frequencies located at

$$\omega_p = \frac{-\left(\dfrac{L}{R_L} + C_LR\right) \pm \left[\left(\dfrac{L}{R_L} + C_LR\right)^2 - 4LC_L\left(\dfrac{R}{R_L} + 1\right)\right]^{1/2}}{2LC_L} \tag{3.10}$$

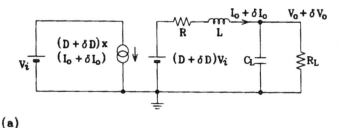

(a)

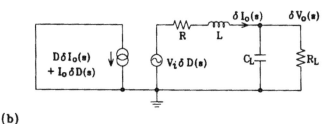

(b)

FIGURE 3.3 (a) Buck converter with perturbation δD and (b) continuous-mode, small-signal equivalent circuit.

If $R \ll R_L$ and $R \ll L/(C_L R_L)$, the two pole frequencies will be approximately equal to

$$\omega_p = \frac{-\dfrac{L}{R_L} \pm [(L/R_L)^2 - 4LC_L]^{1/2}}{2LC_L} \tag{3.11}$$

$$= -\frac{1}{2C_L R_L} \pm \left\{ \left[\frac{1}{2C_L R_L}\right]^2 - \frac{1}{LC_L} \right\}^{1/2} \tag{3.12}$$

for both the $\delta V_o(s)/\delta V_i(s)$ and $\delta V_o(s)/\delta D(s)$ transfer functions.

For the purpose of illustration, the typical shapes of the theoretical $\delta V_o(s)/\delta V_i(s)$ and $\delta V_o(s)/\delta D(s)$ characteristics, plotted according to Eqs. (3.6) and (3.9), are shown in Fig. 3.4 by the solid lines. However, when experimental measurements are carried out on practical circuits, the resultant characteristics will appear more like those shown by the dotted lines. The different shape of the measured characteristics is due to the existence of the effective series resistance (ESR) of the output filtering capacitor C_L. We have not included this effective series resistance in our manual analysis because it makes the analysis too complicated. It will be shown in Chapter 7, however, that such series resistance (and other stray

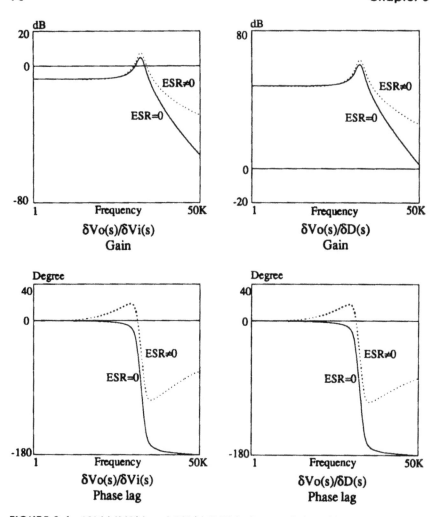

FIGURE 3.4 $\delta V_o(s)/\delta V_i(s)$ and $\delta V_o(s)/\delta D(s)$ characteristics of buck converter in continuous-mode operation. ESR = effective series resistance of C_L.

components as well) can be easily included in computer simulations in order to obtain the true $\delta V_o(s)/\delta V_i(s)$ and $\delta V_o(s)/\delta D(s)$ characteristics of converters.

3.3 THE BUCK CONVERTER IN DISCONTINUOUS-MODE OPERATION

The circuit of the buck converter, together with the various forms of the low-frequency behavior model for discontinuous-mode operation, is shown in Fig. 3.5. These models are derived from the MISSCO model given in Fig. 2.6(a), with the modifications stated in Section 3.1. Such models are included here to illustrate the fact that, although they may appear very different, they actually mean the same thing.

From the model shown in Fig. 3.5(d) or (e), we can find the dc voltage ratio V_o/V_i, denoted as M, for the circuit:

$$V_o = V_i \frac{R_L}{R_L + \dfrac{M}{D^2 G}} \tag{3.13}$$

$$\frac{V_o}{V_i} = M = \frac{R_L}{R_L + M/(D^2 G)} = \frac{D^2 G R_L}{D^2 G R_L + M} \tag{3.14}$$

$$M^2 + M D^2 G R_L - D^2 G R_L = 0 \tag{3.15}$$

$$M = (1/2)\{-D^2 G R_L + [(D^2 G R_L)^2 + 4 D^2 G R_L]^{1/2}\} \tag{3.16}$$

$$M = \frac{D^2 G R_L}{2} \left\{ \left[1 + \frac{4}{D^2 G R_L} \right]^{1/2} - 1 \right\} \tag{3.17}$$

The output voltage V_o is therefore given by

$$V_o = M V_i = V_i \frac{D^2 G R_L}{2} \left\{ \left[1 + \frac{4}{D^2 G R_L} \right]^{1/2} - 1 \right\} \tag{3.18}$$

Or, if M is fixed for a given design, the required duty cycle D can be found from Eq. (3.15) as

$$D = \frac{M}{[(1 - M) G R_L]^{1/2}} \tag{3.19}$$

Now, let us determine the transfer function $\delta V_o(s)/\delta V_i(s)$ for a given duty cycle D. From Eq. (3.18), we have, in the steady state,

$$\delta V_o = M \delta V_i$$

$$\frac{\delta V_o}{\delta V_i} = M$$

However, the behavior models in Figs. 3.5(b) to (f) show that there should be a pole frequency ω_p in the transfer function, such that

$$\frac{\delta V_o(s)}{\delta V_i(s)} = M \frac{1}{1 + (s/\omega_p)} \tag{3.20}$$

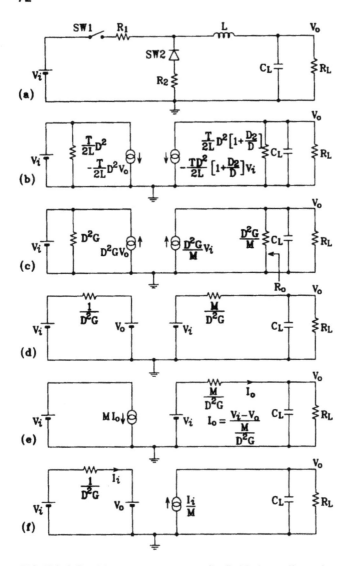

FIGURE 3.5 (a) Buck converter; (b–f) Various discontinuous-mode behavior models.

where

$$\omega_p = \frac{1}{C_L \times [\text{effective small-signal resistance across } C_L]} \tag{3.21}$$

$$= \frac{1}{C_L \dfrac{R_o R_L}{R_o + R_L}} \tag{3.22}$$

The term R_o in Eq. (3.22) is the small-signal output resistance looking into the output terminal of the converter. R_o can be found by referring to Fig. 3.6 [which is derived from Fig. 3.5(c)], where a perturbation δV_o is assumed for evaluation of the resultant change δI_o:

$$I_o + \delta I_o = \frac{-D^2 G}{M + \delta M} V_i + \frac{D^2 G}{M + \delta M} (V_o + \delta V_o) \tag{3.23}$$

where $M = V_o/V_i$, $\delta M = \delta V_o/V_i$.

$$I_o + \delta I_o = -D^2 G V_i \frac{1}{\dfrac{V_o + \delta V_o}{V_i}} + D^2 G (V_o + \delta V_o) \frac{1}{\dfrac{V_o + \delta V_o}{V_i}} \tag{3.24}$$

$$= -D^2 G V_i^2 \frac{1}{V_o \left[1 + \dfrac{\delta V_o}{V_o} \right]} + D^2 G V_i \tag{3.25}$$

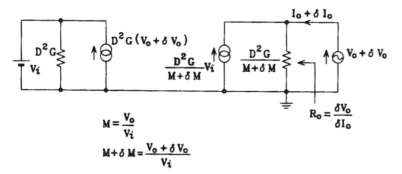

$$M = \frac{V_o}{V_i}$$

$$M + \delta M = \frac{V_o + \delta V_o}{V_i}$$

FIGURE 3.6 Determination of R_o for buck converter in discontinuous-mode operation.

$$\approx -D^2GV_i^2 \frac{1}{V_o}\left[1 - \frac{\delta V_o}{V_o}\right] + D^2GV_i \tag{3.26}$$

$$= \frac{-D^2G}{V_o}V_i^2 + D^2G\left[\frac{V_i}{V_o}\right]^2 \delta V_o + D^2GV_i \tag{3.27}$$

Considering only the changes, we have

$$\delta I_o = D^2G\left[\frac{V_i}{V_o}\right]^2 \delta V_o \tag{3.28}$$

$$\therefore R_o = \frac{\delta V_o}{\delta I_o} = \left[\frac{V_o}{V_i}\right]^2 \frac{1}{D^2G} = M^2 \frac{1}{D^2G} \tag{3.29}$$

Substituting Eq. (3.19) into Eq. (3.29) gives

$$R_o = M^2 \frac{1}{\dfrac{M^2}{(1 - M)GR_L}G} \tag{3.30}$$

$$\therefore R_o = (1 - M)R_L \tag{3.31}$$

By putting Eq. (3.31) into Eq. (3.22), we can determine the pole frequency ω_p:

$$\omega_p = \frac{1}{C_L \dfrac{(1 - M)R_LR_L}{(1 - M)R_L + R_L}} = \frac{1}{C_L \dfrac{1 - M}{2 - M}R_L} \tag{3.32}$$

When Eq. (3.32) is substituted into Eq. (3.20), we finally have the transfer function $\delta V_o(s)/\delta V_i(s)$, found as

$$\frac{\delta V_o(s)}{\delta V_i(s)} = M \frac{1}{1 + sC_L \dfrac{1 - M}{2 - M}R_L} \tag{3.33}$$

Now, let us determine the transfer function $\delta V_o/\delta D$ for a given V_i. Refer back to the model shown in Fig. 3.5(e). By introducing a perturbation δD in the equation

$$I_o = \frac{V_i - V_o}{\dfrac{M}{D^2G}}$$

we have

$$I_o + \delta I_o = \cfrac{V_i - (V_o + \delta V_o)}{\cfrac{M + \delta M}{(D + \delta D)^2 G}} \tag{3.34}$$

$$(I_o + \delta I_o)(M + \delta M) = (V_i - V_o - \delta V_o)(D + \delta D)^2 G \tag{3.35}$$

Substitution of $I_o = V_o/R_L$, $\delta I_o = \delta V_o/R_L$, $\delta M = \delta V_o/V_i$, and $V_o = MV_i$ into Eq. (3.35) gives

$$\left[\frac{V_o}{R_L} + \frac{\delta V_o}{R_L}\right]\left[M + \frac{\delta V_o}{V_i}\right] = V_i$$

$$\times \left[1 - M - \frac{\delta V_o}{V_i}\right](D^2 + 2D\,\delta D + \delta D^2)G \tag{3.36}$$

Considering only the changes and ignoring higher-order terms such as $\delta V_o^2/(R_L V_i)$, δD^2, and $(\delta V_o/V_i)2D\,\delta D$, we have

$$\frac{V_o}{R_L}\frac{\delta V_o}{V_i} + \frac{\delta V_o}{R_L}M = GV_i\left[(1 - M)2D\,\delta D - \frac{\delta V_o}{V_i}D^2\right] \tag{3.37}$$

$$\frac{\delta V_o}{\delta D} = GV_i(1 - M)2D\frac{1}{(2M/R_L) + GD^2} \tag{3.38}$$

Substitution of Eq. (3.19) into Eq. (3.38) yields

$$\frac{\delta V_o}{\delta D} = 2GV_i(1 - M)\frac{M}{[(1 - M)GR_L]^{1/2}}\frac{1}{\cfrac{2M}{R_L} + G\cfrac{M^2}{(1 - M)GR_L}}$$

$$= 2V_i\frac{(1 - M)^2}{2 - M}\left[\frac{GR_L}{1 - M}\right]^{1/2} \tag{3.39}$$

Equation (3.39) gives the dc transfer function between the input δD and the resultant δV_o. If the effect of the filtering capacitance C_L is also considered, we shall have to add the pole frequency found in Eq. (3.32) to the $\delta V_o/\delta D$ transfer function so that

$$\frac{\delta V_o(s)}{\delta D(s)} = 2V_i\frac{(1 - M)^2}{2 - M}\left[\frac{GR_L}{1 - M}\right]^{1/2}\frac{1}{1 + sC_L\cfrac{1 - M}{2 - M}R_L} \tag{3.40}$$

For the purpose of illustration, the typical shapes of the theoretical $\delta V_o(s)/\delta V_i(s)$ and $\delta V_o(s)/\delta D(s)$ characteristics of the buck converter in discontinuous-mode operation, plotted according to Eqs. (3.33) and (3.40), are shown in Fig. 3.7 by the solid lines. However, SPICE simulations that include the effective series resistance (ESR) of C_L give the characteristics shown in Fig. 3.7 by the dotted lines. It is found that when the effective series resistance is included ESR in the computer simulation, the results are much more accurate when compared with the theoretical plot according

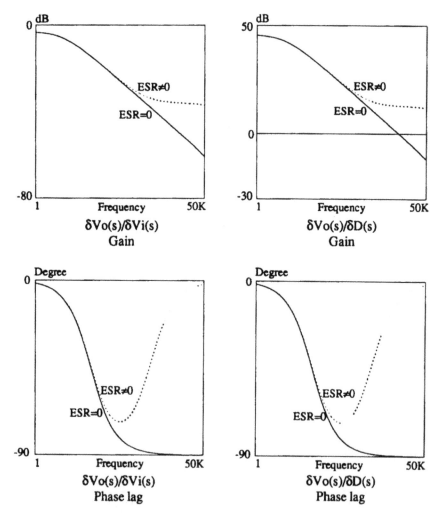

FIGURE 3.7 $\delta V_o(s)/\delta V_i(s)$ and $\delta V_o(s)/\delta D(s)$ characteristics of buck converter in discontinuous-mode operation. ESR = effective series resistance of C_L.

to Eqs. (3.33) and (3.40) (where, for the sake of simplicity in the analysis, the effective series resistance of C_L has been assumed to be zero).

3.4 THE BUCK-BOOST CONVERTER IN CONTINUOUS-MODE OPERATION

The circuit of a buck-boost converter and two forms of the continuous-mode behavior model are shown in Fig. 3.8. The behavior models are obtained from Fig. 2.11, with the modifications stated in Section 3.1.

From the model shown in Fig. 3.8(c), it is found that the dc output voltage V_o is equal to

$$V_o = V_i \frac{D}{(1-D)} \frac{R_L}{R_L + \frac{R}{(1-D)^2}} \tag{3.41}$$

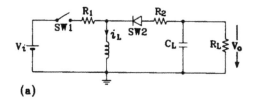

(a)

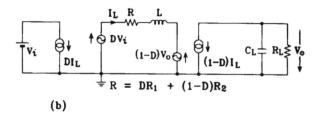

(b)

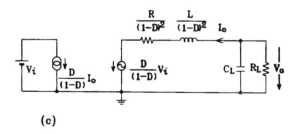

(c)

FIGURE 3.8 (a) Buck-boost converter and (b, c) continuous-mode behavior models.

and

$$V_o = \frac{D}{(1 - D)} V_i \quad \text{if} \quad R_L \gg \frac{R}{(1 - D)^2} \tag{3.42}$$

Based on the model given in Fig. 3.8(b), let us now determine the $\delta V_o(s)/\delta V_i(s)$ transfer function by assuming a perturbation δV_i as shown in Fig. 3.9(a). As a result, a small-signal equivalent circuit is found as shown in Fig. 3.9(b). From this equivalent circuit, we have

$$\delta V_o(s) = (1 - D)\, \delta I_L(s)\, \frac{1}{(1/R_L) + sC_L} \tag{3.43}$$

$$\delta I_L(s) = \left[\frac{D\, \delta V_i(s) - (1 - D)\, \delta V_o(s)}{sL + R} \right] \tag{3.44}$$

Substitution of Eq. (3.44) into Eq. (3.43) gives

$$\delta V_o(s) = (1 - D) \left[\frac{D\, \delta V_i(s) - (1 - D)\, \delta V_o(s)}{sL + R} \right] \frac{R_L}{1 + sC_L R_L} \tag{3.45}$$

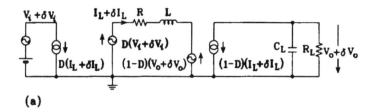

(a)

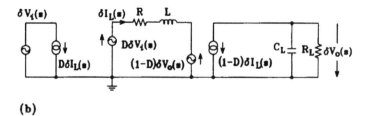

(b)

FIGURE 3.9 (a) Buck-boost converter with perturbation δV, and (b) continuous-mode small-signal equivalent circuit.

$$\delta V_o(s)(sL + R)(1 + sC_LR_L) = (1 - D)D\,\delta V_i(s)R_L$$

$$- (1 - D)^2\,\delta V_o(s)\,R_L \tag{3.46}$$

$$\delta V_o(s) = \frac{D(1 - D)\,\delta V_i(s)\,R_L}{s^2LC_LR_L + s(L + C_LR_LR) + R + (1 - D)^2R_L} \tag{3.47}$$

$$\frac{\delta V_o(s)}{\delta V_i(s)} = \frac{D}{(1 - D)}$$

$$\times \frac{1}{s^2\dfrac{LC_L}{(1 - D)^2} + s\left[\dfrac{L}{(1 - D)^2R_L} + \dfrac{C_LR}{(1 - D)^2}\right] + \dfrac{R}{(1 - D)^2R_L} + 1}$$

$$\tag{3.48}$$

The transfer function obtained in Eq. (3.48) contains two pole frequencies.

To continue the analysis, we now determine the transfer function between the control input δD and the output δV_o. First, we assume that there is a perturbation in the duty cycle D of the equivalent circuit, as shown in Fig. 3.10(a), causing disturbances in I_L and V_o. The resultant incremental changes in the circuit parameters of DI_L, DV_i, $(1 - D)V_o$, $(1 - D)I_L$, and V_o are then found as shown in Table 3.1.

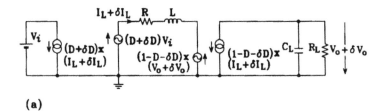

(a)

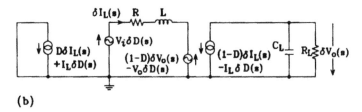

(b)

FIGURE 3.10 (a) Buck-boost converter with perturbation δD and (b) continuous-mode, small-signal equivalent circuit.

Table 3.1 Summary of Incremental Changes Due to δD

Circuit parameter	Parameter with perturbation	Incremental changes[a]
DI_L	$(D + \delta D)(I_L + \delta I_L)$	$D\,\delta I_L + I_L\,\delta D$
DV_i	$(D + \delta D)V_i$	$V_i\,\delta D$
$(1 - D)V_o$	$(1 - D - \delta D)(V_o + \delta V_o)$	$(1 - D)\,\delta V_o - V_o\,\delta D$
$(1 - D)I_L$	$(1 - D - \delta D)(I_L + \delta I_L)$	$(1 - D)\,\delta I_L - I_L\,\delta D$
V_o	$V_o + \delta V_o$	δV_o

[a]Higher-order terms such as $\delta D\,\delta I_L$ and $\delta D\,\delta V_o$ are neglected.

By considering only the incremental changes, a small-signal equivalent circuit of the converter is found, as shown in Fig. 3.10(b). From this circuit, we have

$$\delta V_o(s) = [(1 - D)\,\delta I_L(s) - I_L\,\delta D(s)]\frac{1}{(1/R_L) + sC_L} \tag{3.49}$$

$$\delta I_L(s) = \left[\frac{V_i\,\delta D(s) - (1 - D)\,\delta V_o(s) + V_o\,\delta D(s)}{R + sL}\right] \tag{3.50}$$

Substitution of Eq. (3.50) into Eq. (3.49) gives

$$\delta V_o(s) = \left\{(1 - D)\left[\frac{V_i\,\delta D(s) - (1 - D)\,\delta V_o(s) + V_o\,\delta D(s)}{R + sL}\right]\right.$$

$$\left. - I_L\,\delta D(s)\right\}\frac{R_L}{1 + sC_LR_L} \tag{3.51}$$

$$\delta V_o(s)(R + sL)(1 + sC_LR_L) = [(1 - D)V_i\,\delta D(s) - (1 - D)^2\,\delta V_o(s)$$

$$+ (1 - D)V_o\,\delta D(s) - I_L(R + sL)\,\delta D(s)]R_L \tag{3.52}$$

$$\delta V_o(s) = \delta D(s)\frac{[(1 - D)V_i + (1 - D)V_o - I_L(R + sL)]R_L}{(R + sL)(1 + sC_LR_L) + (1 - D)^2R_L} \tag{3.53}$$

$$\frac{\delta V_o(s)}{\delta D(s)} = \frac{[(1 - D)V_i + (1 - D)V_o - I_L(R + sL)]R_L}{s^2LC_LR_L + s(L + C_LR_LR) + R + (1 - D)^2R_L} \tag{3.54}$$

Therefore, we have the $\delta V_o(s)/\delta D(s)$ transfer function given by

$$\frac{\delta V_o(s)}{\delta D(s)} = \frac{1}{(1 - D)^2}$$

$$\times \frac{(1 - D)V_i + (1 - D)V_o - I_L(R + sL)}{s^2 \dfrac{LC_L}{(1 - D)^2} + s\left[\dfrac{L}{(1 - D)^2 R_L} + \dfrac{C_L R}{(1 - D)^2}\right] + \dfrac{R}{(1 - D)^2 R_L} + 1}$$

$$(3.55)$$

Since, from Eq. (3.42) and Fig. 3.8(b), we have

$$V_o = \frac{D}{(1 - D)} V_i \tag{3.56}$$

$$V_o = (1 - D)I_L R_L \tag{3.57}$$

substitution of Eq. (3.57) into Eq. (3.56) gives

$$I_L = \frac{DV_i}{(1 - D)^2 R_L} \tag{3.58}$$

Further substitution of Eqs. (3.56) and (3.58) into Eq. (3.55) gives a resultant $\delta V_o(s)/\delta D(s)$ transfer function:

$$\frac{\delta V_o(s)}{\delta D(s)} = \frac{1}{(1 - D)^2}$$

$$\times \frac{(1 - D)V_i + (1 - D)\dfrac{D}{(1 - D)} V_i - \dfrac{DV_i}{(1 - D)^2 R_L}(R + sL)}{s^2 \dfrac{LC_L}{(1 - D)^2} + s\left[\dfrac{L}{(1 - D)^2 R_L} + \dfrac{C_L R}{(1 - D)^2}\right] + \dfrac{R}{(1 - D)^2 R_L} + 1}$$

$$(3.59)$$

$$\frac{\delta V_o(s)}{\delta D(s)} = \frac{V_i}{(1 - D)^2}$$

$$\times \frac{-s\dfrac{DL}{(1 - D)^2 R_L} - \dfrac{DR}{(1 - D)^2 R_L} + 1}{s^2 \dfrac{LC_L}{(1 - D)^2} + s\left[\dfrac{L}{(1 - D)^2 R_L} + \dfrac{C_L R}{(1 - D)^2}\right] + \dfrac{R}{(1 - D)^2 R_L} + 1}$$

$$(3.60)$$

Equations (3.48) and (3.60) indicate that both the $\delta V_o(s)/\delta V_i(s)$ and $\delta V_o(s)/\delta D(s)$ transfer functions have two pole frequencies. These pole frequencies are equal to the solutions of the equation

$$s^2 \frac{LC_L}{(1 - D)^2} + s \left[\frac{L}{(1 - D)^2 R_L} + \frac{C_L R}{(1 - D)^2} \right] + \frac{R}{(1 - D)^2 R_L} + 1 = 0$$

$$(3.61)$$

Note that the LHS (left-hand side) of Eq. (3.61) is actually the denominator of the two transfer functions $\delta V_o(s)/\delta V_i(s)$ and $\delta V_o(s)/\delta D(s)$, as given in Eqs. (3.48) and (3.60).

If $R \ll (1 - D)^2 R_L$ and $R \ll L/(C_L R_L)$, then Eq. (3.61) can be simplified to

$$s^2 \frac{LC_L}{(1 - D)^2} + s \left[\frac{L}{(1 - D)^2 R_L} \right] + 1 = 0 \qquad (3.62)$$

We then have the pole frequencies approximately equal to

$$\omega_p = \frac{-\dfrac{L}{(1 - D)^2 R_L} \pm \left\{ \left[\dfrac{L}{(1 - D)^2 R_L} \right]^2 - 4 \dfrac{LC_L}{(1 - D)^2} \right\}^{1/2}}{\dfrac{2LC_L}{(1 - D)^2}} \qquad (3.63)$$

$$= -\frac{1}{2C_L R_L} \pm \left\{ \left[\frac{1}{2C_L R_L} \right]^2 - \frac{(1 - D)^2}{LC_L} \right\}^{1/2} \qquad (3.64)$$

In addition to the two pole frequencies given in Eq. (3.64), the $\delta V_o(s)/\delta D(s)$ transfer function also has a zero frequency, which is equal to the solution of the equation

$$-s \frac{DL}{(1 - D)^2 R_L} - \frac{DR}{(1 - D)^2 R_L} + 1 = 0 \qquad (3.65)$$

Equation (3.65) is actually the numerator of the $\delta V_o(s)/\delta D(s)$ transfer function, as given in Eq. (3.60).

If $R \ll [(1 - D)^2/D]R_L$, the zero frequency ω_z is approximately equal to

$$\omega_z = \frac{(1 - D)^2 R_L}{DL} \qquad (3.66)$$

Equation (3.66) indicates that ω_z is, in fact, an RHP (right half-plane) zero. This results in a maximum phase lag in the $\delta V_o(s)/\delta D(s)$ transfer function approaching $-270°$ at high frequencies. Such a large phase shift is highly undesirable for feedback circuits.

As an example, the theoretical $\delta V_o(s)/\delta V_i(s)$ and $\delta V_o(s)/\delta D(s)$ characteristics, plotted according to Eqs. (3.48) and (3.60), are shown in Fig. 3.11 by the solid lines. The dotted lines in Fig. 3.11 indicate the changed characteristics when the filtering capacitor C_L has a nonzero effective series resistance.

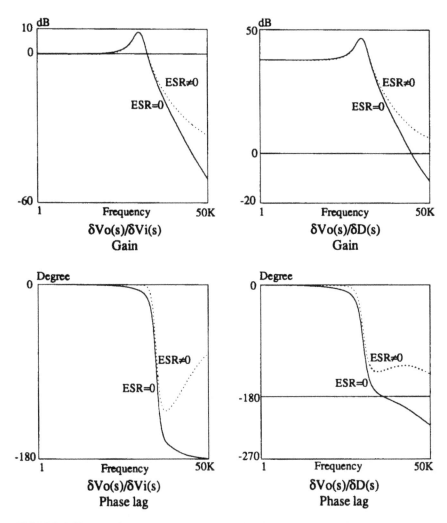

FIGURE 3.11 $\delta V_o(s)/\delta V_i(s)$ and $\delta V_o(s)/\delta D(s)$ characteristics of buck-boost converter in continuous-mode operation. ESR = effective series resistance of C_L.

3.5 THE BUCK-BOOST CONVERTER IN DISCONTINUOUS-MODE OPERATION

The circuit of a buck-boost converter and its low-frequency behavior model for the discontinuous-mode operation are shown in Fig. 3.12. The model in Fig. 3.12(b) is obtained from Fig. 2.13, with the modifications stated in Section 3.1. From Fig. 3.12(b), the voltage ratio V_o/V_i, denoted as M, can be found as follows:

$$V_o = V_i D^2 G \frac{1}{M} R_L \tag{3.67}$$

$$\frac{V_o}{V_i} = \frac{1}{M} D^2 G R_L \tag{3.68}$$

$$M = \frac{1}{M} D^2 G R_L \tag{3.69}$$

$$M = [G R_L]^{1/2} D \tag{3.70}$$

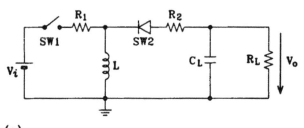

(a)

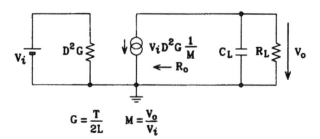

(b)

FIGURE 3.12 (a) Buck-boost converter and (b) discontinuous-mode behavior model.

The output voltage V_o is therefore given by

$$V_o = MV_i = [GR_L]^{1/2} DV_i \tag{3.71}$$

Or, if M is fixed for a given design, the required duty cycle D can be found as

$$D = \frac{M}{[GR_L]^{1/2}} \tag{3.72}$$

Let us now determine the transfer function $\delta V_o(s)/\delta V_i(s)$ for the converter. From Eq. (3.71), we have, in the steady state,

$$\delta V_o = M \, \delta V_i \tag{3.73}$$

$$\therefore \frac{\delta V_o}{\delta V_i} = M \tag{3.74}$$

However, the model in Fig. 3.12(b) indicates that, in a similar way to a buck converter in discontinuous-mode operation, there should be a pole frequency ω_p in the transfer function, so that

$$\frac{\delta V_o(s)}{\delta V_i(s)} = M \frac{1}{1 + (s/\omega_p)} \tag{3.75}$$

where

$$\omega_p = \frac{1}{C_L \times [\text{effective small-signal resistance across } C_L]} \tag{3.76}$$

$$= \frac{1}{C_L[R_o R_L/(R_o + R_L)]} \tag{3.77}$$

The resistance R_o in Eq. (3.77) is the small-signal output resistance of the converter. R_o can be found by referring to Fig. 3.13, where a perturbation δV_o is assumed for determination of the corresponding change δI_o:

$$I_o + \delta I_o = -V_i D^2 G \frac{1}{M + \delta M} \tag{3.78}$$

$$= -V_i D^2 G \frac{V_i}{V_o + \delta V_o} \tag{3.79}$$

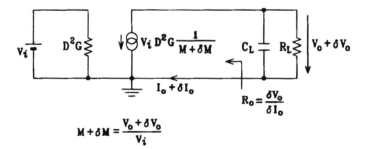

FIGURE 3.13 Determination of R_o for buck-boost converter in discontinuous-mode operation.

$$= -V_i D^2 G \frac{V_i}{V_o} \frac{1}{1 + (\delta V_o / V_o)} \qquad (3.80)$$

$$\approx -V_i D^2 G \frac{1}{M} \left[1 - \frac{\delta V_o}{V_o} \right] \qquad (3.81)$$

$$= -V_i D^2 G \frac{1}{M} + D^2 G \frac{1}{M^2} \delta V_o \qquad (3.82)$$

Considering only the changes, we have

$$\delta I_o = D^2 G \frac{1}{M^2} \delta V_o \qquad (3.83)$$

Substitution of Eq. (3.72) into Eq. (3.83) gives

$$\delta I_o = \frac{M^2}{GR_L} G \frac{1}{M^2} \delta V_o \qquad (3.84)$$

The output resistance is therefore equal to

$$R_o = \frac{\delta V_o}{\delta I_o} = R_L \qquad (3.85)$$

Substituting Eq. (3.85) into Eq. (3.77), and then Eq. (3.77) into Eq. (3.75), gives the resultant transfer function $\delta V_o(s)/\delta V_i(s)$:

$$\frac{\delta V_o(s)}{\delta V_i(s)} = M \frac{1}{1 + sC_L(R_L/2)} \qquad (3.86)$$

Note that the pole frequency of the transfer function (for a buck-boost converter in discontinuous-mode operation) is located at

$$\omega_p = \frac{2}{C_L R_L} \tag{3.87}$$

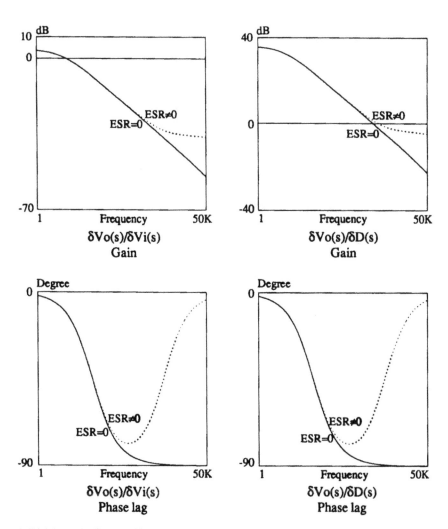

FIGURE 3.14 $\delta V_o(s)/\delta V_i(s)$ and $\delta V_o(s)/\delta D(s)$ characteristics of buck-boost converter in discontinuous-mode operation. ESR = effective series resistance of C_L.

To continue the analysis, we now come to the transfer function $\delta V_o(s)/\delta D(s)$. From Eq. (3.71), we have

$$\delta V_o = [GR_L]^{1/2} \delta DV_i \tag{3.88}$$

$$\therefore \frac{\delta V_o}{\delta D} = [GR_L]^{1/2} V_i \tag{3.89}$$

When the pole frequency is included in the transfer function, the resultant $\delta V_o(s)/\delta D(s)$ is equal to

$$\frac{\delta V_o(s)}{\delta D(s)} = V_i[GR_L]^{1/2} \frac{1}{1 + (1/2)sC_LR_L} \tag{3.90}$$

Note that the RHP zero in the $\delta V_o(s)/\delta D(s)$ transfer function of a continuous-mode operated buck-boost converter no longer exists when the converter enters into the discontinuous-mode operation. The disappearance of the RHP zero reduces the phase lag of the $\delta V_o(s)/\delta D(s)$ transfer function, which simplifies the design of the feedback circuit.

The theoretical plots of the $\delta V_o(s)/\delta V_i(s)$ and $\delta V_o(s)/\delta D(s)$ characteristics according to Eqs. (3.86) and (3.90) are given in Fig. 3.14 by the solid lines. The characteristics shown by the dotted lines are for a converter with a nonzero effective series resistance in the output filtering capacitor C_L.

3.6 THE BOOST CONVERTER IN CONTINUOUS-MODE OPERATION

The circuit of a boost converter and its continuous-mode low-frequency behavior model are shown in Fig. 3.15. The model in Fig. 3.15(b) is obtained from Fig. 2.16, with the modifications stated in Section 3.1.

From the equivalent circuit, we have, in the steady state,

$$I_L = [V_i - (1 - D)V_o] \frac{1}{R} \tag{3.91}$$

$$I_o = (1 - D)I_L \tag{3.92}$$

$$\therefore I_L = I_o \frac{1}{1 - D} = \frac{V_o}{R_L} \frac{1}{1 - D} \tag{3.93}$$

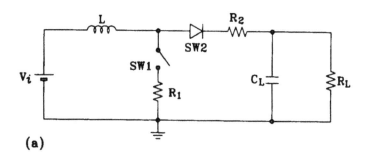

(a)

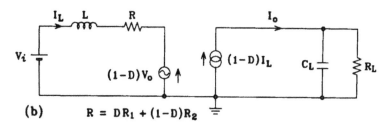

(b) $R = DR_1 + (1-D)R_2$

FIGURE 3.15   (a) Boost converter and (b) continuous-mode behavior model.

By combining Eqs. (3.91) and (3.93), we get

$$V_o \left[\frac{1}{(1-D)R_L} + \frac{(1-D)}{R} \right] = \frac{V_i}{R}$$

$$V_o = \frac{V_i}{(1-D)} \frac{R_L}{R_L + \frac{R}{(1-D)^2}} \tag{3.94}$$

If $R/[(1-D)^2] \ll R_L$, we have

$$V_o = \frac{V_i}{(1-D)} \tag{3.95}$$

Therefore, the steady-state output voltage V_o can be found from Eq. (3.94) or (3.95).

To determine the transfer function $\delta V_o(s)/\delta V_i(s)$, a perturbation in V_i, as shown in Fig. 3.16(a), is now introduced. The resultant small-signal

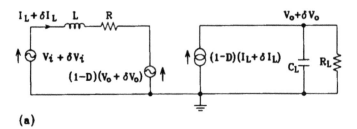

(a)

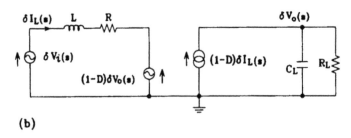

(b)

FIGURE 3.16 (a) Boost converter with perturbation δV, and (b) continuous-mode, small-signal equivalent circuit.

equivalent circuit is shown in Fig. 3.16(b). From the equivalent circuit, we have

$$\delta V_o(s) = (1 - D)\, \delta I_L(s)\, \frac{1}{(1/R_L) + sC_L} \tag{3.96}$$

$$\delta I_L(s) = \left[\frac{\delta V_i(s) - (1 - D)\, \delta V_o(s)}{sL + R}\right] \tag{3.97}$$

Substituting Eq. (3.97) into Eq. (3.96) gives

$$\delta V_o(s) = (1 - D)\left[\frac{\delta V_i(s) - (1 - D)\, \delta V_o(s)}{sL + R}\right]\frac{R_L}{1 + sC_L R_L} \tag{3.98}$$

$$\delta V_o(s)(sL + R)(1 + sC_L R_L) = [(1 - D)\, \delta V_i(s) - (1 - D)^2\, \delta V_o(s)]R_L \tag{3.99}$$

$$\delta V_o(s)[(sL + R)(1 + sC_L R_L) + (1 - D)^2 R_L] = \delta V_i(s)(1 - D)R_L \tag{3.100}$$

$$\frac{\delta V_o(s)}{\delta V_i(s)} = \frac{(1 - D)R_L}{s^2 LC_L R_L + s(L + C_L R_L R) + R + (1 - D)^2 R_L} \tag{3.101}$$

$$\frac{\delta V_o(s)}{\delta V_i(s)} = \frac{1}{(1 - D)}$$

$$\times \frac{1}{s^2 \dfrac{LC_L}{(1 - D)^2} + s\left[\dfrac{L}{(1 - D)^2 R_L} + \dfrac{C_L R}{(1 - D)^2}\right] + \dfrac{R}{(1 - D)^2 R_L} + 1}$$

$$(3.102)$$

The continuous-mode transfer function $\delta V_o(s)/\delta V_i(s)$ obtained in Eq. (3.102) contains two pole frequencies. It is interesting to note that the locations of these poles are identical to those for the buck-boost converter in continuous-mode operation, as indicated by Eq. (3.48).

To determine the $\delta V_o(s)/\delta D(s)$ transfer function, a perturbation in the duty cycle D is now introduced, as shown in Fig. 3.17(a). The resultant small-signal equivalent circuit is shown in Fig. 3.17(b). From this circuit, we have

$$\delta V_o(s) = [(1 - D)\, \delta I_L(s) - I_L\, \delta D(s)] \frac{1}{(1/R_L) + sC_L} \qquad (3.103)$$

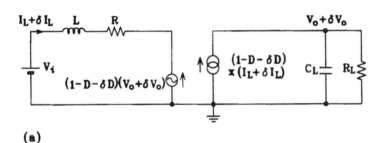

(a)

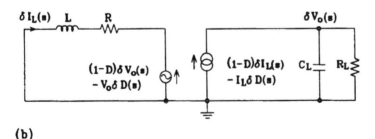

(b)

FIGURE 3.17 (a) Boost converter with perturbation δD and (b) continuous-mode, small-signal equivalent circuit.

$$\delta I_L(s) = \left[\frac{V_o\,\delta D(s) - (1 - D)\delta V_o(s)}{sL + R}\right] \tag{3.104}$$

Substitution of Eq. (3.104) into Eq. (3.103) gives

$$\delta V_o(s) = \left\{(1 - D)\left[\frac{V_o\,\delta D(s) - (1 - D)\,\delta V_o(s)}{sL + R}\right] - I_L\,\delta D(s)\right\}$$

$$\times\,\frac{R_L}{1 + sC_LR_L} \tag{3.105}$$

$$\delta V_o(s)(sL + R)(1 + sC_LR_L) = [(1 - D)V_o\,\delta D(s) - (1 - D)^2\,\delta V_o(s)$$

$$- (sL + R)I_L\,\delta D(s)]R_L \tag{3.106}$$

$$\delta V_o(s)[s^2LC_LR_L + s(L + C_LR_LR) + R + (1 - D)^2R_L]$$

$$= \delta D(s)[(1 - D)V_o - (sL + R)I_L]R_L$$

$$\frac{\delta V_o(s)}{\delta D(s)} = \frac{[(1 - D)V_o - (sL + R)I_L]R_L}{s^2LC_LR_L + s(L + C_LR_LR) + R + (1 - D)^2R_L} \tag{3.107}$$

Since

$$V_o = \frac{V_i}{(1 - D)} \qquad \text{[from Eq. (3.95)]} \tag{3.108}$$

$$I_L = \frac{V_o}{R_L}\frac{1}{1 - D} \qquad \text{[from Eq. (3.93)]}$$

we have

$$I_L = \frac{V_i}{(1 - D)^2R_L} \tag{3.109}$$

Substitution of Eqs. (3.108) and (3.109) into Eq. (3.107) gives the resultant $\delta V_o(s)/\delta D(s)$ transfer function:

$$\frac{\delta V_o(s)}{\delta D(s)} = \frac{\left[(1 - D)\dfrac{V_i}{(1 - D)} - (sL + R)\dfrac{V_i}{(1 - D)^2R_L}\right]R_L}{s^2LC_LR_L + s(L + C_LR_LR) + R + (1 - D)^2R_L} \tag{3.110}$$

$$\frac{\delta V_o(s)}{\delta D(s)} = \frac{V_i}{(1 - D)^2}$$

$$\times \frac{-s\dfrac{L}{(1 - D)^2 R_L} - \dfrac{R}{(1 - D)^2 R_L} + 1}{s^2 \dfrac{LC_L}{(1 - D)^2} + s\left[\dfrac{L}{(1 - D)^2 R_L} + \dfrac{C_L R}{(1 - D)^2}\right] + \dfrac{R}{(1 - D)^2 R_L} + 1}$$

$$(3.111)$$

By comparing Eq. (3.111), which is the $\delta V_o(s)/\delta D(s)$ transfer function for a boost converter in continuous-mode operation, with Eq. (3.60), the $\delta V_o(s)/\delta D(s)$ transfer function for a buck-boost converter also in continuous-mode operation, it can be seen that the two transfer functions are actually similar. Both have two pole frequencies located approximately at

$$\omega_p = -\frac{1}{2C_L R_L} \pm \left\{\left[\frac{1}{2C_L R_L}\right]^2 - \frac{(1 - D)^2}{LC_L}\right\}^{1/2} \tag{3.112}$$

Equation (3.111) also has an RHP zero frequency, which is located approximately at

$$\omega_z = \frac{(1 - D)^2 R_L}{L} \tag{3.113}$$

In a similar way to a buck-boost converter, the extra phase lag in a boost converter due to the RHP zero can easily cause an instability problem in a regulator if the feedback circuit is not designed properly.

3.7 THE BOOST CONVERTER IN DISCONTINUOUS-MODE OPERATION

The circuit of a boost converter and its discontinuous-mode behavior model are shown in Fig. 3.18. The model in Fig. 3.18(b) is obtained from Fig. 2.19, with the modifications stated in Section 3.1.

From the model shown in Fig. 3.18(b), we have, in the steady state,

$$V_o = V_i D^2 G \frac{1}{M - 1} R_L \tag{3.114}$$

$$\frac{V_o}{V_i}(M - 1) = D^2 G R_L \tag{3.115}$$

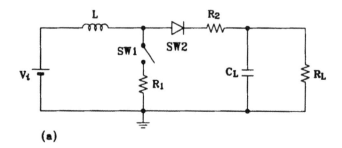

(a)

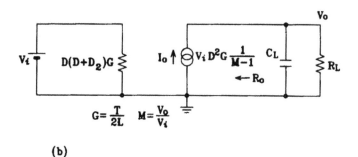

(b)

FIGURE 3.18   (a) Boost converter and (b) discontinuous-mode behavior model.

$$M^2 - M - D^2GR_L = 0 \tag{3.116}$$

$$M = \frac{1 + [1 + 4D^2GR_L]^{1/2}}{2} \tag{3.117}$$

Therefore, the dc output voltage V_o is given by

$$V_o = MV_i = V_i \left\{ \frac{1 + [1 + 4D^2GR_L]^{1/2}}{2} \right\} \tag{3.118}$$

Or, for a given value of M, the required value of D can be found from Eq. (3.116):

$$D = \left[\frac{M^2 - M}{GR_L} \right]^{1/2} \tag{3.119}$$

Let us now determine the small-signal transfer function $\delta V_o(s)/\delta V_i(s)$. In the steady state, we have, from Eq. (3.118),

$$\delta V_o = M\delta V_i \tag{3.120}$$

But the equivalent circuit in Fig. 3.18(b) indicates that there should be a pole frequency ω_p in the $\delta V_o(s)/\delta V_i(s)$ transfer characteristic and that we should have

$$\frac{\delta V_o(s)}{\delta V_i(s)} = M \frac{1}{1 + (s/\omega_p)} \tag{3.121}$$

where

$$\omega_p = \frac{1}{C_L \times [\text{small-signal resistance across } C_L]} \tag{3.122}$$

$$= \frac{1}{C_L[R_L R_o/(R_L + R_o)]} \tag{3.123}$$

The value of R_o can be determined by assuming a perturbation in V_o and by finding the corresponding change in I_o, as shown in Fig. 3.19. From there, we have

$$I_o + \delta I_o = -V_i D^2 G \frac{1}{M + \delta M - 1} \tag{3.124}$$

$$= -D^2 G V_i \frac{1}{\dfrac{V_o + \delta V_o}{V_i} - 1} \tag{3.125}$$

$$= -D^2 G V_i^2 \frac{1}{(V_o - V_i)\left[1 + \dfrac{\delta V_o}{V_o - V_i}\right]} \tag{3.126}$$

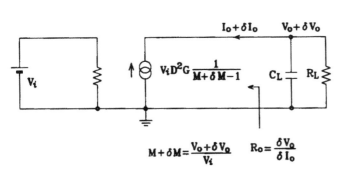

$$M + \delta M = \frac{V_o + \delta V_o}{V_i} \qquad R_o = \frac{\delta V_o}{\delta I_o}$$

FIGURE 3.19 Determination of R_o for boost converter in discontinuous-mode operation.

$$\approx -D^2GV_i^2 \frac{1}{V_o - V_i}\left[1 - \frac{\delta V_o}{V_o - V_i}\right] \tag{3.127}$$

$$= -D^2GV_i^2 \frac{1}{V_o - V_i} + D^2GV_i^2 \frac{1}{(V_o - V_i)^2}\delta V_o \tag{3.128}$$

Considering only the changes, we have

$$\delta I_o = D^2G \frac{V_i^2}{(V_o - V_i)^2}\delta V_o \tag{3.129}$$

$$\therefore R_o = \frac{\delta V_o}{\delta I_o} = \frac{1}{D^2G}\frac{(V_o - V_i)^2}{V_i^2} = \frac{1}{D^2G}(M - 1)^2 \tag{3.130}$$

Substituting Eq. (3.119) into Eq. (3.130) gives

$$R_o = \frac{1}{\dfrac{M^2 - M}{GR_L}G}(M - 1)^2 \tag{3.131}$$

$$= \frac{M - 1}{M}R_L \tag{3.132}$$

By combining Eqs. (3.132), (3.123), and (3.121), we get the $\delta V_o(s)/\delta V_i(s)$ transfer function:

$$\frac{\delta V_o(s)}{\delta V_i(s)} = M\frac{1}{1 + sC_L\dfrac{R_L[(M - 1)/M]R_L}{R_L + [(M - 1)/M]R_L}} \tag{3.133}$$

$$= M\frac{1}{1 + sC_L[(M - 1)/(2M - 1)]R_L} \tag{3.134}$$

The resultant $\delta V_o(s)/\delta V_i(s)$ transfer function found in Eq. (3.134) indicates that the pole frequency ω_p is equal to

$$\omega_p = \frac{2M - 1}{M - 1}\frac{1}{C_LR_L} \tag{3.135}$$

To determine the transfer function $\delta V_o/\delta D$, we now refer back to Fig. 3.18(b). The output voltage V_o in the steady state is

$$V_o = I_oR_L = V_iD^2G\frac{1}{M - 1}R_L \tag{3.136}$$

When there is a perturbation δD, we have

$$V_o + \delta V_o = (D + \delta D)^2 \frac{1}{M + \delta M - 1} GR_L V_i \tag{3.137}$$

$$(V_o + \delta V_o)(M - 1 + \delta M) = (D^2 + 2D\,\delta D + \delta D^2)GR_L V_i \tag{3.138}$$

If we consider only the incremental changes and neglect the higher-order terms such as δD^2 and $\delta V_o\,\delta M$, we have

$$V_o\,\delta M + \delta V_o(M - 1) = 2D\,\delta DGR_L\,V_i \tag{3.139}$$

$$V_o \frac{\delta V_o}{V_i} + \delta V_o(M - 1) = 2D\,\delta DGR_L\,V_i \tag{3.140}$$

Substitution of $V_o = MV_i$ into Eq. (3.140), followed by a rearrangement, gives

$$\frac{\delta V_o}{\delta D} = 2DGR_L V_i \frac{1}{2M - 1} \tag{3.141}$$

Further substitution of Eq. (3.119) into Eq. (3.141) then yields

$$\frac{\delta V_o}{\delta D} = 2\left[\frac{M^2 - M}{GR_L}\right]^{1/2} GR_L V_i \frac{1}{2M - 1} \tag{3.142}$$

$$= V_i \frac{2}{2M - 1}\left[(M^2 - M)GR_L\right]^{1/2} \tag{3.143}$$

When the pole frequency found in Eq. (3.135) is incorporated into Eq. (3.143), the resultant $\delta V_o(s)/\delta D(s)$ transfer function is equal to

$$\frac{\delta V_o(s)}{\delta D(s)} = V_i \frac{2}{2M - 1}\left[(M^2 - M)GR_L\right]^{1/2}$$

$$\times \frac{1}{1 + sC_L[(M - 1)/(2M - 1)]R_L} \tag{3.144}$$

It should be noted that all three basic converters (buck, buck-boost, and boost), when operated in the discontinuous mode, will have single-pole characteristics in their $\delta V_o(s)/\delta V_i(s)$ and $\delta V_o/\delta D(s)$ transfer functions.

3.8 ELECTRONIC TRANSFORMERS AND THE ĆUK CONVERTER

The basic circuit of an electronic transformer is shown in Fig. 3.20. As explained in Section 2.6, the same circuit may also be interpreted as an electronic dc-to-dc transformer or as a Ćuk converter working in the continuous-mode operation. The characteristics of all these circuits are therefore similar, provided that they operate within their respective operation ranges.

Figure 3.21 shows the low-frequency behavior model of the electronic transformer [5], as found in Fig. 2.23. To help determine the dc output voltage V_o, we can write the following equations for the equivalent circuit:

$$0 = (1 - D)I_A - DI_B \tag{3.145}$$

$$V_i = (1 - D)V_C + (I_A + I_B)R \tag{3.146}$$

$$0 = DV_C - (I_A + I_B)R - I_BR_L \tag{3.147}$$

The three unknowns in Eqs. (3.145–3.147), namely, I_A, I_B, and V_C, can be found as

$$I_A = \frac{D^2}{(1 - D)^2} V_i \frac{1}{R_L + \dfrac{R}{(1 - D)^2}} \tag{3.148}$$

$$I_B = \frac{D}{(1 - D)} V_i \frac{1}{R_L + \dfrac{R}{(1 - D)^2}} \tag{3.149}$$

$$V_C = \frac{V_i}{(1 - D)} \frac{R_L + \dfrac{R}{(1 - D)}}{R_L + \dfrac{R}{(1 - D)^2}} \tag{3.150}$$

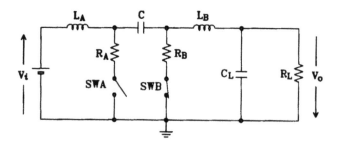

FIGURE 3.20   Basic circuit of electronic transformer and Ćuk converter.

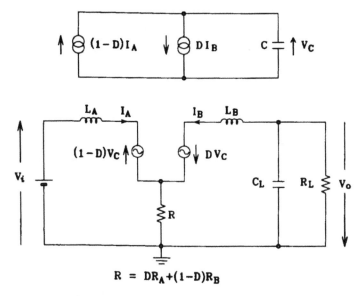

R = DR$_A$+(1−D)R$_B$

FIGURE 3.21 Low-frequency behavior model of electronic transformer.

From Eq. (3.149), the dc output voltage can also be determined:

$$V_o = I_B R_L = \frac{D}{(1 - D)} V_i \frac{R_L}{R_L + \frac{R}{(1 - D)^2}} \tag{3.151}$$

If $R/[(1 - D)^2] \ll R_L$, Eqs. (3.151) and (3.148–3.150) can be approximated as

$$V_o = \frac{D}{(1 - D)} V_i \tag{3.152}$$

$$I_A = \frac{D^2}{(1 - D)^2} \frac{V_i}{R_L} \tag{3.153}$$

$$I_B = \frac{D}{1 - D} \frac{V_i}{R_L} \tag{3.154}$$

$$V_C = \frac{V_i}{1 - D} \tag{3.155}$$

Let us now determine the transfer function $\delta V_o(s)/\delta V_i(s)$. Assume that a small perturbation δV_i in series with V_i is introduced, as shown in Fig.

3.22(a). This results in a small-signal equivalent circuit of the electronic transformer, as shown in Fig. 3.22(b). From this small-signal equivalent circuit, it can be found that

$$\delta V_C(s) = [(1 - D)\, \delta I_A(s) - D\, \delta I_B(s)]\, \frac{1}{sC} \tag{3.156}$$

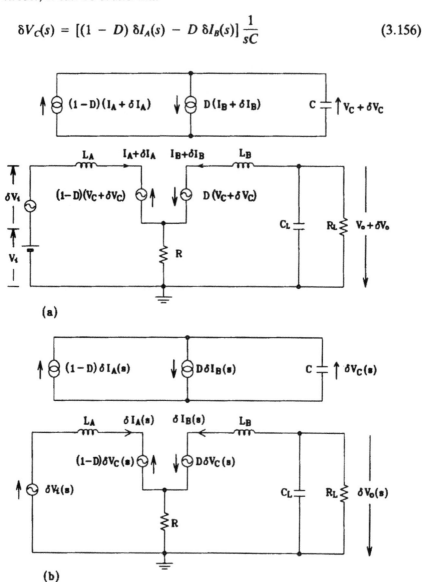

(a)

(b)

FIGURE 3.22 (a) Electronic transformer with perturbation δV, and (b) small-signal equivalent circuit.

$$\delta V_i(s) = \delta I_A(s)[sL_A + R] + (1 - D)\, \delta V_C(s) + R\, \delta I_B(s) \qquad (3.157)$$

$$0 = -D\, \delta V_C(s) + \delta I_B(s)\left[sL_B + \frac{1}{(1/R_L) + sC_L} + R\right] + R\, \delta I_A(s)$$

$$(3.158)$$

After some lengthy but straightforward algebraic manipulations, the three unknowns—$\delta I_A(s)$, $\delta I_B(s)$, and $\delta V_C(s)$—in Eqs. (3.156–3.158) can be found. $\delta I_B(s)$ can be determined as

$$\delta I_B(s) =$$

$$\frac{\delta V_i(s)\left[\dfrac{(1 - D)D}{sC} - R\right]}{\left[sL_A + \dfrac{(1 - D)^2}{sC} + R\right]\left[sL_B + R + \dfrac{1}{(1/R_L) + sC_L} + \dfrac{D^2}{sC}\right] - \left[\dfrac{(1 - D)D}{sC} - R\right]^2}$$

$$(3.159)$$

The transfer function $\delta V_o(s)/\delta V_i(s)$ is then found as

$$\frac{\delta V_o(s)}{\delta V_i(s)} = \frac{\delta I_B(s)}{\delta V_i(s)}\, \frac{1}{(1/R_L) + sC_L} \qquad (3.160)$$

$$= \frac{D}{(1 - D)}\, \frac{-s\,\dfrac{CR}{D(1 - D)} + 1}{K_4 s^4 + K_3 s^3 + K_2 s^2 + K_1 s + K_0} \qquad (3.161)$$

where

$$K_4 = \frac{L_A L_B C C_L}{(1 - D)^2} \qquad (3.162)$$

$$K_3 = \frac{L_A L_B C}{(1 - D)^2 R_L} + \frac{(L_A + L_B) C R C_L}{(1 - D)^2} \qquad (3.163)$$

$$K_2 = \frac{(L_A + L_B) C R}{(1 - D)^2 R_L} + \frac{D^2 L_A C_L + L_A C}{(1 - D)^2} + L_B C_L \qquad (3.164)$$

$$K_1 = \frac{D^2 L_A}{(1 - D)^2 R_L} + \frac{L_B}{R_L} + \frac{CR + C_L R}{(1 - D)^2} \qquad (3.165)$$

$$K_0 = \frac{R}{(1 - D)^2 R_L} + 1 \qquad (3.166)$$

Equation (3.161) shows that the transfer function $\delta V_o(s)/\delta V_i(s)$ contains four poles and one RHP zero.

To find the $\delta V_o(s)/\delta D(s)$ transfer function, we now assume that a perturbation δD is introduced, as shown in Fig. 3.23(a). The small-signal equivalent circuit is then determined, as shown in Fig. 3.23(b).

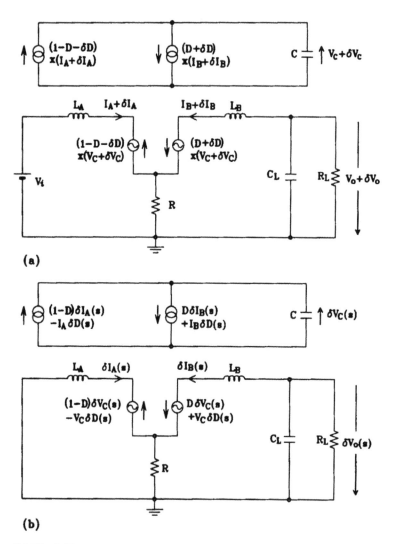

(a)

(b)

FIGURE 3.23 (a) Electronic transformer with perturbation δD and (b) small-signal equivalent circuit.

For the small-signal changes, the following equations can be obtained directly from the equivalent circuit in Fig. 3.23(b):

$$\delta V_C(s) = [(1 - D)\delta I_A(s) - I_A \,\delta D(s) - D \,\delta I_B(s) - I_B \,\delta D(s)]\frac{1}{sC}$$

$$(3.167)$$

$$0 = \delta I_A(s)[sL_A + R] + (1 - D)\delta V_C(s) - V_C \,\delta D(s) + R \,\delta I_B(s)$$

$$(3.168)$$

$$0 = -D \,\delta V_C(s) - V_C \,\delta D(s)$$

$$+ \,\delta I_B(s) \left[sL_B + \frac{1}{(1/R_L) + sC_L} + R \right] + R \,\delta I_A(s)$$

$$(3.169)$$

Again, after some lengthy but straightforward manipulation of Eqs. (3.167–3.169), the three unknowns—$\delta V_C(s)$, $\delta I_A(s)$, and $\delta I_B(s)$—can be determined, and the transfer function $\delta V_o(s)/\delta D(s)$ can be found as

$$\frac{\delta V_o(s)}{\delta D(s)} = \frac{\delta I_B(s)}{\delta D(s)} \frac{1}{(1/R_L) + sC_L}$$

$$(3.170)$$

$$= V_i \frac{1}{(1 - D)^2} \frac{R_L}{R_L + \dfrac{R}{(1 - D)^2}}$$

$$\times \frac{H_2 s^2 + H_1 s + H_0}{K_4 s^4 + K_3 s^3 + K_2 s^2 + K_1 s + K_0}$$

$$(3.171)$$

where

$$H_2 = L_A C \left[\frac{1}{(1 - D)} + \frac{R}{(1 - D)^2 R_L} \right]$$

$$(3.172)$$

$$H_1 = -\frac{D^2 L_A}{(1 - D)^2 R_L}$$

$$(3.173)$$

$$H_0 = \frac{(1 - 2D)R}{(1 - D)^2 R_L} + 1$$

$$(3.174)$$

K_4, K_3, K_2, K_1, and K_0 have the same definitions as those for Eq. (3.161).

Equation (3.171) indicates that the transfer function $\delta V_o(s)/\delta D(s)$ contains four poles and two zeros.

Figure 3.24 shows a typical example of the $\delta V_o(s)/\delta V_i(s)$ and $\delta V_o(s)/\delta D(s)$ transfer characteristics of an electronic transformer (or of a Ćuk converter in continuous-mode operation). The solid lines are for circuits in which the output filtering capacitor C_L is assumed to have zero effective series resistance. The dotted lines (which are much closer to practical measurements) take into account the effective series resistance of C_L.

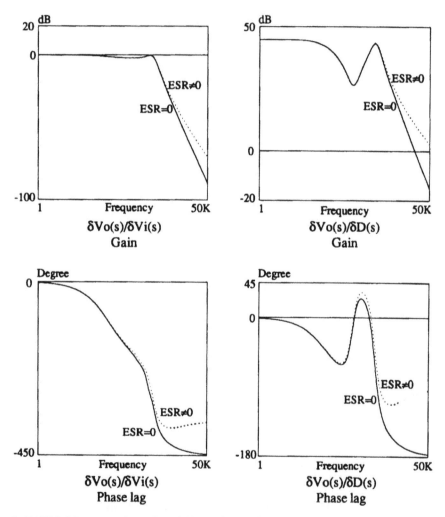

FIGURE 3.24 Typical $\delta V_o(s)/\delta V_i(s)$ and $\delta V_o(s)/\delta D(s)$ characteristics of electronic transformer. ESR = effective series resistance of C_L.

Note that while the nonzero effective series resistance of C_L can introduce much complexity in the manual analysis, it can be easily included in the SPICE simulation, as will be shown in Chapter 7.

3.9 COMPARISON OF THE VARIOUS CONVERTERS IN DIFFERENT MODES OF OPERATION

Based on the materials presented above, some important characteristics of converters will be summarized in this section. The implications of these characteristics on the design of the feedback loops for regulators will also be briefly discussed.

In the discussions that follow, it is assumed that the effective series resistance of the filtering capacitor C_L is zero.

3.9.1 $\delta V_o(s)/\delta V_i(s)$ and $\delta V_o(s)/\delta D(s)$ Transfer Functions for Discontinuous-Mode Operation

For buck, buck-boost, and boost converters in discontinuous-mode operation, the $\delta V_o(s)/\delta V_i(s)$ and $\delta V_o(s)/\delta D(s)$ transfer functions are all single-pole functions. The design of feedback circuits for regulators using such converters (in discontinuous-mode operation) should be straightforward, and the resultant transient responses are usually very good. However, in the discontinuous-mode operation, the current ratings of switching devices have to be larger for a given dc output current.

3.9.2 $\delta V_o(s)/\delta V_i(s)$ and $\delta V_o(s)/\delta D(s)$ Transfer Functions for Continuous-Mode Operation

For a buck converter in continuous-mode operation, both the $\delta V_o(s)/\delta V_i(s)$ and $\delta V_o(s)/\delta D(s)$ transfer functions have a fixed pair of poles located near $[1/(LC_L)]^{1/2}$ (rad/sec). For a buck-boost or boost converter in continuous-mode operation, both the $\delta V_o(s)/\delta V_i(s)$ and $\delta V_o(s)/\delta D(s)$ transfer functions have a variable pair of poles located near $[(1 - D)^2/(LC_L)]^{1/2}$ (rad/sec). The existence of such variable poles in the $\delta V_o(s)/\delta D(s)$ transfer function, together with the RHP zero, makes the design of the feedback circuits more difficult for buck-boost or boost converters when compared with buck converters.

Note that, for regulators using a continuous-mode operated buck-boost or boost converter, the stability problem is more likely to occur when the duty cycle D is large. This is due to the higher gain and larger phase lag in the $\delta V_o(s)/\delta D(s)$ transfer function for a larger duty cycle.

3.9.3 RHP Zero in the $\delta V_o(s)/\delta D(s)$ Transfer Functions of Buck-Boost and Boost Converters

Both buck-boost and boost converters have an RHP zero in their $\delta V_o(s)/\delta D(s)$ transfer functions. The existence of this RHP zero has the following effects on the $\delta V_o(s)/\delta D(s)$ characteristics:

1. It adds a $+20$ dB/dec upward slope at the high-frequency portion of the amplitude response.
2. It introduces at the same time an additional $-90°$ phase lag at the high-frequency portion of the phase response.

Because of the higher gain and larger phase shift at high frequency, the existence of the RHP zero can easily cause oscillation in a feedback loop if it is not handled carefully.

3.9.4 Four-Pole, Two-Zero $\delta V_o(s)/\delta D(s)$ Transfer Functions of Electronic Transformers and Ćuk Converters

The $\delta V_o(s)/\delta D(s)$ transfer function of electronic transformers, as well as of Ćuk converters in continuous-mode operation, contains four poles and two zeros, which are functions of the duty cycle D. The complex nature of the transfer function makes the optimization of feedback circuits (for regulators) difficult. However, in the worst case, an external dominant pole can still be added in the feedback circuit to achieve a dominant-pole compensation. The resultant transient response of the regulator will then not, of course, be an optimized one.

3.10 SUMMARY AND FURTHER REMARKS

Based on the low-frequency behavior models examined in Chapter 2, the low-frequency characteristics of various square-wave converters and electronic transformers have been analyzed and compared. In the course of the analysis, it was found that manual analysis is very helpful in understanding converter characteristics. However, it is difficult to include in the manual analysis stray elements such as the effective series resistance of the output filtering capacitor or the internal impedance of the dc supply. For this and other reasons, computer simulation is therefore necessary if accurate and detailed analyses are required. The computer simulation of the low-frequency behaviors of square-wave converters will be covered in detail in Chapter 7.

Chapters 1 to 3 are designed to familiarize the reader with the principles of operation and the low-frequency behaviors of basic square-wave converters. The next chapter will focus on the operation and characteristics of converters with transformer isolation.

EXERCISES

1. Explain why assumptions and approximations are often made in the manual analysis of power converters.
2. Compare the advantages and disadvantages of the continuous-mode operation with those of the discontinuous-mode operation for
   a. a buck converter
   b. a buck-boost converter
   c. a boost converter
3. Explain the shapes of the $\delta V_o(s)/\delta V_i(s)$ and $\delta V_o(s)/\delta D(s)$ characteristics shown in Figs. 3.4, 3.7, 3.11, 3.14, and 3.24 for
   a. ESR $= 0$
   b. ESR $\neq 0$
   (ESR $=$ effective series resistance of output filtering capacitor C_L.) Comment on the uses of these characteristics in the design of regulators.
4. Sketch the shapes of the $\delta V_o(s)/\delta V_i(s)$ and $\delta V_o(s)/\delta D(s)$ characteristics of a boost converter, assuming
   a. ESR $= 0$
   b. ESR $\neq 0$
5. Explain why the existence of
   a. RHP zeros
   b. multiple poles in the $\delta V_o(s)/\delta D(s)$ characteristics makes the design of regulator feedback circuits difficult.

4
Power Converters with Transformer Isolation

All the square-wave converters examined in previous chapters have a common feature in that no high-frequency power transformer is used. This lack of a transformer severely limits the uses of such converters. The following are some typical applications in which transformers are essential or highly desirable:

1. For off-line switch-mode power supplies (which are connected directly to the ac mains), transformer isolation is necessary to protect the user from the ac mains.
2. For power supplies with isolated multiple outputs, a transformer with multiple output windings can eliminate the need for multiple converters.
3. For converters that are required to provide voltage step-up or step-down, the use of a transformer offers a simple and easy solution.

The use of a high-frequency transformer would not significantly alter the low-frequency behavior of a converter (except by stepping the output voltage up or down). However, it does change the high-frequency switching waveforms and may produce other undesirable side effects. The following are some examples of such side effects:

1. The existence of primary inductance results in an inductive shunting current. This may cause problems in certain types of circuits.
2. The unavoidable leakage inductances of transformers may result in large voltage spikes during switching periods. This problem becomes particularly serious in circuits operating at high switching frequencies.

3. Additional losses such as eddy-current loss, copper loss, and hysteresis loss are introduced.

It will be shown in Section 4.1 that the existence of primary inductance in a single-ended forward converter will impose an operational limit on the usable range of the duty cycle D.

The transformer leakage inductance and heavy losses at high frequencies, on the other hand, are important factors limiting the upper switching frequency of power conversion circuits.

However, this chapter is not concerned with the effects of transformer leakage inductance and losses. Rather, it provides an introduction to the conversion principles without going into unnecessary detail. (The detailed analysis is much better performed by computer simulation, as discussed in Chapters 6, 8, and 9.)

The power converter circuits to be studied in this chapter include the following types: forward, flyback, Ćuk, electronic transformers, transformer-coupled push-pull, half-bridge, and full-bridge. Note that a forward converter is a buck converter with an isolation transformer and that a flyback converter is a buck-boost converter with an isolation transformer.

4.1 FORWARD CONVERTER

This section explains the operation of the forward converter, using the example of a direct off-line (mains-operated) switching regulator that employs a multiple-output forward converter as the power conversion engine. Figure 4.1 shows the circuit of such a regulator. Figure 4.2 shows the idealized waveforms of the circuit.

In the circuit shown in Fig. 4.1, the bridge rectifier BR and the filtering capacitor C_f convert the ac mains into an unregulated dc (V_{in}) to provide power to the forward converter. The negative feedback through the pulse-width modulator (PWM) regulates the output voltage V_o. The high-frequency power transformer T_1 provides the functions of voltage step-down and galvanic isolation between the regulator outputs and the ac mains input. The pulse transformer T_2 magnetically couples the output of the PWM to the power transistor.

Referring to the waveforms shown in Fig. 4.2, the operation of the forward converter can be explained as follows:

1. It is assumed that the switching action of the converter has reached a steady state and that it operates in the continuous-mode operation.
2. For $0 < t < DT$, the MOS switch SW is turned on by the gate

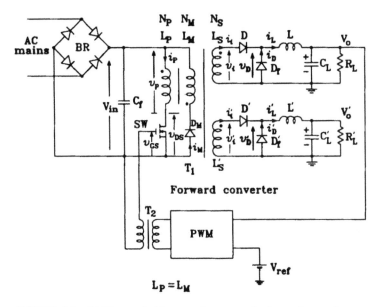

FIGURE 4.1   Off-line switching regulator employing a forward converter.

voltage v_{GS}. We then have (refer to Fig. 4.1 for the circuit and to Fig. 4.2 for the waveforms)

$$v_p = V_{in} \tag{4.1}$$

$$v_i = V_{in} \frac{N_S}{N_P} = v_D \tag{4.2}$$

$$\frac{di_L}{dt} = (v_i - V_o) \frac{1}{L} \qquad (v_i \text{ assists } i_L \text{ to build up})$$

$$= \left(V_{in} \frac{N_S}{N_P} - V_o \right) \frac{1}{L} \tag{4.3}$$

$$i_i = i_L \tag{4.4}$$

$$i_D = 0 \qquad (D_f \text{ being reversely biased}) \tag{4.5}$$

At the same time, the magnetic flux in the magnetic core of T_1 builds up linearly.

3. At $t = DT$, the MOS switch SW is turned off by the gate voltage v_{GS}. Since the primary current i_p can no longer exist, the collapse of the magnetic flux in the magnetic core of T_1 induces back emf's

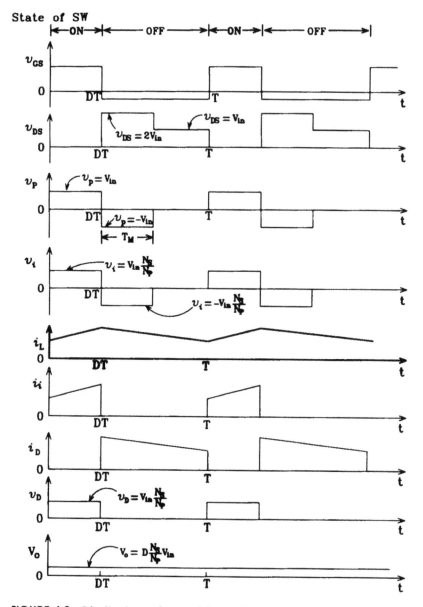

FIGURE 4.2   Idealized waveforms of forward converter.

in the windings L_M and L_S. The back emf in L_S reversely biases the diode D, causing i_i to drop to zero. In the absence of the reset winding L_M, the back emf can become very large and cause damage to the MOS switch SW and the output rectifiers.

4. The inclusion of the reset winding L_M provides a path along which the magnetizing current can continue to flow, through the diode D_M back to the supply, when SW is turned off. This clamps the back emf to a safe value. Then the energy stored in the magnetic flux is also able to return to the power source. The time taken by the magnetic flux to reset to zero is defined as T_M in the waveform v_p, as shown in Fig. 4.2.

5. For $DT < t < T$, the switch SW remains in the off state. While the back emf is causing i_M to flow through D_M and L_M (within T_M), the voltage across L_M is clamped to the value V_{in}. If L_P and L_M have the same number of turns (which is usually the case), the amplitude of the primary voltage v_p will also be clamped to V_{in}. In mathematical terms, we have

$$N_P = N_M \tag{4.6}$$

$$v_p = -V_{in} \text{ within the reset time } T_M$$

$$= 0 \text{ after reset} \tag{4.7}$$

$$v_{DS} = V_{in} - v_p = 2V_{in} \text{ within } T_M$$

$$= V_{in} \text{ after reset} \tag{4.8}$$

$$v_i = -V_{in} \frac{N_S}{N_P} \text{ within } T_M$$

$$= 0 \text{ after reset} \tag{4.9}$$

$$i_i = 0 \tag{4.10}$$

$$i_L = i_D \tag{4.11}$$

$$\frac{di_L}{dt} = \frac{-V_o}{L} \tag{4.12}$$

$$v_D = 0 \tag{4.13}$$

6. The reset time T_M can be found by equating the increase of the magnetizing current during $0 < t < DT$ to the decrease of the magnetizing current during T_M:

$$\frac{V_{in}}{L_P} DT = \frac{V_{in}}{L_M} T_M \tag{4.14}$$

$$\therefore T_M = DT \qquad \text{(assuming } L_P = L_M) \qquad\qquad (4.15)$$

7. At $t = T$, SW is turned on again. This initiates the next switching cycle, repeating the actions described above.
8. Since the output voltage V_o is equal to the averaged value of v_D, we have

$$V_o = D \frac{N_S}{N_P} V_{in} \qquad\qquad (4.16)$$

It is obvious from Eq. (4.16) that the output voltage V_o can be controlled by either the duty cycle D or the turns ratio N_S/N_P.

9. Note that although the waveforms of v_i', i_L', i_i', i_D', v_D', and V_o' are not shown in Fig. 4.2, they are actually similar to those of v_i, i_L, i_i, i_D, v_D, and V_o, respectively.

The waveforms of i_L, i_i, i_D, v_D, and V_o shown in Fig. 4.2 for a forward converter should be compared with those given in Fig. 1.5 for a buck converter. It can be found that they are similar.

A forward converter also enters into the discontinuous-mode operation when

$$R_L > \frac{2L}{(1-D)T} \qquad \text{(which is similar to a buck converter)} \qquad (4.17)$$

To make the situation a little more complex, in a multiple-output forward converter, different outputs may operate in different modes of operation.

As an interim remark, it may be said that, as far as the low-frequency behavior is concerned, a forward converter is functionally a buck converter followed by an ideal transformer that operates down to zero frequency (dc).

It is found, however, that because of the existence of the primary inductance L_P, the duty cycle D of the forward converter must be limited to below 0.5 in order to prevent excessive magnetizing current. The cause of such a large magnetizing current for D larger than 0.5 can be explained as follows:

1. Referring to Fig. 4.1, it can be imagined that, during the time the electronic switch SW is turned on, there is a magnetizing current i_{MAG} in L_P (note that i_{MAG} is a component of i_P and that, for the

purpose of avoiding confusion, i_{MAG} is not shown in Fig. 4.1), increasing at the rate of

$$\frac{di_{MAG}}{dt} = \frac{V_{in}}{L_P} \qquad (4.18)$$

Assuming that i_{MAG} is initially zero, the waveform of i_{MAG} can be drawn as shown in Fig. 4.3. (It is assumed that $D > 0.5$.)

2. At $t = DT$, the switch SW is turned off. i_P is forced to zero. However, the magnetizing current i_{MAG} (in L_P) now switches to the reset winding L_M and appears as i_M. If we assume that $L_P = L_M$, then

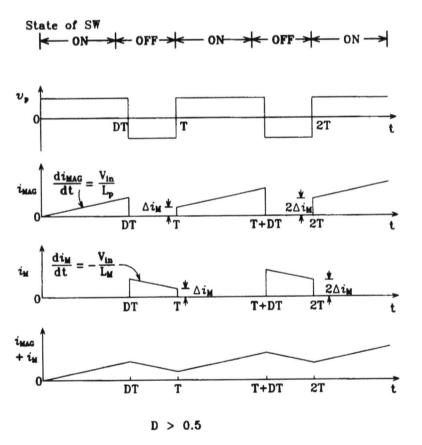

D > 0.5

FIGURE 4.3 Magnetizing currents in forward converter with $D > 0.5$.

the amplitude of i_M (in L_M) just after switching will be equal to the amplitude of i_{MAG} (in L_P) just before switching. As far as the magnetizing current is concerned, i_M may therefore be considered as a continuation of i_{MAG}.

3. For $DT < t < T$, the magnetizing current i_M decreases at the rate of

$$\frac{di_M}{dt} = \frac{-V_{in}}{L_M} \tag{4.19}$$

4. At $t = T$, there will be a net gain in the magnetizing current if the increase of i_{MAG} during $0 < t < DT$ is larger than the decrease of i_M during $DT < t < T$. This situation occurs when

$$\frac{V_{in}}{L_P} DT > \frac{V_{in}}{L_M} (1 - D)T \tag{4.20}$$

or

$$D > 0.5 \quad \text{(assuming } L_M = L_P) \tag{4.21}$$

When the switch SW is turned on again in the next switching cycle, the residual magnetizing current Δi_M will be switched back to the primary winding L_P to form an initial value for i_{MAG}.

5. If, for each switching cycle, there is a net gain of magnetizing current,

$$\Delta i_M = \frac{V_{in}}{L_P} DT - \frac{V_{in}}{L_M} (1 - D)T \tag{4.22}$$

the magnetizing current will eventually grow so large that it will drive the transformer into saturation and cause the circuit to malfunction.

The duty cycle D of a single-ended forward converter, therefore, cannot exceed 0.5 if $L_P = L_M$.

4.2 FLYBACK CONVERTER

This section considers the operation of the flyback converter. Figure 4.4 shows the basic circuit of a flyback converter. Figure 4.5 shows the idealized current and voltage waveforms of the circuit.

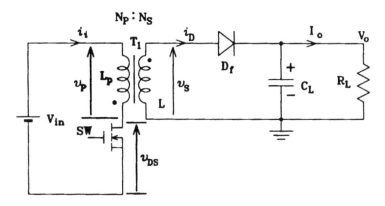

FIGURE 4.4 Flyback converter.

Referring to the waveforms shown in Fig. 4.5, the operation of the flyback converter can be explained as follows:

1. It is assumed that the switching action of the converter has reached a steady state and that the circuit operates in the continuous-mode operation.
2. For $0 < t < DT$, SW is turned to the on state:

$$v_{DS} = 0 \qquad (4.23)$$

$$v_p = V_{in} \qquad (4.24)$$

$$v_S = -v_p \frac{N_S}{N_P} = -V_{in} \frac{N_S}{N_P} \qquad (4.25)$$

$i_D = 0$ (because D_f is reversely biased by a negative v_s)

$$ \qquad (4.26)$$

$$\frac{di_i}{dt} = \frac{V_{in}}{L_P} \qquad (4.27)$$

Magnetic energy is being stored in the primary inductance, so that

Energy stored $= (1/2)L_P i_i^2$ (4.28)

3. At $t = DT$, SW is turned off. Since i_i is no longer able to exist, the collapse of the magnetic flux in the magnetic core (actually, it is only a decrease in the magnetic flux) induces a back emf to force the current i_D to flow in the secondary winding and charge up the output filtering capacitor C_L.

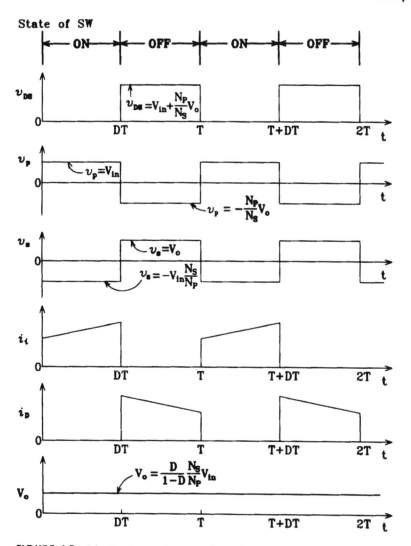

FIGURE 4.5 Idealized waveforms of flyback converter.

4. For $DT < t < T$, SW remains off. The inductive current i_D continues to flow through D_f to charge up the output filtering capacitor C_L. Since the flow of i_D is opposed by the output voltage V_o, i_D decreases linearly.

$$\frac{di_D}{dt} = \frac{-V_o}{L} \qquad (4.29)$$

where L is the inductance of the secondary winding of T_1 (when the primary winding is open-circuited). At the same time, we have

$$v_s = V_o \tag{4.30}$$

$$v_p = -v_s \frac{N_P}{N_S} = -V_o \frac{N_P}{N_S} \tag{4.31}$$

$$v_{DS} = V_{in} - v_p \tag{4.32}$$

$$= V_{in} + \frac{N_P}{N_S} V_o \tag{4.33}$$

5. At $t = T$, SW is turned on once more, so that

$$v_{DS} = 0 \tag{4.34}$$
$$v_p = V_{in} \tag{4.35}$$

$$v_s = -v_p \frac{N_S}{N_P} = -V_{in} \frac{N_S}{N_P} \tag{4.36}$$

Since v_s is negative and i_D drops to zero, the primary current i_i now replaces the secondary current i_D in order to maintain the magnetic flux in the magnetic core of the transformer T_1. The actions described in steps 2–4 will then repeat.

When voltage and current waveforms of the flyback converter (shown in Fig. 4.5) are compared with those of the buck-boost converter (shown in Fig. 1.11), it will be found that they are similar except for the following:

1. The use of a transformer in the flyback converter divides the inductor current i_L in Fig. 1.11 into a primary current i_i and a secondary current i_D, as shown in Fig. 4.5.
2. The output transformer T_1 of the flyback converter steps the output voltage down or up by a factor of N_S/N_P. (At the same time, it also steps the secondary current up or down by a factor of N_P/N_S.)
3. Instead of having

$$V_o = \frac{D}{1-D} V_i \qquad \text{for a buck-boost converter} \tag{4.37}$$

we have

$$V_o = \frac{N_S}{N_P} \frac{D}{1 - D} V_i \qquad \text{for a flyback converter} \qquad (4.38)$$

A flyback converter, like a buck-boost converter, enters into discontinuous-mode operation when

$$R_L > \frac{2L}{(1 - D)^2 T} \qquad \text{[which is essentially similar to Eq. (1.44)]} \qquad (4.39)$$

It should be noted that, as far as the low-frequency behavior is concerned, a flyback converter is functionally a buck-boost converter followed by an ideal transformer that operates down to zero frequency (dc).

Both flyback and forward converters may be designed to provide multiple outputs. This can be achieved for a flyback converter by simply adding multiple secondary windings, output rectifiers, and filtering capacitors to the circuit of Fig. 4.4

4.3 ĆUK CONVERTER WITH TRANSFORMER ISOLATION

Figure 4.6(a) shows the circuit of a Ćuk converter with a one-to-one isolation transformer T. The circuit shown in Fig. 4.6(b) is its equivalent circuit if the following assumptions can be made:

1. The transformer T has zero leakage inductance and infinite primary inductance.
2. The lack of galvanic isolation in the equivalent circuit shown in Fig. 4.6(b) is not considered as a violation of the equivalence.

Note that the transformer T within the dashed rectangle in Fig. 4.6(a) is equivalent to the circuit T' shown within the dashed rectangle in Fig. 4.6(b).

So, as far as circuit operation is concerned, the Ćuk converter shown in Fig. 4.6(a) (with an isolation transformer) is the same as that shown in Fig. 1.19 (without a transformer) if

$$\frac{C_A C_B}{C_A + C_B} = C \qquad (4.40)$$

A Ćuk converter having a transformer with turns ratio N_S/N_P is equivalent to a Ćuk converter without a transformer but followed by an ideal trans-

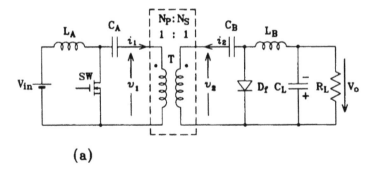

(a)

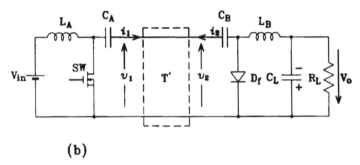

(b)

FIGURE 4.6 (a) Ćuk converter with transformer and (b) equivalent circuit.

former with a transforming ratio equal to N_S/N_P. For a Ćuk converter with a real transformer, such as the one shown in Fig. 4.6(a), we have the dc output voltage given by

$$V_o = \frac{N_S}{N_P} \frac{D}{(1 - D)} V_{in} \tag{4.41}$$

4.4 ELECTRONIC TRANSFORMERS WITH TRANSFORMER ISOLATION

Figure 4.7 shows the circuit of an electronic transformer with a real (magnetically coupled) transformer T. Functionally, the circuit is equivalent to an electronic transformer having a transform ratio equal to (N_S/N_P) $[D/(1 - D)]$.

The circuit shown in Fig. 4.8 is a unipolar (dc-to-dc) electronic transformer whose effective transform ratio is equal to $(N_S/N_P)[D/(1 - D)]$.

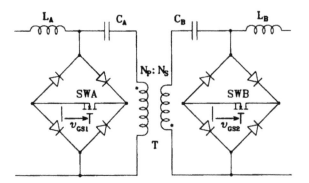

FIGURE 4.7   Electronic transformer with isolation transformer. *SWA* and *SWB* are in complementary states.

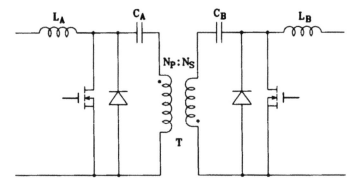

FIGURE 4.8   Electronic dc-to-dc transformer with isolation transformer.

4.5   TRANSFORMER-COUPLED PUSH-PULL CONVERTER

The transformer-coupled push-pull converter is functionally a push-pull version of the forward converter. Figure 4.9 shows the circuit of such a converter. Figure 4.10 gives the idealized waveforms of the circuit, assuming a perfect transformer *T* (with zero leakage inductance and infinite primary inductance).

The operation of the circuit can be briefly explained as follows:

1.   It is assumed that the switching action of the converter has reached a steady state and that it operates in the continuous mode.

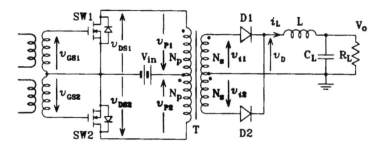

FIGURE 4.9 Transformer-coupled push-pull converter.

2. Within $0 < t < t_1$, the electronic switch $SW1$ is turned on and $SW2$ is turned off.

$$v_{DS1} = 0, \qquad v_{P1} = V_{in}, \qquad v_{P2} = -V_{in}$$

$$v_{DS2} = V_{in} - v_{p2} = 2V_{in} \tag{4.42}$$

$$v_{i1} = -v_{p1} \frac{N_S}{N_P} = -V_{in} \frac{N_S}{N_P} \tag{4.43}$$

$$v_{i2} = v_{p1} \frac{N_S}{N_P} = V_{in} \frac{N_S}{N_P} \tag{4.44}$$

$$v_D = v_{i2} = V_{in} \frac{N_S}{N_P} \tag{4.45}$$

$$\frac{di_L}{dt} = (v_D - V_o)\frac{1}{L} = \left(V_{in}\frac{N_S}{N_P} - V_o\right)\frac{1}{L} \tag{4.46}$$

Under this condition, $SW1$ is active to enable the power from V_{in} to be delivered to the loading circuit. Equations (4.45) and (4.46) should be compared with Eqs. (1.7) and (1.4), which are for a buck converter.

3. Within $t_1 < t < t_2$, $SW1$ is turned off. At the same time, since $SW2$ is also in the off state, we have

$$v_{p1} = 0, \qquad v_{p2} = 0 \tag{4.47}$$

$$v_{i1} = 0, \qquad v_{i2} = 0 \tag{4.48}$$

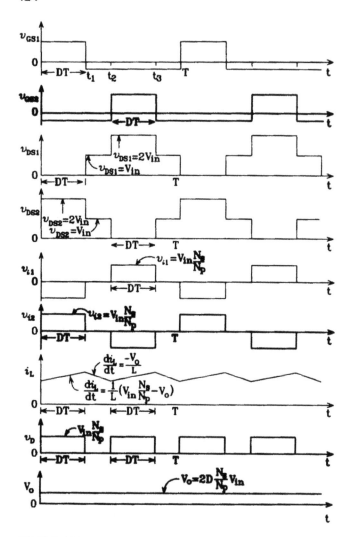

FIGURE 4.10 Idealized waveforms of transformer-coupled push-pull converter.

But the inductive current i_L will continue to flow through both D_1 and D_2 against the output voltage V_o.

$$\therefore \frac{di_L}{dt} = \frac{-V_o}{L} \tag{4.49}$$

Equation (4.49) should be compared with Eq. (1.8).

4. Within $t_2 < t < t_3$, $SW2$ is turned to the on state, while $SW1$ remains off. As a result, we have

$$v_{DS2} = 0, \qquad v_{P2} = V_{in}, \qquad v_{P1} = -V_{in} \qquad (4.50)$$

$$v_{DS1} = V_{in} - v_{p1} = 2V_{in} \qquad (4.51)$$

$$v_{i1} = v_{p2} \frac{N_S}{N_P} = V_{in} \frac{N_S}{N_P} \qquad (4.52)$$

$$v_{i2} = -v_{p2} \frac{N_S}{N_P} = -V_{in} \frac{N_S}{N_P} \qquad (4.53)$$

$$v_D = v_{i1} = V_{in} \frac{N_S}{N_P} \qquad (4.54)$$

$$\frac{di_L}{dt} = (v_D - V_o) \frac{1}{L} = \left(V_{in} \frac{N_S}{N_P} - V_o \right) \frac{1}{L} \qquad (4.55)$$

Under this condition, $SW2$ is active to enable the power from V_{in} to be delivered to the loading circuit. (The switches $SW1$ and $SW2$ together act as a push-pull circuit to carry out the power conversion process.)

5. Within $t_3 < t < T$, both $SW1$ and $SW2$ are turned off.

$$v_{p1} = 0, \qquad v_{p2} = 0 \qquad (4.56)$$

$$v_{i1} = 0, \qquad v_{i2} = 0 \qquad (4.57)$$

The inductive current i_L continues to flow through both D_1 and D_2, but with its amplitude falling at the rate of

$$\frac{di_L}{dt} = \frac{-V_o}{L} \qquad (4.58)$$

6. When $t = T$, $SW1$ is turned on again to initiate the next cycle of switching in the converter.

From the waveform of v_D, it can be found that the output voltage V_o is given by

$$V_o = \text{Average value of } v_D = 2D \frac{N_S}{N_P} V_{in} \qquad (4.59)$$

where D is the duty cycle of $SW1$ or $SW2$.

From the explanation given above, it should be obvious that a trans-former-coupled push-pull converter is functionally a push-pull version of the forward converter. The waveforms given in Fig. 4.10 for a transformer-coupled push-pull converter should be compared with those shown in Fig. 4.2 for a single-ended forward converter and with those shown in Fig. 1.5 for a buck converter.

4.6 HALF-BRIDGE CONVERTER

Like a transformer-coupled push-pull converter, a half-bridge converter is effectively also a push-pull version of the forward converter. But the output coupling method is different. Figure 4.11 shows the circuit of a half-bridge converter and Fig. 4.12 the idealized waveforms of the circuit for contin-uous-mode operation.

When the waveforms of the transformer-coupled push-pull converter (Fig. 4.10) are compared with those of the half-bridge converter (Fig. 4.12), it will be found that they are very similar. However, the following should be noted:

1. Both C_1 and C_2 in Fig. 4.11 are assumed to have large and equal capacitance, so that the voltage across each capacitor is steady and equal to $(1/2)V_{in}$.

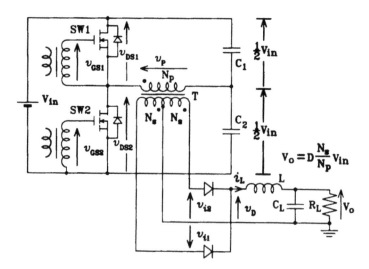

FIGURE 4.11 Half-bridge converter.

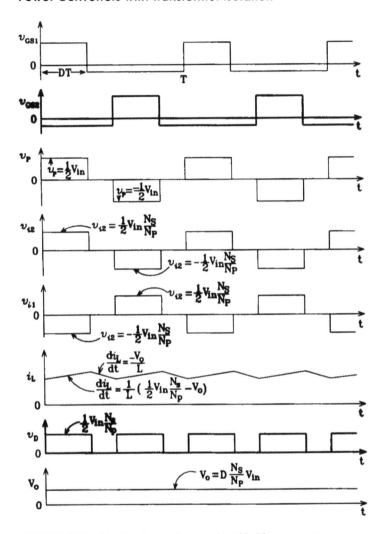

FIGURE 4.12 Idealized waveforms of half-bridge converter.

2. The voltage waveform of the primary winding for the half-bridge circuit is shown as v_p in Fig. 4.12. The method for finding the waveform of v_p is quite straightforward because we have

$$v_p = (1/2)V_{in} \quad \text{when } SW1 \text{ is on and } SW2 \text{ is off} \qquad (4.60)$$

$$v_p = 0 \quad \text{when both } SW1 \text{ and } SW2 \text{ are off} \qquad (4.61)$$

$$v_p = -(1/2)V_{in} \quad \text{when } SW1 \text{ is off and } SW2 \text{ is on} \qquad (4.62)$$

The dc output voltage of the half-bridge converter is equal to

$$V_o = 2D \frac{N_S}{N_P} \frac{1}{2} V_{in} = D \frac{N_S}{N_P} V_{in} \tag{4.63}$$

where D is the duty cycle of $SW1$ or $SW2$.

4.7   FULL-BRIDGE CONVERTER

Effectively, the full-bridge converter is also a push-pull version of the forward converter. Figure 4.13 shows the circuit of a full-bridge converter.

A comparison of the full-bridge circuit (Fig. 4.13) and the half-bridge circuit (Fig. 4.11) shows that they are similar except for the following:

1. In the full-bridge circuit, four transistors are used.
2. The swing of v_p is from $+V_{in}$ to $-V_{in}$ for a full-bridge circuit instead of from $(+1/2) V_{in}$ to $(-1/2) V_{in}$ for a half-bridge circuit. The output voltage of the full-bridge circuit is therefore equal to

$$V_o = 2D \frac{N_S}{N_P} V_{in} \tag{4.64}$$

which is twice as large as that of a half-bridge circuit.

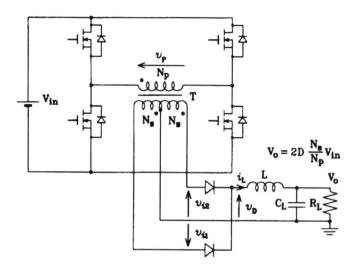

FIGURE 4.13   Full-bridge converter.

4.8 COMPARISON OF THE VARIOUS CONVERTERS

In this section, the characteristics of commonly used dc-to-dc converters, all of which are assumed to have an isolation transformer, will be compared, using the forward converter as a reference. The comparison is given in Table 4.1.

In this comparison, the following assumptions are made:

1. All the converters have the same input supply voltage V_{in}.
2. All the power transistors have the same current rating.
3. All the converters except the flyback converter operate only in the continuous mode, with a reasonably large energy-storage inductor L, so that the inductor current may be assumed to be smooth.
4. There are no leakage inductances in the transformers.

Although both the input and output currents of a flyback converter remain in the pulse form for either the continuous or the discontinuous mode of

Table 4.1 Comparison of Various Converters with Transformer Isolation

Type of converter	Voltage stress of transistor	Need for reset (clamp) winding	Limit on duty cycle D	Power output (normalized with respect to forward converter)
Forward	$2 V_{in}$	Needed	$D < 0.5$	Normalized to 1
Flyback	$2 V_{in}$ for $D = 0.5$	Optional		1 for continuous mode; 0.5 for discontinuous mode
Cuk	$2 V_{in}$ for $D = 0.5$	Not needed		1
Transformer-coupled push-pull	$2 V_{in}$	Not needed	$D < 0.5$ for each transistor	2 (using 2 transistors)
Half-bridge	V_{in}	Not needed	$D < 0.5$ for each transistor	1 (using 2 transistors with voltage rating = V_{in})
Full-bridge	V_{in}	Not needed	$D < 0.5$ for each transistor	2 (using 4 transistors with voltage rating = V_{in})

operation, the flyback converter has the distinct advantage of very good cross-load regulation for multiple outputs because the outputs are tied closely together, through the power transformer, without filtering inductors between them.

4.9 LOW-FREQUENCY BEHAVIOR MODELS OF CONVERTERS WITH ISOLATION TRANSFORMERS

Once the principles of operation of converters with isolation transformers have been understood, it is important for the circuit designer to realize that the use of a high-frequency power transformer does not actually alter the low-frequency behavior of a converter substantially. This means that the low-frequency behavior models of converters developed in Chapter 2 need only minor modifications to enable them to represent converters with isolation transformers. Details about the modeling of the low-frequency behaviors of high-frequency transformers and the simulation of converters with multiple transformer windings will be discussed in Section 7.6.

4.10 SUMMARY AND FURTHER REMARKS

In this chapter, the operation principles of various types of square-wave power converters with transformer isolation have been examined. The characteristics of commonly used converters (with isolation transformers) have also been compared.

Although the addition of a transformer may significantly affect the high-frequency switching waveforms, it does not alter substantially the low-frequency behavior of the converter. The modeling and analysis techniques developed in Chapters 2 and 3 for the low-frequency behaviors of basic converters can therefore be applied also to converters with transformers.

Now that the background to power converters has been covered, the next chapter will examine the operation of switching regulators.

EXERCISES

1. For what purposes are high-frequency transformers used in power converters?
2. What are the problems a high-frequency transformer can cause for a power converter?
3. Sketch the waveforms of v_1, v_2, i_1, and i_2 for the Ćuk converter circuit shown in Fig. 4.6(a).
4. Almost all the waveforms given in this chapter are idealized, based on the assumption that power transformers do not have leakage inductances.

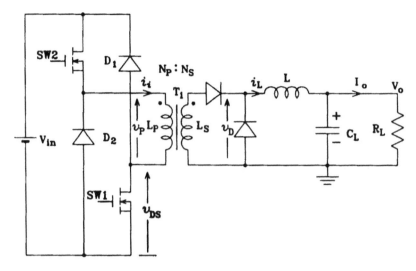

FIGURE 4.14   Two-transistor forward converter.

Assuming that power transformers do have small leakage inductances, sketch the new waveforms of:
a. v_{DS} in Fig. 4.4
b. v_{DS1} in Fig. 4.9
c. v_P in Fig. 4.11

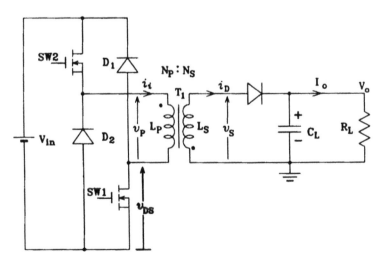

FIGURE 4.15   Two-transistor flyback converter.

5. Why doesn't the high-frequency power transformer substantially affect the low-frequency behavior of a converter?
6. By assuming that the transformer T in Fig. 4.6(a) has a primary to secondary turns ratio N_p/N_s not equal to 1, develop a low-frequency behavior model of the complete Ćuk converter for continuous-mode operation.
7. With the aid of waveform diagrams, explain the operation of:
   a. the two-transistor forward converter circuit shown in Fig. 4.14
   b. the two-transistor flyback converter circuit shown in Fig. 4.15

5
Voltage-Mode and Current-Mode-Controlled Switching Regulators

In Chapters 1 to 4, the operation and characteristics of various square-wave converters have been examined. Such converters are often used in switching regulators to provide regulated dc output voltages.

In order to regulate the output voltage, a control feedback loop is required. Two control schemes are commonly used: voltage-mode control and current-mode control. In a traditional voltage-mode-controlled regulator, the feedback voltage is used to control the duty cycle of the converter directly. In a current-mode-controlled regulator, however, the feedback voltage is used to control only the output current of the converter. (The feedback voltage controls the duty cycle only indirectly.) Since the output current is directly controlled by the input, the converter used in a current-mode-controlled regulator is effectively a current-controlled converter.

This chapter covers the following topics of study:

1. The principle of operation of voltage-mode-controlled and current-mode-controlled regulators
2. The characteristics of voltage-mode-controlled and current-mode-controlled regulators
3. The principle of operation and characteristics of current-controlled converters
4. The design of feedback and compensation circuits for switching regulators
5. The special design considerations for current-mode-controlled regulators

Since current-controlled converters differ significantly from the ordinary converters studied in previous chapters, much attention will be focused on the operation and characteristics of current-controlled converters.

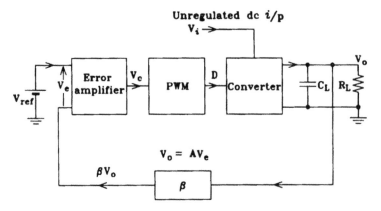

FIGURE 5.1   Voltage-mode-controlled regulator.

5.1   OPERATION OF VOLTAGE-MODE-CONTROLLED REGULATORS

Figure 5.1 is a block diagram of a voltage-mode-controlled regulator arranged in the form of a feedback circuit.

Assuming that $V_o = AV_e$, we have

$$V_o = A(V_{ref} - \beta V_o) \tag{5.1}$$

$$V_o = V_{ref} \frac{A}{1 + A\beta} \tag{5.2}$$

If $A\beta \gg 1$, the output voltage will be approximately equal to

$$V_o = V_{ref} \frac{1}{\beta} \tag{5.3}$$

V_o is therefore a regulated voltage, quite independent of the dc input voltage V_i and the load current in R_L.

5.2   OPERATION OF CURRENT-MODE-CONTROLLED REGULATORS

Figure 5.2 shows the block diagram of a current-mode-controlled regulator using a current-controlled dc-to-dc converter. The output current I_o in Fig. 5.2 is used to mean the averaged output current of the current-controlled converter. It is assumed here that I_o is directly proportional to the control voltage V_c, so that

$$I_o = GV_c \tag{5.4}$$

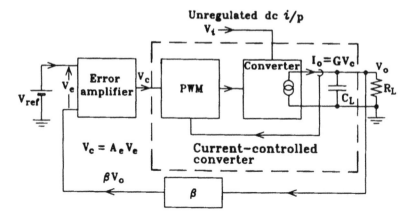

FIGURE 5.2 Current-mode-controlled regulator. (I_o is the averaged value of the converter output current.)

The operation of the regulator can be explained as follows:

1. Assume that, for some reason (such as a drop in V_i or a decrease in R_L), V_o decreases. This increases the error voltage V_e and there-fore also the feedback control voltage V_c, which is equal to A_eV_e.
2. The increase in V_c results in an increase in the dc output current GV_c, which in turn raises the output V_o so as to compensate for the drop assumed in point 1.
3. If, for any reason, V_o increases, the opposite of what is described in points 1 and 2 will occur to regulate the output.

It can be found that, under the steady state, there must exist an error voltage V_e that is just sufficient to maintain the existing I_o in the circuit. This amplitude of error voltage can be determined by considering the following equations, which are derived from Fig. 5.2:

$$V_e = V_{ref} - \beta V_o \tag{5.5}$$
$$= V_{ref} - \beta G V_c R_L \tag{5.6}$$
$$= V_{ref} - \beta G A_e V_e R_L \tag{5.7}$$

where A_e is the voltage gain (V_c/V_e) of the error amplifier. (Note that this A_e is different from the A of the voltage-mode-controlled regulator de-scribed in Section 5.1.)

$$\therefore V_e = \frac{V_{ref}}{1 + \beta A_e GR_L} \tag{5.8}$$

The voltages βV_o and V_o can then be found as

$$\beta V_o = V_{ref} - V_e \tag{5.9}$$

$$= V_{ref} - \frac{V_{ref}}{1 + \beta A_e GR_L} = V_{ref} \frac{\beta A_e GR_L}{1 + \beta A_e GR_L} \tag{5.10}$$

$$V_o = V_{ref} \frac{A_e GR_L}{1 + \beta A_e GR_L} \tag{5.11}$$

It is seen from Eq. (5.11) that if $\beta A_e GR_L$ is much larger than 1, V_o will be very close to V_{ref}/β and should be quite independent of V_i and R_L, thus implementing the function of a voltage regulator.

5.3 CHARACTERISTICS OF VOLTAGE-MODE-CONTROLLED REGULATORS AND CURRENT-MODE-CONTROLLED REGULATORS

From the current-mode-controlled regulator circuit shown in Fig. 5.2, it can be found that the output circuit of a current-controlled converter is a first-order single-pole system. The maximum phase shift between the control input V_c and the output V_o of the current-controlled converter is limited to $-90°$. The feedback circuit is therefore very stable, and the regulator can be easily designed to give very good dynamic regulation against transient changes in V_i and R_L. This compares favorably with voltage-mode-controlled regulators using continuous-mode-operated converters, whose output circuits are second- or higher-order systems.

However, when ordinary dc-to-dc converters, such as buck, buck-boost, boost, or their variants, are operated in the discontinuous-mode operation, their output equivalent circuits are also first-order single-pole systems (consisting effectively of a current source in parallel with a resistance and a filtering capacitance). Good stability and transient regulation can therefore also be obtained in these circuits, just as well as in circuits using current-controlled converters. There is, however, one important difference: while current-controlled converters can maintain their single-pole characteristics for either the continuous- or discontinuous-mode operation, ordinary converters lose their single-pole characteristics when they enter into the continuous-mode operation.

The disadvantages of current-mode-controlled regulators are the need to build stable current-controlled converters and their poor immunity to noise and interference. Current-controlled converters will be examined in greater detail in the next section.

5.4 OPERATION AND CHARACTERISTICS OF CURRENT-CONTROLLED dc-TO-dc CONVERTERS

In this section, the principle of operation, characteristics, and problems of current-controlled converters are discussed. Two typical examples of current-controlled converters will be covered. They are:

1. The hysteretic type
2. The constant-frequency type

The hysteretic type is closer to an ideal current-controlled converter than the constant-frequency type, but it has the disadvantage of requiring a variable switching frequency to control the output current. This is undesirable for circuits that are sensitive to variable-frequency interferences, such as video display units and telecommunication receivers. The constant-frequency type, on the other hand, operates at a constant switching frequency. However, it has a problem of instability when the duty cycle D of the electronic switch is larger than 0.5, as will be explained in Subsection 5.4.3.

5.4.1 Hysteretic Type of Current-Controlled Converter

Figure 5.3(a) shows the circuit of a hysteretic type of current-controlled buck converter. Figures 5.3(b) and 5.3(c) show the associated waveforms. The resistance R_f in Fig. 5.3(a) is a small sampling resistance to produce a sample voltage $i_L R_f$, representing the amplitude of the current i_L. This sample voltage is amplified by the amplifier A_1 and then applied to the input of the amplifier A_2, which is connected as a Schmitt trigger. The output of the Schmitt trigger is used to control the on-off of the electronic switch SW.

The operation of the converter can be explained as follows:

1. Assume that the operation of the converter has already reached a steady state and that the switch SW is driven to the on state by the output of the Schmitt trigger at $t = 0$. As a result, the inductor current i_L increases almost linearly, as shown in Fig. 5.3(c), with a slope equal to

$$\frac{di_L}{dt} = \frac{V_i - V_o}{L} \quad \text{(assuming } R_f = 0\text{)} \tag{5.12}$$

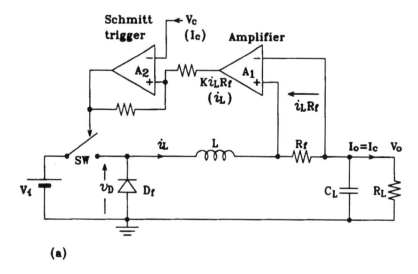

(a)

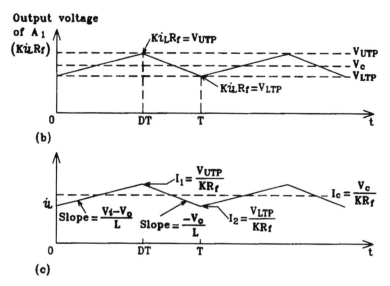

(b)

(c)

FIGURE 5.3 (a) Hysteretic type of current-controlled buck converter; (b) voltage waveform; and (c) current waveform.

2. When i_L increases to such a value that the amplified sample voltage $Ki_L R_f$ reaches the upper trigger point voltage V_{UTP} of the Schmitt trigger, as shown in Fig. 5.3(b) at $t = DT$, the output of the Schmitt trigger toggles to the low state and turns the electronic switch SW off. The inductor current i_L is then forced to pass through the fly-

wheel diode D_f. As a result, the amplitude of i_L decreases almost linearly, with a slope equal to

$$\frac{di_L}{dt} = \frac{-V_o}{L} \quad \text{(assuming } R_f = 0, v_D = 0) \tag{5.13}$$

3. When i_L decreases to such a value that the amplified sample voltage $Ki_L R_f$ falls to the lower trigger point voltage V_{LTP} of the Schmitt trigger, as shown in Fig. 5.3(b) at $t = T$, the output of the Schmitt trigger toggles back to the high state and turns the switch SW on again. The stages described in entries 1 and 2 will then repeat.

For the convenience of analysis, the voltage waveform shown in Fig. 5.3(b) can be converted into the current waveform shown in Fig. 5.3(c). The currents in Fig. 5.3(c) are obtained simply by dividing the voltages in Fig. 5.3(b) by KR_f. From these waveforms, it is observed that we can actually adjust I_c to any value by setting the control voltage to

$$V_c = KI_c R_f \tag{5.14}$$

This implements the function of a current-controlled converter, in which

$$I_o = I_c = \frac{V_c}{KR_f} = GV_c \tag{5.15}$$

5.4.2 Constant-Frequency Type of Current-Controlled Converter

Figure 5.4(a) shows the circuit of a constant-frequency type of current-controlled buck converter. Figure 5.4(b) shows the associated waveforms. The operation of the converter can be explained as follows:

1. Assume that the operation of the converter has already reached a steady state and that, at $t = 0$, the constant-frequency set-pulse generator generates a pulse v_{set} to set the Q output of the R–S flip-flop to high, to turn on the electronic switch SW. As a result, the inductor current i_L increases almost linearly, with a slope m_1 equal to

$$m_1 = \frac{di_L}{dt} = \frac{V_i - V_o}{L} \quad \text{(assuming } R_f = 0) \tag{5.16}$$

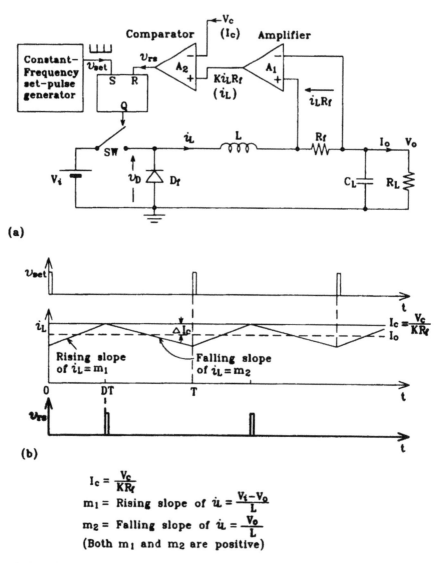

$$I_c = \frac{V_c}{KR_f}$$

$m_1 = \text{Rising slope of } \dot{u} = \frac{V_i - V_o}{L}$

$m_2 = \text{Falling slope of } \dot{u} = \frac{V_o}{L}$

(Both m_1 and m_2 are positive)

FIGURE 5.4 (a) Constant-frequency type of current-controlled buck converter and (b) voltage and current waveforms.

2. When i_L grows to such a value that the amplified sample voltage $Ki_L R_f$ reaches (actually, is slightly larger than) the control voltage V_c, shown in Fig. 5.4(a), the output of the comparator rises and resets the R–S flip-flop. This causes the output Q of the flip-flop to fall and to turn off the switch SW. The inductor current i_L then

decreases, with a falling slope m_2 equal to

$$m_2 = \frac{V_o}{L} \quad \text{(assuming } R_f = 0, v_D = 0) \tag{5.17}$$

(m_2 is defined to have a positive value.)

3. At $t = T$ (the end of a switching cycle), the set-pulse generator generates the next set pulse to turn on the switch SW again, thus restarting the switching cycle described in points 1 and 2.

Assuming that the converter is operating in the continuous-mode operation (i.e., i_L is always larger than zero), we have, from Fig. 5.4(b),

$$I_o = \frac{V_c}{KR_f} - \Delta I_C \tag{5.18}$$

$$= \frac{V_c}{KR_f} - \frac{1}{2} m_2 (1 - D)T$$

$$= \frac{V_c}{KR_f} - \frac{1}{2} \frac{V_o}{L} (1 - D)T \tag{5.19}$$

Although Eq. (5.19) is not in the ideal form of $I_o = GV_c$, we can still control the amplitude of I_o by varying V_c so as to implement the function of a current-controlled converter.

An advantage of the converter described above is its ability to operate at a constant switching frequency. But it has the disadvantage that, under the condition of duty cycle of $SW > 0.5$, the circuit can become unstable. This phenomenon will be explained in the following subsection.

5.4.3 Instability Problem in Constant-Frequency Current-Controlled Converters

When the duty cycle of the constant-frequency current-controlled converter shown in Fig. 5.4 is larger than 0.5, the circuit will become unstable. To find out how this happens, let us consider again the converter circuit shown in Fig. 5.4(a), but this time assuming a duty cycle D larger than 0.5. The new waveform of i_L is shown in Fig. 5.5. The solid line in Fig. 5.5 is the idealized steady-state waveform of i_L for a duty cycle D larger than 0.5. The dashed line shows how the waveform of i_L may change when a very small perturbation (disturbance) ΔI_1 is introduced in the initial value of i_L. Such a perturbation may be due to noise, interference, or other changes in the operating environment. The following analysis will show that, no

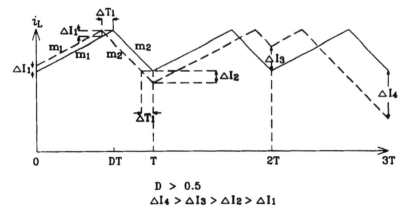

D > 0.5

$$\Delta I_4 > \Delta I_3 > \Delta I_2 > \Delta I_1$$

FIGURE 5.5 Waveform of i_L with perturbation ΔI_1 for $D > 0.5$.

matter how small ΔI_1 is, it will eventually cause an oscillation in the current i_L:

1. When ΔI_1 is introduced at $t = 0$, as shown in Fig. 5.5, the resultant change in ΔT_1 can be found by referring to the small triangle formed by ΔI_1 and ΔT_1:

$$\frac{\Delta I_1}{\Delta T_1} = m_1 \tag{5.20}$$

$$\therefore \Delta T_1 = \frac{\Delta I_1}{m_1} \tag{5.21}$$

2. As a result of the change ΔT_1, the change in i_L at $t = T$, denoted as ΔI_2, can be found by referring to the triangle formed by ΔT_1 and ΔI_2:

$$\frac{\Delta I_2}{\Delta T_1} = m_2 \tag{5.22}$$

$$\therefore \Delta I_2 = m_2 \, \Delta T_1 \tag{5.23}$$

Substitution of Eq. (5.21) into Eq. (5.23) gives

$$\Delta I_2 = \frac{m_2}{m_1} \, \Delta I_1 \tag{5.24}$$

3. Equation (5.24) shows that if m_2/m_1 is larger than 1, the perturbation will be amplified by a factor of m_2/m_1 each time the converter undergoes a switching cycle. Such repeated amplification for each cycle will eventually result in an oscillation in the waveform of i_L. The condition for a stable operation is therefore

$$\frac{m_2}{m_1} < 1 \qquad (5.25)$$

4. Since we know from Fig. 5.5 that, in the steady state,

$$m_1 DT = m_2(1 - D)T \qquad (5.26)$$

or

$$\frac{m_2}{m_1} = \frac{D}{1 - D} \qquad (5.27)$$

the condition for a stable operation may alternatively be stated as

$$\frac{m_2}{m_1} = \frac{D}{(1 - D)} < 1 \qquad (5.28)$$

$$D < 1 - D \quad \text{or} \quad D < 0.5 \qquad (5.29)$$

In order to maintain a stable operation, the duty cycle of the converter shown in Fig. 5.4(a) must therefore be kept below 0.5. As an example illustrating how a perturbation ΔI_1 may eventually die out in a converter operating at a duty cycle D of less than 0.5, Fig. 5.6 shows how ΔI_2 becomes

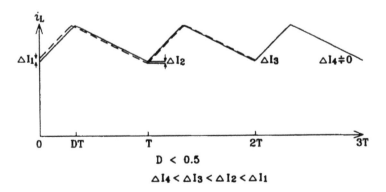

FIGURE 5.6 Waveform of i_L with perturbation ΔI_1 for $D < 0.5$.

smaller than ΔI_1 after the first switching cycle, then ΔI_3 becomes smaller than ΔI_2 after the second switching cycle, and so on.

A method that can be used to stabilize the operation of the converter circuit, which allows a duty cycle larger than 0.5, is to introduce into the control voltage V_c a sawtooth compensation wave, as explained in the next subsection.

5.4.4 Constant-Frequency Current-Controlled Converter with Compensation Slope

The stability problem mentioned in Subsection 5.4.3 for constant-frequency current-controlled converters with a duty cycle larger than 0.5 can be solved by adding a so-called compensation voltage v_{saw} to the control voltage V_c, as shown in Fig. 5.7(a). This compensation voltage v_{saw} is, in fact, a sawtooth wave with a linearly falling slope, the starting point of which is synchronized to the set-pulse output of the set-pulse generator. The compensation voltage v_{saw} and the control voltage V_c are added together to form v_c before it is applied to the input of the comparator A_2.

The introduction of a compensation slope in the control voltage produces the new current waveforms as shown in Fig. 5.7(b).

In this waveform diagram, we have

$$m_1 \text{ (the rising slope of } i_L) = \frac{V_i - V_o}{L} \quad \text{(assuming } R_f = 0)$$

$$m_2 \text{ (the falling slope of } i_L) = \frac{V_o}{L} \quad \text{(assuming } R_f = 0, v_D = 0)$$

$$m_c = \text{Compensation slope}$$

$$i_c = I_c - m_c t \quad \text{(control input for } 0 < t < T)$$

(Note that m_1, m_2, and m_c are all defined to have positive values in order to avoid confusion.)

To consider the stability of the converter, it is assumed that, in Fig. 5.7(b), a perturbation ΔI_1 in i_L is introduced at $t = 0$. In the following analysis, it will be found that if we can maintain

$$m_c > (1/2)m_2$$

then the resultant change of current ΔI_2 at the end of the switching cycle will be smaller than ΔI_1. This will ensure stable operation of the converter.

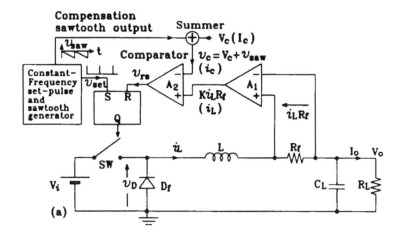

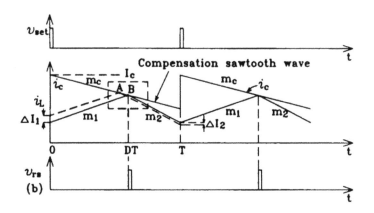

(a) Current-controlled converter with compensation slope and (b) voltage and current waveforms.

$$i_c = I_c - m_c t \quad \text{(for } 0 < t < T)$$

Compensation slope $m_c = |\text{Slope of } i_c|$

FIGURE 5.7 (a) Current-controlled converter with compensation slope and (b) voltage and current waveforms.

To enable us to determine the relationship between ΔI_1 and ΔI_2, the current waveforms within the dashed-line rectangle in Fig. 5.7(b) are expanded, as shown in Fig. 5.8. From Fig. 5.8, it is found that

$$\Delta I_2 = CA = CD - AD \tag{5.30}$$

$$= m_2 \Delta T_1 - m_c \Delta T_1 \tag{5.31}$$

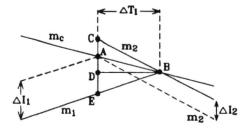

FIGURE 5.8   Expanded current waveform.

$$\Delta I_1 = AE = AD + DE \tag{5.32}$$
$$= m_c \, \Delta T_1 + m_1 \, \Delta T_1 \tag{5.33}$$
$$\therefore \frac{\Delta I_2}{\Delta I_1} = \frac{m_2 - m_c}{m_c + m_1} \tag{5.34}$$

In order to have a stable operation, we need to have

$$\frac{\Delta I_2}{\Delta I_1} < 1 \tag{5.35}$$

or

$$\frac{\Delta I_2}{\Delta I_1} = \frac{m_2 - m_c}{m_c + m_1} < 1 \tag{5.36}$$
$$m_2 - m_c < m_c + m_1 \tag{5.37}$$
$$2m_c > m_2 - m_1 \tag{5.38}$$

Knowing that, in the worst case, m_1 may be equal to zero, the condition for stability is therefore

$$m_c > (1/2)m_2 \tag{5.39}$$

where

$$m_2 = \frac{V_o}{L} \tag{5.40}$$

With a compensation slope m_c larger than half m_2, the converter circuit will be unconditionally stable for all values of the duty cycle (from 0 to 1).

Note that, in the preceding discussion of the stability problem, although only the buck converter has been considered, the same problem actually occurs in other types of constant-frequency current-controlled converters too. The techniques for solving the problem are also similar.

5.5 DESIGN OF FEEDBACK AND COMPENSATION CIRCUITS

This section is a qualitative discussion on the design of feedback and compensation circuits that can enable dc-to-dc converters (whether ordinary or current-controlled) to operate as regulators. The emphasis here is on understanding the fundamental principles that are essential for the computer-aided design work to be discussed in later chapters.

Consider the regulator block diagram, Fig. 5.9, which is valid for both voltage-mode- and current-mode-controlled regulators. For the voltage-mode-controlled circuit, the output V_o is directly controlled by V_c. For the current-mode-controlled circuit, the output I_o is directly controlled by V_c. However, for either case, it is possible to obtain a transfer function relating the output voltage $V_o(s)$ to the control input $V_c(s)$:

$$A_{pc}(s) = \frac{V_o(s)}{V_c(s)} \qquad (5.41)$$

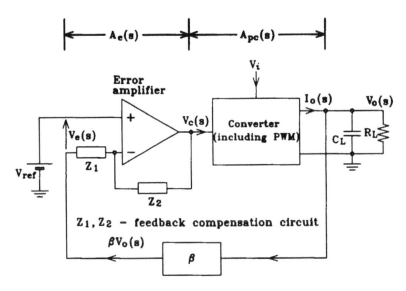

FIGURE 5.9 Regulator block diagram. Loop gain $A_L(s) = A_e(s)A_{pc}(s)\beta$.

In Fig. 5.9, the feedback circuit β is a simple resistive potential divider, and $A_e(s)$ is the frequency-response characteristic of the error amplifier (together with Z_1 and Z_2):

$$A_e(s) = \frac{V_c(s)}{V_e(s)} \approx \frac{Z_2}{Z_1} \tag{5.42}$$

Assume that the impedances Z_1 and Z_2 in Fig. 5.9 are the feedback and compensation circuit to be designed. Either Z_1 or Z_2 may consist of a number of components. In selecting the components (and component values) for the impedances Z_1 and Z_2, our objective is to ensure that the regulator:

1. Operates in a stable manner (without oscillation)
2. Meets the steady-state line and load regulation specifications
3. Meets the transient-response specifications (with acceptable overshoot and settling time, etc.)

In an idealized design process, the required performance specifications (such as line and load regulations and transient response) should first be transformed into a set of objective frequency-domain requirements for the loop-gain $A_L(s)$ of the feedback loop to achieve. The compensation circuit in the feedback loop should then be designed to provide the necessary frequency characteristic.

In practice, it is not a straightforward matter to translate a set of time-domain performance specifications into a set of open-loop frequency-domain requirements. This is particularly difficult for nonlinear circuits working under large signal conditions. Instead, experienced circuit designers often assume a set of objective dc gain, bandwidth, and phase margin requirements and then design the feedback circuit to fit these requirements.

As a guideline for the less experienced designer, it may be assumed that the objective loop-gain characteristic $A_L(s)$ to be achieved should look like a single-pole function, as shown in Fig. 5.10, with suitable dc gain, unity-gain bandwidth, and phase margin. The required frequency-response characteristic of the error amplifier (together with the feedback ratio β), $A_e(s)\beta$, is then the difference between the objective loop-gain characteristic $A_L(s)$ and the converter (and PWM) characteristic $A_{pc}(s)$, as shown in Fig. 5.11.

In selecting the objective gain, bandwidth, and phase margin, the following are the basic requirements/considerations:

1. Since the dc loop gain $A_L(o)$ directly controls the steady-state line and load regulations, it should be selected to satisfy at least the

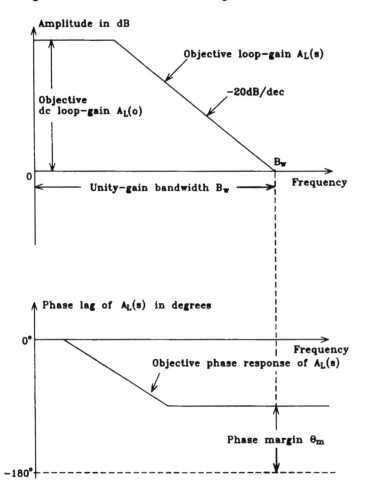

FIGURE 5.10   Bode plot of objective loop-gain characteristic $A_L(s)$.

minimum dc line and load regulation requirements. From negative feedback theory, it can be found that the closed-loop steady-state line and load regulations would be improved by a factor of $[1 + A_L(o)]$ when compared with the open-loop performance. Given any set of open-loop characteristics, the required dc loop gain $A_L(o)$ can therefore be determined.

2. Even in the worst case, the phase margin θ_m should be larger than 45°.

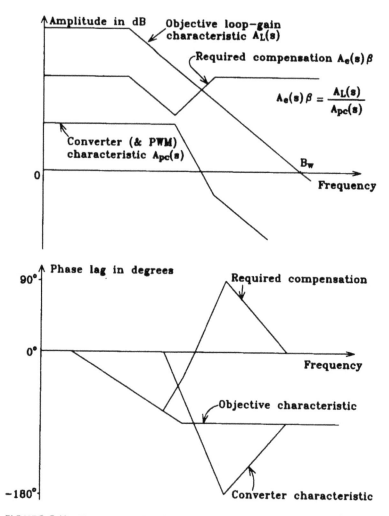

FIGURE 5.11 Bode plot of objective loop-gain characteristic $A_L(s)$ and required compensation.

3. It should be well understood that a wider open-loop bandwidth, together with a larger phase margin, should give a better closed-loop transient response.

4. Because of the need to remove the switching-frequency spikes in the feedback voltage, the maximum unity-gain bandwidth, B_w, of the loop-gain characteristic should be smaller than a quarter of the

converter switching frequency f_s. If there are additional constraints, such as those due to the existence of RHP (right-half-plane) zeros, the unity-gain bandwidth may have to be further reduced to ensure a sufficient phase margin.

When designing a feedback and compensation circuit, the designer should also be aware that the equivalent circuit of a converter may depend on its operating condition. For example, the continuous-mode equivalent circuit is normally different from the discontinuous-mode equivalent circuit. For some converters, such as the boost, buck-boost, flyback, and Ćuk converters, even within the continuous-mode operation, the effective inductance of the output equivalent circuit can also vary as a function of the duty cycle. (Effective inductance $= L/[(1 - D)^2]$)

In designing the feedback compensation circuit, we therefore need to consider all the worst-case combinations. For example, when we specify the minimum dc loop-gain requirement, we must ensure that it is large enough to provide not only the required line regulation for the continuous-mode operation but also the required load regulation for the discontinuous-mode operation (if the converter is expected to operate in both modes of operation). Similarly, when we consider the phase margin for a regulator employing a boost, buck-boost, flyback, or Ćuk converter, we need to ensure that an acceptable phase margin is maintained even for the maximum duty cycle condition (when the gain and phase lag in the $\delta V_o(s)/\delta D(s)$ transfer function are at their maximum).

5.6 SPECIAL DESIGN CONSIDERATIONS FOR CURRENT-MODE-CONTROLLED REGULATORS

A current-mode-controlled regulator has, in fact, two feedback loops. These are:

1. The inner current feedback loop within the current-controlled converter
2. The outer voltage feedback loop

While the inner current feedback loop is used to implement only the function of a controlled dc current source, the outer voltage feedback loop is used to regulate the output voltage. In order to achieve a stable operation of the regulator, both feedback loops must be maintained stable.

Since the outer voltage feedback loop for a current-mode-controlled regulator is basically a first-order feedback system, it is relatively easy to design, and the resultant regulator can also have good transient line and load regulations.

Because of its current-control nature, the function of current limit (for protection purposes) in a current-mode-controlled regulator can be achieved by simply limiting the control voltage V_c. Hence, no extra current detector or current-limiting circuitry is required.

Parallel operation of current-mode-controlled regulators can be implemented easily in order to increase the output current rating of such regulators. An example scheme is shown in Fig. 5.12. In this circuit, a single error amplifier A_e is used to provide a common control voltage V_c to all three current-controlled converters. Appropriate adjustment of the gain parameters G_1, G_2, and G_3 ensures that the output loading current is shared properly by the three regulators.

In the design of current-mode-controlled regulators, special care should be taken to minimize the noise and interference in both the current feedback loop and the voltage feedback loop.

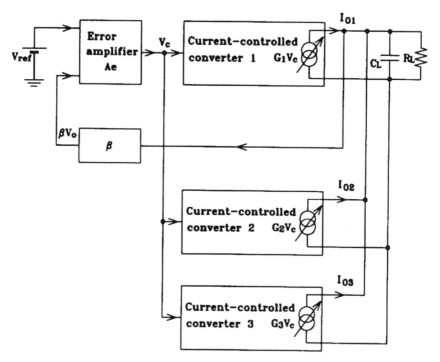

FIGURE 5.12 Parallel operation of current-mode-controlled regulators.

5.7 SUMMARY AND FURTHER REMARKS

In this chapter, the principles of operation and characteristics of voltage-mode-controlled and current-mode-controlled regulators have been examined and compared. The implementation and the stability problem of current-controlled converters have been discussed. The principles of designing feedback and compensation circuits have also been introduced.

The objective of Chapters 1 to 5 is to familiarize the reader with the principles of operation and the characteristics of square-wave power converters and regulators. The emphasis is on understanding the fundamental principles. Such a background is essential for the simulation and computer-aided design work to be discussed in Chapters 6 to 9.

EXERCISES

1. Compare voltage-mode-controlled regulators and current-mode-controlled regulators. Consider aspects of their
    a. principle of operation
    b. performance
    c. relative advantages and disadvantages
2. What are the relative advantages and disadvantages of the constant-frequency type of current-controlled converter compared with the hysteretic type of current-controlled converter?
3. With the aid of circuit and waveform diagrams, explain the operation of a constant-frequency, current-controlled
    a. forward converter
    b. flyback converter
    Does the problem of instability discussed in Subsection 5.4.3 occur in these converters?
4. Explain the general principles and essential requirements in the design of a feedback and compensation circuit for a regulator.
5. Explain why current-mode-controlled regulators are suitable for parallel operation to increase the output current whereas voltage-mode-controlled regulators are not.

6
Cycle-by-Cycle Simulation of Power Converter Circuits

Direct and manual analysis of power converters and switching regulators can be very difficult when exact solutions are required. This is particularly so when various secondary effects, such as the nonperfect switching behavior of transistors, losses of inductors and capacitors, and nonlinear characteristics of behavior models, must be included in the analysis. Under such circumstances, computer-aided design (CAD) tools will become indispensable to the circuit designer.

In this and the following three chapters, extensive use will be made of the computer simulation program SPICE (Simulation Program with Integrated Circuit Emphasis) as a tool for the analysis and design of converter and regulator circuits.

Two levels of SPICE simulation can be used for the analysis and design of converters and regulators:

1. The direct cycle-by-cycle simulation of the switching action of the converter circuit. This kind of time-domain simulation is referred to as transient simulation in SPICE.
2. The simulation based on the low-frequency behavior model derived from the converter or regulator circuit. This includes the dc, ac (frequency-domain), and transient simulations in SPICE.

Whereas the direct cycle-by-cycle simulation of the converter circuit is useful for the prediction of detailed switching waveforms, the simulation based on the low-frequency behavior model is more appropriately used for the design of feedback circuits and for the prediction of the overall performance of the complete regulator, such as line and load regulations.

This chapter describes the cycle-by-cycle simulation of the converter circuit and the modeling of discrete circuit components such as transistors, inductors, transformers, and capacitors. Chapter 7 will deal with simulations based on the low-frequency behavior models and the modeling of various types of converters.

Readers unfamiliar with the SPICE program are referred to Appendix A, "An Introduction to SPICE," which explains the nature, format, and applications of SPICE. For the more advanced SPICE users, Appendix B, "Summary of Commonly Used SPICE Statements," and Appendix C, "Nonconvergence and Related Problems in SPICE," are included as references to help them solve practical simulation problems.

It should be mentioned that the graphic outputs of the examples given in Chapters 6–8 have actually been processed by the Probe program of PSpice, a commercial version of SPICE, to improve their appearance. More information about PSpice can be found in Appendix D, "About PSpice."

6.1 USES OF CYCLE-BY-CYCLE SIMULATION

A very useful application of SPICE in the analysis and design of switch-mode power supplies is the cycle-by-cycle simulation of the switching action of converters. Such simulations can provide very useful data, either directly or indirectly, for analysis and design purposes. These data include:

1. The voltage and current waveforms and stresses
2. The switching trajectory of devices
3. The switching losses of components

A problem that must be solved in the cycle-by-cycle simulation is the modeling of power switches, diodes, resistors, capacitors, inductors, and transformers. In Sections 6.2–6.5, various techniques for modeling such components will be discussed. The practical problems and examples of simulation will be studied in Section 6.6.

6.2 MODELING A NEAR-IDEAL SWITCH

In the simulation of circuits such as that shown in Fig. 6.1, a model has to be given to the switching transistor. The SPICE model parameters of the transistor can often be obtained from transistor manufacturers or software vendors. If such parameters are not available, however, we may have to develop a switch model to replace the transistor.

An almost ideal switch is not difficult to model in the SPICE program. However, the idealized infinite off-state resistance or zero on-state resist-

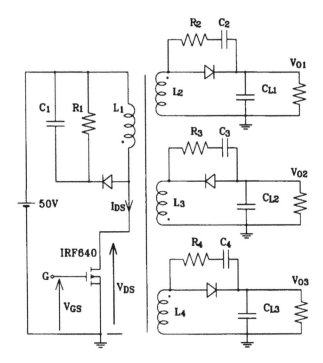

FIGURE 6.1 Flyback converter.

ance often results in nonconvergence problems in simulations that involve the switching of inductive or capacitive components because such circuits may produce, in theory, infinitely large voltages or currents under transient conditions.

To make a switch model practically useful, the maximum off-state resistance and minimum on-state resistance should be limited to practically acceptable values. By using a voltage-controlled current source, a near-ideal switch can be modeled as shown in Fig. 6.2(a). In this example, the switch is modeled as a voltage-controlled current source, named G89 (for more details about such controlled sources, see Sections 3.9–3.12 of Appendix B):

```
G89  8  9  POLY(2)  8  9  1  0  0  1E-6  0  0  1000
```

which means that the current G89 is given by

$$G89 = V(8,9)[1000 \ V(1,0) + 10^{-6}] \tag{6.1}$$

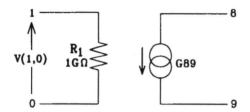

```
R1  1  0  1G
G89  8  9  POLY(2)  8  9  1  0  0  1E-6  0  0  1000
```

(a)

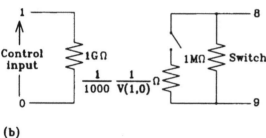

(b)

FIGURE 6.2 Near-ideal switch: (a) SPICE model, and (b) equivalent circuit.

where $V(8,9)$ represents the voltage at node 8 with respect to node 9, and $V(1,0)$ represents the control voltage at node 1 with respect to node 0. (R_1 is a dummy resistance to satisfy the SPICE requirement that there are at least two circuit elements connected to each node.)

If the control voltage $V(1,0)$ is zero, current G89 will be reduced to $V(8,9)10^{-6}$, meaning that the off-state resistance of the switch is $10^6\ \Omega$. However, if the control voltage $V(1,0)$ is assumed to be 1 V, the current G89 will be equal to

$$G89 = V(8,9)\ [1000 + 10^{-6}] \tag{6.2}$$

Equation (6.2) indicates that the conductance between nodes 8 and 9 is now approximately 1000 S (equivalent to 1 mΩ). This results in a practically closed switch between nodes 8 and 9. The equivalent circuit of the switch model is shown in Fig. 6.2(b).

Based on the switch model given in Fig. 6.2(a), simulations can be carried out to find the voltage and current waveforms of the converter circuit.

6.3 MODELING SEMICONDUCTOR DEVICES

The development of a full and accurate SPICE model for a semiconductor
device for all applications can be difficult when information is insufficient.
However, if such devices are to be used as switches, only two characteristics
are of prime importance: the switching speed and the on-state voltage drop.
Given the data sheet of a semiconductor device, it should not be difficult
for a circuit designer to develop a useful SPICE model for switching applications.

In the following three subsections, the modeling techniques for MOS
transistors, bipolar transistors, and diodes will be discussed.

6.3.1 Modeling MOS Switching Transistors

Depending on the type and amount of information available, an MOS
switch can be modeled using one of the following two approaches:

1. If all the necessary SPICE model parameters are available, they can
    be entered directly into the model file of the transistor. The follow-
    ing is an example:

```
M1  1  2  0  0  IRF640  W=0.66  L=2U
* The preceding statement means the following:
* 1. The drain, gate, source, and substrate of
*    the MOS transistor M1 are connected to
*    the nodes 1, 2, 0, and 0, respectively.
* 2. The transistor's model name is IRF640,
*    whose parameters are given under the
*    .MODEL IRF640 statement, as shown below.
* 3. The effective channel width of M1 is 0.66 m.
* 4. The effective channel length of M1 is 2 µm.
* Note that, while in SPICE the parameters W=
* 0.66 and L=2U must follow the device statement
* (as shown above), such parameters may either
* follow the device statement or be included in
* the .MODEL statement in PSpice.
```

```
.MODEL        IRF640        NMOS
+ (LEVEL=3    GAMMA=0       DELTA=0
+ ETA=0       THETA=0       KAPPA=0
+ VMAX=0      XJ=0          TOX=100N
+ UO=600      PHI=0.6       RS=19.61M
```

```
+ KP=20.73U       VTO=3.788       RD=95.58M
+ CBD=1.872N      PB=0.8          MJ=0.5
+ FC=0.5          CGSO=1.745N     CGDO=334.7P
+ IS=16.39P)
```

The meaning of the symbols used in the above .MODEL statement can be found from SPICE user's guides [57,125].

2. If SPICE model parameters are not available but transistor data sheets are given, we can still produce a good model of the transistor by starting with an almost ideal transistor model and then adding elements to model the imperfections of the transistor.

Consider again the example of a power MOS transistor. This time, we assume that the SPICE parameters are not given, but we know from data sheets the following key parameters that affect the switching behavior of the transistor:

Turn-on threshold voltage V_{TO} = 3.8 V
On-state resistance R_{DS} = 0.085 Ω
Gate to drain capacitance C_{GD} = 400 pF
Gate to source capacitance C_{GS} = 2500 pF
Drain to source capacitance C_{DS} = 1000 pF

Based on the parameters given above, a model can be developed as shown in Fig. 6.3. In this model, the transistor M1 is assumed to be an almost ideal MOS transistor, with an effective channel width W of 2 m and an effective channel length L of 2 μm:

```
M1   4   2   3   3   MOSMOD   W=2   L=2U
```

It should be noted that the channel width W of 2 m does not mean that the transistor is physically 2 m wide. What it implies is: The sum of the effective widths of the many parallel transistors on the transistor chip area is 2 m. This effective channel width, together with the channel length and other default parameters, determines the drain current for any given gate and drain voltages:

$$I_{DS} = K_p \frac{W}{L} (V_{GS} - V_{TO})V_{DS} \tag{6.3}$$

where I_{DS} is the drain current, $K_p(W/L)$ the transconductance, V_{GS} the gate to source voltage, V_{TO} the threshold voltage, and V_{DS} the drain to source voltage, of the MOS transistor.

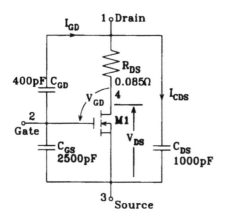

```
* SPICE SUBCIRCUIT OF MOS TRANSISTOR
.SUBCKT MOS   1 2 3
RDS  1 4   0.085
CGD  1 2   400PF
CGS  2 3   2500PF
CDS  1 3   1000PF
M1  4 2 3 3   MOSMOD W=2 L=2U
.MODEL MOSMOD NMOS (VTO=3.8)
.ENDS MOS
```

FIGURE 6.3   MOS model.

In the model given in Fig. 6.3, it is not always necessary to specify an accurate value of W for M1, except that it should be large enough. However, the default value of W in SPICE, 100 μm, is unsuitable for modeling power MOS transistors. The complete transistor circuit shown in Fig. 6.3 can be more conveniently modeled as a subcircuit. The required subcircuit SPICE listing is also given in Fig. 6.3. (More about the definition and uses of SPICE subcircuits can be found in Section 3.13 of Appendix B.)

6.3.2   Modeling of Bipolar Switching Transistors

In the modeling of a bipolar transistor, if all necessary SPICE model parameters are available, they can be entered directly into the model file. The following is an example:

```
.MODEL Q2N3055 NPN (IS=974.4F XTI=3
+ EG=1.11 VAF=50 BF=99.49 NE=1.941
```

```
+  ISE=902.5P  IKF=4.029  XTB=1.5  BR=2.949
+  NC=2  ISC=0  IKR=0  RC=0.1  CJC=276P
+  VJC=0.75  MJC=0.3333  FC=0.5  CJE=569.1P
+  VJE=0.75  MJE=0.3333  TR=971.7N  TF=39.11N
+  ITF=20  VTF=10  XTF=2  RB=0.1)
```

The meaning of the terms used can be found from SPICE user's guides [57,125].

If SPICE model parameters, such as those given above, are not all available, we can still model a bipolar transistor reasonably accurately if we have the following data:

1. The forward current gain β, which is called B_F in the SPICE program
2. The zero-bias collector-to-base depletion capacitance C_{JC} and the zero-bias base-to-emitter depletion capacitance C_{JE}
3. The collector ohmic resistance R_C
4. The forward transit time T_F

(Note that since subscripts are not allowed in SPICE files, the parameters B_F, C_{JC}, C_{JE}, R_C, and T_F, are denoted as BF, CJC, CJE, RC, and TF, respectively in the SPICE file.)

Of the parameters mentioned above, B_F, C_{JC}, and C_{JE} can be found directly from data sheets, or estimated by experience. The collector ohmic resistance R_C can be determined from the relationship

$$R_C = \frac{\text{collector to emitter saturation voltage}}{\text{saturation collector current}} \qquad (6.4)$$

The forward transit time T_F can also be calculated from the relationship

$$T_F = \frac{1}{2\pi f_T} \qquad (6.5)$$

where f_T is the gain-bandwidth product of the transistor. f_T can normally be found from data sheets.

When the parameters BF, CJC, CJE, RC, and TF are found, they can be entered into the .MODEL statement as shown in the preceding example. For all unspecified parameters, the default values will be used.

In the worst case, if we know very little about the bipolar transistor except that it is a fast switching transistor with a very small saturation resistance, a default bipolar transistor can still be used. The required model statement is then simply:

.MODEL QMOD1 NPN

for an NPN transistor model called QMOD1, or

.MODEL QMOD2 PNP

for a PNP transistor model called QMOD2. The default bipolar transistor has

$$B_F = 100$$
$$C_{JC} = 0$$
$$C_{JE} = 0$$
$$R_C = 0$$
$$T_F = 0$$

implying that it is assumed almost perfect.

6.3.3 Modeling of Diodes

In the modeling of a switching diode the following characteristics should be included:

1. The zero-bias junction capacitance C_{JO}
2. The forward voltage drop due to ohmic resistance R_S
3. The saturation current I_s in the diode current equation

$$I = I_s\{e^{[Vq/(nkT)]} - 1\} \tag{6.6}$$

   (I_s is one of the key parameters that determines the forward voltage drop for a given current.)
4. The transit time T_T (a major factor that determines the turnoff recovery time)

The zero-bias junction capacitance C_{JO} can be found from data sheets. The ohmic resistance R_S of the diode can also be estimated by measuring the slope of

$$\frac{\delta V}{\delta I} = R_S \tag{6.7}$$

at the high-current end of the $V - I$ characteristic of the diode.

The determination of the saturation current I_s, however, requires some calculation based on the theoretical equation

$$I = I_s\{e^{[Vq/(nkT)]} - 1\} \tag{6.8}$$

where $q/(KT) = 1/25$ mV and $n = 1$ for silicon or Schottky diodes. The circuit designer will need to sample a point at the low-current end of the diode $V - I$ characteristic. This pair of voltage and current readings should satisfy Eq. (6.8). Substitution of these values into the equation

$$I_s = \frac{I}{\{e^{[Vq/(nkT)]} - 1\}} \tag{6.9}$$

will give the value of I_s, which can then be put into the SPICE model file.

Now, let us consider the turnoff reverse recovery time. For Schottky diodes, it can be assumed to be zero because practically only major carriers are used to conduct current. (It is only the junction capacitance that limits the switching speed of a Schottky diode.) For other types of junction diodes, the nonzero reverse recovery time can be modeled by entering a suitable transit time T_T into the SPICE model file.

Unfortunately, diode data sheets usually do not give the transit time T_T directly. Instead, the measured reverse recovery time t_{rr} under specified testing conditions is normally given. The circuit designer therefore needs to know how t_{rr} is measured and how T_T is related to t_{rr}.

A method commonly used to determine the reverse recovery time is shown in Fig. 6.4. The pulse generator v_g in Fig. 6.4(a) suddenly switches the voltage v from a positive value to a negative value, as shown in Fig. 6.4(b). For the idealized current waveform shown in Fig. 6.4(c), the theoretical transit time T_T can be found from the relationship (assuming $C_{JO} = 0$)

$$T_T I_F = t_{rr} I_R \tag{6.10}$$

or

$$T_T = \frac{I_R}{I_F} t_{rr} \tag{6.11}$$

However, in the practical current waveform shown in Fig. 6.4(d), manufacturers often define t_{rr} as the time required for the reverse current to fall

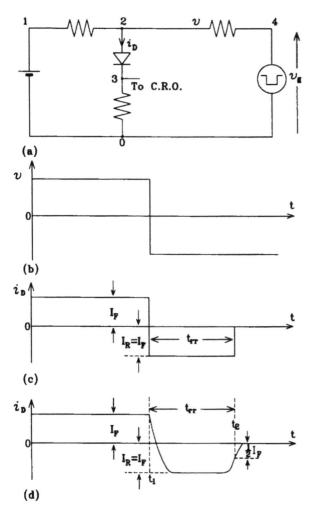

FIGURE 6.4 Measurement of diode reverse recovery time: (a) testing circuit; (b) voltage waveform; (c) idealized current waveform; (d) practical current waveform.

to one-half of the amplitude of the forward current. In the measurement of t_{rr}, the maximum reverse current I_R is normally also limited to I_F or $2I_F$. (The measured values of t_{rr} for different values of I_R are, of course, different.) Approximately, it may be assumed that

$$T_T \approx t_{rr} \qquad \text{if } I_R = I_F \tag{6.12}$$

$$T_T \approx 2t_{rr} \qquad \text{if } I_R = 2I_F \tag{6.13}$$

Based on Eqs. (6.12) and (6.13), the value of T_T, which needs to be entered into the SPICE model file, can be estimated from the t_{rr} specified in the data sheets.

When all the parameters C_{JO}, R_S, I_S, and T_T are found, a diode model can be prepared as shown in the following example:

```
.MODEL DMOD D (CJO=100P RS=0.5 IS=1E-13
+ TT=1E-8 N=1)
```

where N is the emission coefficient n in Eq. (6.8). Note that a diode can be modeled to have a very small forward voltage drop by simply specifying a small emission coefficient n (0.001, for example). Diodes modeled in this way are extremely useful for the simulation of near-ideal clipping or clamping circuits.

6.4 MODELING OF LINEAR CAPACITORS, INDUCTORS, AND TRANSFORMERS

In modeling an output filtering capacitor for a converter, it is desirable to include its effective series resistance in the model. This series resistance can significantly increase the ripple component of the output voltage. It can also change significantly the high-frequency characteristic of the transfer function between the control input and the output of the converter. The effective series resistances of capacitors can often be obtained from their specification sheets or be determined through measurements. Typical values may range from 0.001 to 0.5 Ω.

Sometimes, it may also be necessary to add a large resistance (e.g., 1 GΩ) across a capacitor to avoid the "floating node" problem encountered in SPICE simulation.

In the modeling of an inductor, it is a good practice to place a small resistance in series with the inductance to represent the dc resistance and high-frequency losses of the inductor. This not only improves the accuracy of the simulation but also eliminates other potential problems (e.g., SPICE does not allow a voltage source to be connected directly across an inductor with zero series resistance because this would imply that the circuit can have any value of dc current). While the dc resistance can be measured directly using an ohmmeter, the total effective series resistance at high frequency can be estimated from a Q-factor measurement.

In SPICE, transformers can be modeled as coupled inductors. For example, the transformer in Fig. 6.5 may be described in the SPICE format as

```
L1  1   2   0.5
L2  3   4   1
K   L1  L2  0.98
```

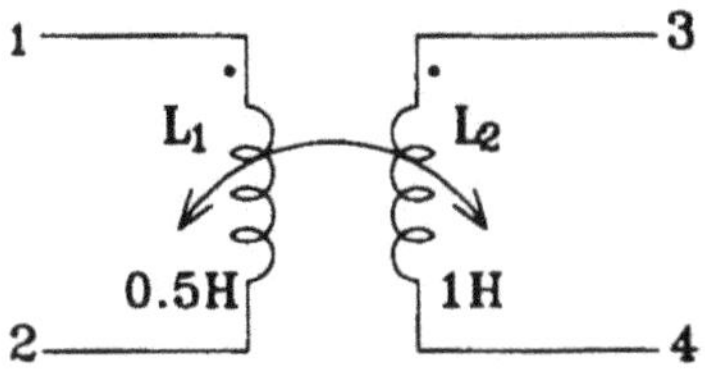

Coupling coefficient K=0.98

```
L1 1 2  0.5
L2 3 4  1
K L1 L2 0.98
```

FIGURE 6.5 SPICE model of linear transformer.

To ensure that the transformer model is reasonably accurate, the lossy components of the transformer may also be included in the model. The use of a series resistance to represent the lossy component in an inductor, as described above, can be equally applied to a transformer.

6.5 MODELING OF NONLINEAR RESISTORS, INDUCTORS, TRANSFORMERS, AND CAPACITORS

In the simulation of converter circuits, it may be necessary to take into account the nonlinear characteristics of various circuit components. Typical examples include the nonlinear resistances and capacitances of switching devices and inductors/transformers working near magnetic saturation. Although such nonlinear characteristics are readily available from data books, including them in the SPICE model is not so straightforward.

The objectives of this section are:

1. To develop a general approach to the SPICE simulation of arbitrarily nonlinear characteristics (such as those obtained from data books) of resistors, inductors, transformers, and capacitors, using the curve-fitting method.
2. To identify and solve potential problems that may be encountered in the simulation of the nonlinear characteristics of circuit components.

In Subsection 6.5.1, the basic principles of the modeling method, based on the example of a nonlinear resistance, will be introduced. The modeling of nonlinear inductors, transformers, and capacitors will be described in Subsections 6.5.2, 6.5.3, and 6.5.4, respectively. The applications and limitations of the modeling method will be discussed in Subsection 6.5.5. A simulation example with nonlinear resistance, capacitance, and transformer will be given in Attachment 6.2 at the end of this chapter.

6.5.1 Basic Principles

In the simulation of the nonlinear characteristic of a circuit component, the method we use here consists of

1. A curve-fitting process in which the nonlinear characteristic of the circuit component is expressed mathematically as a polynomial series
2. A modeling process in which the behavioral response of the nonlinear component that is due to any excitation is modeled as the response of a linearized component scaled by the polynomial series obtained in step 1.

Consider, for example, the nonlinear resistance shown in Fig. 6.6(a). Assume that we can use a curve-fitting program (such as MATLAB) to find an expression for the resistance R, so that

$$R = R_o(1 + K_{R1}I + K_{R2}I^2 + K_{R3}I^3 + \cdots) \tag{6.14}$$

The voltage drop due to a current I is then given by

$$\begin{aligned} V &= IR \\ &= IR_o(1 + K_{R1}I + K_{R2}I^2 + K_{R3}I^3 + \cdots) \\ &= SIR_o \end{aligned} \tag{6.15}$$

where

$$S = 1 + K_{R1}I + K_{R2}I^2 + K_{R3}I^3 + \cdots \tag{6.16}$$

The parameter S in Eq. (6.15) will be referred to as the scaling factor for the voltage drop of the nonlinear resistance.

In order to simulate the voltage drop across the nonlinear resistance, we shall first develop a scaling factor circuit to produce a voltage representing the scaling factor S. A nonlinearly scaled voltage source, whose amplitude is made equal to SIR_o (the product of S and IR_o) will then be used as the equivalent circuit of the nonlinear resistance R. Note that while nonlinear resistances are not supported directly by SPICE, nonlinear controlled sources (current or voltage), such as SIR_o, are easily simulated in the program. (For details, see Sections 3.9–3.12 of Appendix B.)

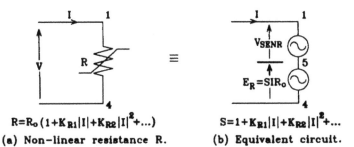

$R = R_0 (1 + K_{R1}|I| + K_{R2}|I|^2 + ...)$

(a) Non-linear resistance R.

$S = 1 + K_{R1}|I| + K_{R2}|I|^2 + ...$

(b) Equivalent circuit.

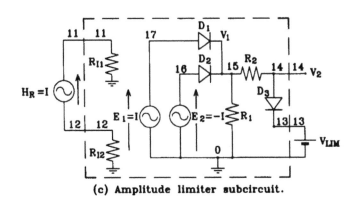

(c) Amplitude limiter subcircuit.

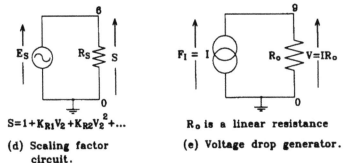

$S = 1 + K_{R1}V_2 + K_{R2}V_2^2 + ...$

(d) Scaling factor circuit.

R_0 is a linear resistance

(e) Voltage drop generator.

FIGURE 6.6 Modeling nonlinear resistance.

In the actual simulation of Eq. (6.15), special attention should, however, be paid to the following potential problems:

1. In many cases, the nonlinear resistance

$$R = R_o(1 + K_{R1}I + K_{R2}I^2 + K_{R3}I^3 + \cdots)$$

   is used to mean

$$R = R_o(1 + K_{R1}|I| + K_{R2}|I|^2 + K_{R3}|I|^3 + \cdots) \tag{6.17}$$

   where $|I|$ is the absolute amplitude of the current.

2. When using a polynomial series to approximate a nonlinear function, there is a limited range within which good accuracy can be maintained. Outside this range, the error in Eq. (6.14) or (6.17) can easily increase to absurdly large values and possibly cause nonconvergence problems in simulations.

To solve problem 1 mentioned above, it is necessary that the scaling factor S be generated as a function of $|I|$ (instead of I):

$$S = 1 + K_{R1}|I| + K_{R2}|I|^2 + K_{R3}|I|^3 + \cdots \tag{6.18}$$

To solve problem 2, the value of $|I|$ in Eq. (6.18) needs to be limited within the range for which curve fitting has been done to obtain the polynomial expression (6.14) or (6.17).

Figure 6.6 gives the detailed modeling arrangements. There, the nonlinear resistance R in Fig. 6.6(a) is modeled as the voltage-controlled voltage source E_R in Fig. 6.6(b). (The dummy voltage source V_{SENR} is used only to sense the current I.) Figures 6.6(c) and 6.6(d) show the arrangement used to generate a range-limited scaling factor S to satisfy Eq. (6.18). In this example, we use a full-wave rectifier, shown as D_1 and D_2 in Fig. 6.6(c), to obtain a voltage V_1 representing the current $|I|$. The amplitude limiter (consisting of the resistor R_2, diode D_3, and voltage source V_{LIM}) is used to limit the value of V_1, so as to confine its value to the curve-fitting range of Eq. (6.14) or (6.17). Only the range-limited voltage V_2 (representing $|I|$) is used to generate the scaling factor S, based on Eq. (6.18), as shown in Fig. 6.6(d). The circuit in Fig. 6.6(e) is used to produce the voltage IR_o, to enable an $E_R = SIR_o$ to be generated in Fig. 6.6(b).

An example of the SPICE listing for a nonlinear resistance model can be found in the simulation file given in Attachment 6.2, "Cycle-by-Cycle Simulation of Flyback Converter with Nonlinear RCL" (at the end of this chapter). In this example, the listing is prepared in the form of a subcircuit called R_{DS}, which represents the on-state drain resistance of the MOS transistor in Fig. A6.2.1. The actual characteristic of R_{DS} is shown by the curve AB in Fig. A6.2.2(a). When the subcircuit listing in Section A6.2 is examined, the following should be noted:

1. The R_{DS} subcircuit listing starts with the statement

    ```
    .SUBCKT RDS 1 4
    ```

    and the ends with the statement

    ```
    .ENDS RDS
    ```

2. The node numbers used in the subcircuit listing are the same as those shown in Fig. 6.6. (Only the term R is changed to R_{DS}.)
3. The R_{DS} subcircuit listing contains a nested subcircuit call XLIMR for a LIMITER, the listing of which is given (at the beginning of the input file) under the statement

    ```
    .SUBCKT LIMITER 11 12 13 14
    ```

    The node numbers used in the LIMITER subcircuit are also the same as those used in Fig. 6.6(c).
4. The voltage-controlled voltage source

    ```
    ES 6 0
    + POLY(1) 14 0 1 1.61082543625219E-2
    + . . .
    ```

    is used to simulate the scaling factor S. The meaning of the coefficients can be found by referring to the voltage-controlled current source examples given in Section 3.9 of Appendix B. These coefficients are obtained by curve fitting using MATLAB.

6.5.2 Modeling of Nonlinear Inductances

In the modeling of a nonlinear inductance, the method of nonlinearly scaling the voltage drop of the linearized component, as described in Subsection 6.5.1, can also be used.

Assume that an inductive voltage V is given by

$$V = L \frac{dI}{dt} \tag{6.19}$$

$$V = L_o(1 + K_{L1}I + K_{L2}I^2 + K_{L3}I^3 + \cdots) \frac{dI}{dt} \tag{6.20}$$

where the nonlinear coefficients K_{L1}, K_{L2}, $K_{L3} \cdots$ of the inductor L can be found from an inductance-measuring and curve-fitting exercise.

By rearranging Eq. (6.20), we can have V expressed as

$$V = (1 + K_{L1}I + K_{L2}I^2 + K_{L3}I^3 + \cdots) \left[L_o \frac{dI}{dt} \right] \tag{6.21}$$

$$V = S \left[L_o \frac{dI}{dt} \right] \tag{6.22}$$

where

$$S = 1 + K_{L1}I + K_{L2}I^2 + K_{L3}I^3 + \cdots \tag{6.23}$$

S is then the required nonlinear scaling factor for the back emf of the linearized inductor. However, in order to enable the nonlinear inductor model to work for both current directions, Eq. (6.23) should be modified to

$$S = 1 + K_{L1}|I| + K_{L2}|I|^2 + K_{L3}|I|^3 + \cdots \tag{6.24}$$

Based on Eqs. (6.22) and (6.24), the nonlinear inductor can be modeled as the controlled voltage source E_L shown in Fig. 6.7(b).

6.5.3 Modeling of Nonlinear Transformers

A nonlinear transformer may be regarded as effectively a nonlinear inductor with an additional tightly coupled winding. Based on this concept, the SPICE model of a nonlinear transformer can be developed as shown in Fig. 6.8.

Compared with the nonlinear inductor model shown in Fig. 6.7, the nonlinear transformer model in Fig. 6.8 has the following special features:

1. There are two coupled windings in the transformer model, namely, the original winding L and the additional secondary winding L_1.
2. The current that is now used to determine the nonlinear scaling factor S for the back emf is the sum of two currents, $(I + NI_1)$,

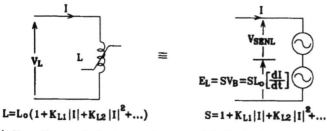

$L = L_0 (1 + K_{L1}|I| + K_{L2}|I|^2 + \ldots)$

(a) Non-linear inductance L.

$S = 1 + K_{L1}|I| + K_{L2}|I|^2 + \ldots$

(b) Equivalent circuit.

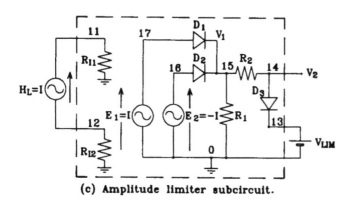

(c) Amplitude limiter subcircuit.

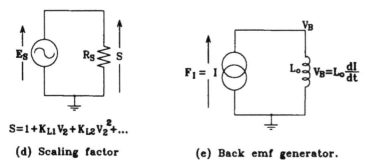

$S = 1 + K_{L1}V_2 + K_{L2}V_2^2 + \ldots$

(d) Scaling factor circuit.

(e) Back emf generator.

FIGURE 6.7   Modeling nonlinear inductance.

where I is the primary current, I_1 the secondary current, and N the secondary to primary turns ratio of the transformer.
3. Two back emfs, V_B and V_{B1}, are generated from the two coupled linear inductors L_o and N^2L_o, respectively, as shown in Fig. 6.8(e).
4. Two nonlinearly scaled voltages, SV_B and SV_{B1}, are used to model the back emfs of the two windings of the transformer.

6.5.4 Modeling of Nonlinear Capacitances

In the simulation of a nonlinear capacitance, the method of nonlinearly scaling the voltage drop of the linearized component can also be used. However, here we use an alternative method of nonlinearly scaling the current response for a given voltage excitation. There is a subtle difference between these two simulation methods: In the first method, the analysis of the voltage response involves integration of the current and division by the capacitance while, in the second method, the analysis of the current response involves only differentiation of the voltage and multiplication by the capacitance:

$$V = \frac{1}{C} \int_0^t I \, dt$$

$$I = C \frac{dV}{dt}$$

Since integration and division are less easily dealt with than differentiation and multiplication, the second method is used here for the modeling of nonlinear capacitors.

As an example, Fig. 6.9 shows the SPICE model of a nonlinear capacitance C. In this model we assume that

$$I = C \frac{dV}{dt} \tag{6.25}$$

$$I = C_o(1 + K_{C1}V + K_{C2}V^2 + K_{C3}V^3 + \cdots) \frac{dV}{dt} \tag{6.26}$$

where the polynomial coefficients K_{C1}, K_{C2}, K_{C3} $\cdots$ can be found by a capacitance-measuring and curve-fitting exercise. Rearranging Eq. (6.26), we have

$$I = (1 + K_{C1}V + K_{C2}V^2 + K_{C3}V^3 + \cdots) \left[C_o \frac{dV}{dt} \right] \tag{6.27}$$

$$= S \left[C_o \frac{dV}{dt} \right] \tag{6.28}$$

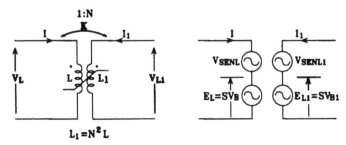

(a) Non-linear transformer.   (b) Equivalent circuit.

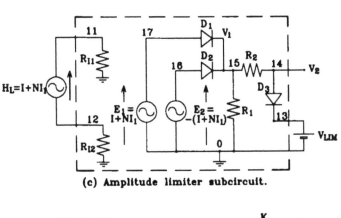

(c) Amplitude limiter subcircuit.

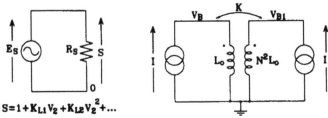

$$S = 1 + K_{L1}V_2 + K_{L2}V_2^2 + \cdots$$

(d) Scaling factor circuit.   (e) Back emf generator.

FIGURE 6.8 Modeling nonlinear transformer.

where

$$S = 1 + K_{C1}V + K_{C2}V^2 + K_{C3}V^3 + \cdots \tag{6.29}$$

The parameter S is then the required nonlinear scaling factor for the capacitive current. The nonlinearly scaled current source F_C in Fig. 6.9(b) is the equivalent circuit of the nonlinear capacitor C.

If the nonlinear capacitance C shown in Fig. 6.9 is controlled not by the voltage across the capacitor itself but by another voltage, V_C, the current expression given in Eq. (6.26) will have to be modified to

$$I = C_o(1 + K_{C1}'V_C + K_{C2}'V_C^2 + K_{C3}'V_C^3 + \cdots) \frac{dV}{dt} \qquad (6.30)$$

where the K_{C1}', K_{C2}', $K_{C3}' \cdots$ are the new polynomial coefficients relating the nonlinear capacitance C and the control voltage V_C. Rearranging Eq. (6.30) yields

$$I = (1 + K_{C1}'V_C + K_{C2}'V_C^2 + K_{C3}'V_C^3 + \cdots) \left[C_o \frac{dV}{dt} \right] \qquad (6.31)$$

$$= S \left[C_o \frac{dV}{dt} \right] \qquad (6.32)$$

where

$$S = 1 + K_{C1}'V_C + K_{C2}'V_C^2 + K_{C3}'V_C^3 + \cdots \qquad (6.33)$$

Equation (6.33) is then the new expression for the voltage scaling factor.

6.5.5 Applications and Limitations of the Curve-Fitting Method

A major application of the curve-fitting method is to help circuit designers develop nonlinear component models based on the information obtained from data books or experimental measurements. As far as circuit designers are concerned, the reasons for nonlinear characteristics may not be so relevant; however, such characteristics have to be taken into account in the design work. The method of curve fitting and nonlinear scaling provides a general solution to the modeling of such apparently arbitrary nonlinear characteristics.

A problem of the curve-fitting method is the limited range for which curve fitting can be performed to obtain the scaling factor, as discussed in Subsection 6.5.1. If this range is exceeded during the simulation, absurdly large errors or nonconvergence problems may result. A limiter circuit is therefore required to limit the signal level used to determine the scaling

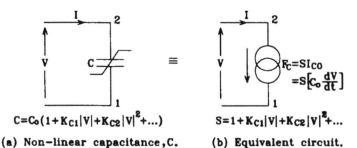

$$C=C_0(1+K_{C1}|V|+K_{C2}|V|^2+...)$$

$$S=1+K_{C1}|V|+K_{C2}|V|^2+...$$

(a) Non-linear capacitance, C.   (b) Equivalent circuit.

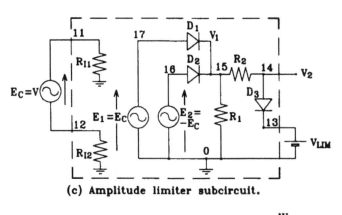

(c) Amplitude limiter subcircuit.

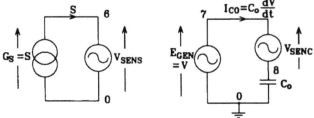

$$S=1+K_{C1}V_2+K_{C2}V_2^2+...$$ C_0 is a linear capacitance

(d) Scaling factor   (e) Capacitive current generator.
circuit.

FIGURE 6.9   Modeling nonlinear capacitance.

factor. It may appear that such large-error or nonconvergence problems can be eliminated by carefully confining the current or voltage swings. This is actually not always possible because the swings occupied by the internal Newton–Raphson iteration process in SPICE can still exceed the curve-fitting range and cause nonconvergence problems.

Even when a limiter is used, care should be taken not to allow the current or voltage swing to exceed the curve-fitting range because, outside this range, the model is no longer valid.

It should also be noted that when the nonlinear characteristics are included in the model, the simulation time will be increased.

6.6 PRACTICAL SIMULATION PROBLEMS AND EXAMPLES

In this section some typical problems frequently encountered in the practical cycle-by-cycle simulations of converters, as well as possible solutions to these problems, will be discussed. Practical simulation examples will also be given to illustrate the modeling and simulation techniques described in previous sections.

6.6.1 Inappropriate Initial Conditions

For a normal transient simulation, SPICE starts with a biasing point (operating point) analysis before the transient analysis. This fact is often overlooked by circuit designers. For example, if designers use the gate driving voltage shown in Fig. 6.10(a) for a SPICE transient simulation of the circuit shown in Fig. 6.1 (but without specifying an initial condition), they will find unexpectedly large initial currents in the converter circuit. The reason is actuallly simple: In the calculation of the biasing point, SPICE assumes that a V_{GS} of 15 V has been applied to the gate of the MOS transistor for an infinitely long time before $t = 0$, so that

$$\text{Initial } I_{DS} = \frac{50V}{\text{on-state resistance of IRF640}} \approx 280A \qquad (6.34)$$

The following methods can be used to solve this problem:

1. Use an alternative gate drive which is 0 V at $t = 0$, such as that shown in Fig. 6.10(b).
2. Use a .TRAN statement with UIC (use initial conditions) option to request SPICE to perform the simulation based on the initial conditions specified in the IC = ... or .IC statement. When UIC

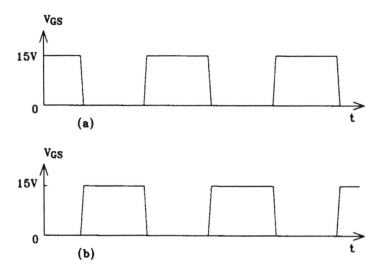

FIGURE 6.10 Gate drive voltage for MOS switch in Fig. 6.1: (a) with 15-V initial value, and (b) with 0 V initial value.

option is used, no dc analysis will be carried out to determine the dc biasing point of the circuit before the transient analysis. The following is an example:

```
L1  9  99  8U  IC=14
CL  990  0  3500U  IC=5
.IC  V(33)=50
.TRAN  0.1U  60U  10U  UIC
* The above statements (extracted from At-
* tachment 6.1) request SPICE to carry out
* a transient analysis for 60 µs in steps of
* 0.1 µs, using the following initial con-
* ditions:
* 1. Initial current in L1=14A (in the di-
*      rection from node 9, through L1, to
*      node 99)
* 2. Initial voltage across CL=5V
* 3. Initial V(33)=50V
* Note that
```

```
*  1. If no initial conditions are speci-
*     fied,   the   initial   voltages   and
*     currents are assumed zero.
*  2. If there is conflict between the IC=
*     ... and .IC statements, the IC=...
*     parameter will take precedence over
*     the .IC value.
```

When a .TRAN statement with UIC option is used, all initial voltages across capacitors and initial currents in inductors should be specified, if they are known.

6.6.2 Long Simulation Time to Reach the Steady State

A problem often encountered in the cycle-by-cycle simulation of converters is the length of time a simulation may take to reach the steady state. Consider, for example, the simulation result given in Fig. A6.1.2 of Attachment 6.1 (at the end of this chapter) for the circuit shown in Fig. A6.1.1. The simulation given there requires 1 min to complete (using an IBM PC with a 33-MHz 80486 CPU). Yet the output voltage V(99) shown in Fig. A6.1.2(h) (which is the output voltage V_o) is still very far from the steady-state value of 5 V. This rather slow rise of V_o is mainly due to the large filtering capacitance C_L, and the large filtering inductance L_1 of the output circuit. One way of reducing the time required for the circuit to reach a near-steady state is to assign proper initial voltages to capacitors, especially large capacitors, and initial currents to inductors, especially large inductors. An example showing how the initial conditions are set can be found in the last subsection. When proper initial conditions are specified in the example given in Fig. A6.1.1, the steady-state waveforms, as shown in Figs. A6.1.3 and A6.1.4, can be obtained also in about 1 min. (Note that the input listing given in Attachment 6.1 actually includes these initial conditions.)

The initial voltages and currents can often be estimated by rough calculations. The accuracy of such initial conditions need not be very high and may be improved through repeated simulations (intelligent trials). In the worst case, they may also be found by a long simulation with zero initial conditions.

If the estimated initial conditions are accurate, a transient simulation for two cycles of the switching operation would be sufficient to give all the detailed switching waveforms of the converter in the steady state.

6.6.3   Nonconvergence Problems

When cycle-by-cycle simulations are carried out for a converter, nonconvergence problems may occur because of the large di/dt in an inductor (causing a large voltage) or large dv/dt across a capacitor (causing a large current). Based on the example of a flyback converter, some typical causes of such problems and methods of solution will be described in this subsection.

Consider the flyback converter shown in Fig. 6.11(a), in which it is assumed that

1.   The electronic switch SW has zero on-state resistance and infinite off-state resistance
2.   The control voltage V_c is so sharp that the switch SW will take zero time to turn on or turn off

The first problem we encounter is that when the switch is turned off (after it has been turned on for some time), a nonconvergence problem will occur. This is due to the infinite di/dt in the leakage inductance of the transformer, which produces an infinitely large voltage. Possible methods of solving this problem are:

1.   Include a capacitance C_p to represent the effective capacitance of the electronic switch SW, as shown in Fig. 6.11(b).
2.   Include a parallel resistance R_p to represent the actual off-state resistance of the electronic switch SW, also as shown in Fig. 6.11(b).
3.   Slow down the falling edge of the control voltage V_c of the electronic switch SW to a practical value.

Note that the suggestions given above are actually requests to ask the designer to provide a more realistic model of the converter. If the nonconvergence problem continues to exist even after the model is modified, it would probably mean that a snubber circuit across the switch SW may be required to reduce the amplitude of the voltage spikes.

The addition of a parallel capacitance C_p across the switch SW, as shown in Fig. 6.11(b), may solve the nonconvergence problem encountered during turn-off switching. It may, however, also create a new problem for the turn-on switching because the capacitive discharging current through C_p (and the switch SW) can now become infinitely large. Possible methods of solving this new problem are:

1.   Add a small resistance R_s in series with the switch SW to model its on-state resistance.

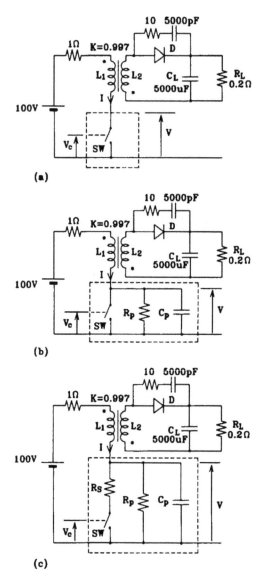

(a)

(b)

(c)

FIGURE 6.11 Simulation of flyback converter.

2. Slow down the rising edge of the control voltage V_c to a practical value.

Again, these suggestions are actually requests for a more realistic model of the converter. After modifications, the resultant model of the flyback converter is as shown in Fig. 6.11(c).

The nonconvergence problems encountered in simulations involving nonlinear components have been discussed in Subsection 6.5.5.

A more general description of the nonconvergence problems in SPICE and methods of solution are given in Appendix C at the end of this book.

6.6.4 Simulation Examples

For the purpose of illustration, some typical examples of cycle-by-cycle simulations are given in Attachments 6.1 and 6.2 at the end of this chapter.

Attachment 6.1 is a simulation file for a forward converter with linear resistors, capacitors, and transformer. The converter circuit is shown in Fig. A6.1.1. The simulation result given in Fig. A6.1.2 assumes a zero initial condition, while the ones in Figs. A6.1.3 and A6.1.4 assume the following initial conditions:

Initial current in L_1 = 14 A

Initial voltage across C_L = 5 V

Initial voltage at node 33 = 50 V

From these results, the power of computer simulation for gathering design information, such as voltage stress, current stress, power dissipation (peak and average values), and output ripple, is quite obvious.

Attachment 6.2 is a simulation file for a three-output flyback converter, the circuit of which is shown in Fig. A6.2.1. In this circuit, the following assumptions are made:

1. The near-ideal MOS transistor M_1, together with nonlinear components R_{DS}, C_{DS}, C_{GD}, and C_{GS}, forms a model of the switching transistor. The nonlinear characteristics of R_{DS}, C_{DS}, C_{GD}, and C_{GS} are given by the curves AB, CD, EF, and GH in Figs. A6.2.2(a), A6.2.3(a), A6.2.4(a), and A6.2.5(a), respectively.
2. The transformer T has a nonlinear characteristic as given by the curve JK in Fig. A6.2.6(a).

Using the techniques described in Section 6.5, the nonlinear components R_{DS}, C_{DS}, C_{GD}, and C_{GS} are modeled as subcircuits RDS, CDS, CGD, and

CGS, respectively, in the SPICE listing. The nonlinear transformer T is modeled by the statements between the two comment statements "*MODELING OF TRANSFORMER" and "*END OF TRANS-FORMER MODEL" in the listing.

As mentioned in Section 6.5, limiters have to be employed to limit the signal levels used to determine the scaling factors of the nonlinear models. Because of such limiters, the simulated nonlinear characteristics of R_{DS}, C_{DS}, C_{GD}, C_{GS}, and L_A would have the shapes shown in Figs. A6.2.2(b), A6.2.3(b), A6.2.4(b), A6.2.5(b), and A6.2.6(b). It is interesting to compare these characteristics with those shown in dotted lines in Figs. A6.2.2(a), A6.2.3(a), A6.2.4(a), A6.2.5(a), and A6.2.6(a), which are the characteristics if limiters are not used. The comparison shows clearly the need for such amplitude limiters in order to avoid the larger-error or nonconvergence problems mentioned in Section 6.5.

Figures A6.2.7 and A6.2.8 are samples of the simulated waveforms and power dissipation characteristics of the flyback converter.

The input files shown in Attachments 6.1 and 6.2 can be run by either SPICE or PSpice. (When run by SPICE, the .PROBE statement should be deleted.)

In the examples given in the attachments, although both the print and plot outputs are requested in the input file (as indicated by the .PRINT and .PLOT statements), only the graphic outputs are shown in the simulation results. This is purely for the purpose of saving space. These graphic outputs have also been processed by the PROBE program in PSpice to improve their appearance. (Examples of the printed outputs directly from SPICE can be found in Appendix A at the end of this book.)

6.7 SUMMARY AND FURTHER REMARKS

In this chapter, the techniques of modeling the circuit components in power converters for cycle-by-cycle simulations have been introduced. The modeling of the nonlinear characteristics, using the curve-fitting method, has also been studied. Typical simulation examples were given to illustrate the power of such simulations.

A problem often encountered in the cycle-by-cycle simulation is the possible long simulation time required for the switching operation to reach a steady state. One method of reducing this simulation time is to assign initial voltages to capacitors, especially large capacitors, and initial currents to inductors, especially large inductors, of the converter.

The inclusion of the nonlinear characteristics of components into the converter model improves the accuracy of simulation. But it will also increase the simulation time, the model complexity, and the probability of nonconvergence. In the course of simulation, designers should therefore make good judgments about how many of the nonlinear properties should be included and make wise compromises among speed, model complexity, and accuracy according to their needs and objectives.

EXERCISES

1. Develop a SPICE model for a switch whose on-state resistance is 0.1 Ω and off-state resistance is 2 MEGΩ.
2. Why is it not desirable to model a switch with unrealistically low on-state resistance or unrealistically high off-state resistance?
3. Develop a SPICE model for a fast-switching NPN bipolar transistor whose on-state resistance is 0.05 Ω.
4. Develop a SPICE model for a fast-switching N-channel enhancement-mode MOS transistor that has
   a. An on-state resistance of 0.2 Ω.
   b. A turn-on threshold voltage of 3 V.
5. Comment on the advantages and disadvantages of including the nonlinear characteristics of the circuit components into a converter model.
6. For the flyback converter circuit shown in Fig. 6.11(c), prepare a SPICE input file for a cycle-by-cycle simulation for ten switching cycles.
   Make the following assumptions in your input file:
   a. Switching frequency = 100 kHz.
   b. Duty cycle of $SW = 0.5$.
   c. $L_1 = 65$ μH.
   d. $L_2 = 0.2$ μH.
   e. $R_s = 0.1$ Ω.
   f. $R_p = 2 \times 10^6$ Ω.
   g. $C_p = 900$ pF.
   Include initial-condition statements in your input file, so that the simulation waveforms are close to the steady-state waveforms.
7. Repeat question 6 as many times as necessary (through iterations) until a reasonable set of "steady-state" switching waveforms is obtained.

ATTACHMENT 6.1: CYCLE-BY-CYCLE SIMULATION OF FORWARD CONVERTER

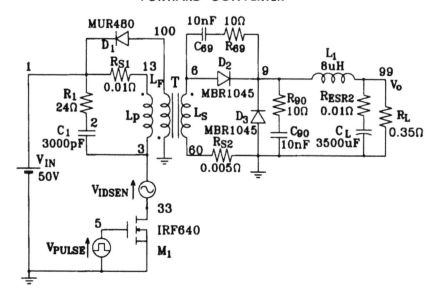

Fig. A6.1.1   Circuit of Forward converter.

A6.1   CYCLE-BY-CYCLE SIMULATION OF FORWARD CONVERTER

```
CYCLE-BY-CYCLE SIMULATION OF FORWARD CONVERTER
*THIS SIMULATION ASSUMES AN INITIAL CONDITION OF
*CURRENT IN L1 = 14A
*VOLTAGE ACROSS CL = 5V
*VOLTAGE AT NODE 33 = 50V
*THE SIMULATION RESULTS ARE SHOWN IN FIGS. A6.1.3 AND A6.1.4.
*
VIN 1 0 DC 50
R1 1 2 24
C1 2 3 3000P
RS1 1 13 0.01
D1 100 1 MUR480
*
*
VIDSEN 3 33
M1 33 5 0 0 IRF640 W=0.66 L=2U
*
*
VPULSE 5 0 PULSE(0 15 0 0.1U 0.1U 2.56U 6.3U)
*
*
RS2 60 0 0.005
D2 6 9 MBR1045
```

```
C69 6 69 10N
R69 69 9 10
D3 0 9 MBR1045
R90 9 90 10
C90 90 0 10N
L1 9 99 8U IC=14
RESR2 99 990 0.01
CL 990 0 3500U IC=5
RL 99 0 .35
*
*
LF 0 100 0.385M
LP 13 3 0.385M
LS 6 60 0.036M
KFP LF LP 0.998
KPS LP LS 0.995
KFF LF LS 0.995
*THE FOLLOWING ARRANGEMENT IS FOR MEASUREMENT OF TRANSISTOR
*POWER DISSIPATION ONLY
FAV 0 501 POLY(1) VIDSEN 0 1
RFAV 501 0 1
*V(501) = DRAIN CURRENT OF M1
GAV 0 500 POLY(2) 33 0 501 0 0 0 0 0 1
*V(33,0) = DRAIN VOLTAGE OF M1
*GAV = INSTANTANEOUS POWER DISSIPATION OF M1
CAV 500 0 1
*V(500) = ENERGY
RAV 500 0 1G
*AVERAGED POWER DISSIPATION OF M1 = V(500)/TIME
*
.MODEL IRF640    NMOS(LEVEL=3 GAMMA=0 DELTA=0 ETA=0 THETA=0
+                KAPPA=0 VMAX=0 XJ=0 TOX=100N UO=600 PHI=.6
+                RS=19.61M KP=20.73U VTO=3.788 RD=95.58M
+                CBD=1.872N PB=.8 MJ=.5 FC=.5 CGSO=1.745N
+                CGDO=334.7P IS=16.39P)
.MODEL MUR480    D(IS=92.91F RS=58.12M N=1 XTI=5 EG=1.11
+                CJO=66.41P M=.5022 VJ=.75 FC=.5 TT=185N)
.MODEL MBR1045   D(IS=168.4N RS=8.013M N=1 XTI=0 EG=1.11
+                CJO=888.9P M=0.4639 VJ=0.75 FC=0.5)
.OPTIONS ITL5=0 LIMPTS=1000
.IC V(33)=50
.TRAN 0.1U 60U 10U UIC
.PRINT TRAN I(VIDSEN) V(33) V(99)
.PLOT TRAN I(VIDSEN) V(33) V(99)
*MORE VARIABLES CAN BE ADDED TO THE .PRINT AND
*.PLOT LIST AT THE DISCRETION OF THE DESIGNER
.PROBE
*THE .PROBE STATEMENT IS FOR PSPICE ONLY.
*DELETE THE .PROBE STATEMENT IF RUN BY SPICE.
.END
```

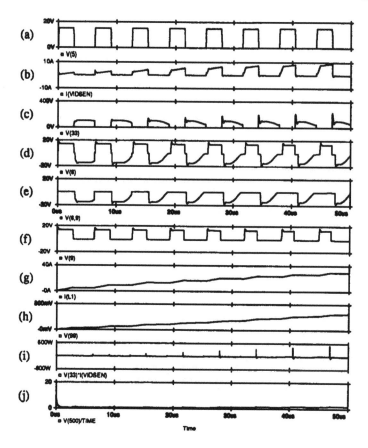

(a) V(5) = Gate voltage
(b) I(VIDSEN) = Drain current
(c) V(33) = Drain voltage
(d) V(6) = Secondary voltage
(e) V(6,9) = Voltage across D_2
(f) V(9) = Voltage across D_3
(g) I(L1) = Current in L_1
(h) V(99) = Output voltage V_o
(i) V(33)*I(VIDSEN) = Instantaneous MOSFET power dissipation
(j) V(500)/TIME = Average MOSFET power dissipation

Fig. A6.1.2   Simulation results of Forward converter with linear RCL and
              zero initial conditions.

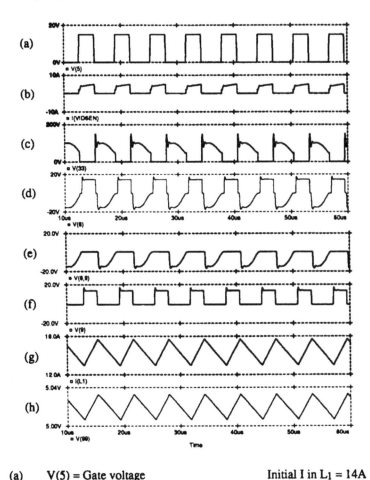

(a)	V(5) = Gate voltage	Initial I in L_1 = 14A
(b)	I(VIDSEN) = Drain current	Initial V across C_L = 5V
(c)	V(33) = Drain voltage	Initial V(33) = 50V
(d)	V(6) = Secondary voltage	
(e)	V(6,9) = Voltage across D_2	
(f)	V(9) = Voltage across D_3	
(g)	I(L1) = Current in L_1	
(h)	V(99) = Output voltage V_o	

Fig. A6.1.3 Simulation results of Forward converter with assigned initial conditions.

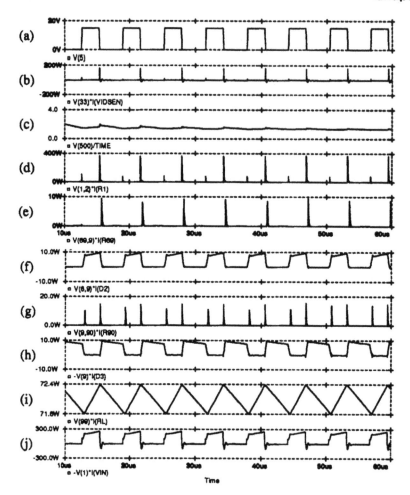

(a)   V(5) = Gate voltage
(b)   V(33)*I(VIDSEN) = Instantaneous MOSFET power dissipation
(c)   V(500)/TIME = Averaged MOSFET power dissipation
(d)   V(1,2)*I(R1) = Instantaneous power loss in R_1
(e)   V(59,9)*I(R69) = Instantaneous power loss in R_{69}
(f)   V(6,9)*I(D2) = Instantaneous power loss in D_2
(g)   V(9,90)*I(R90) = Instantaneous power loss in R_{90}
(h)   -V(9)*I(D3) = Instantaneous power loss in D_3
(i)   V(99)*I(RL) = Instantaneous power output
(j)   -V(1)*I(VIN) = Instantaneous power input

Fig. A6.1.4   Power dissipations of Forward converter.

ATTACHMENT 6.2: CYCLE-BY-CYCLE SIMULATION OF FLYBACK CONVERTER WITH NONLINEAR RCL

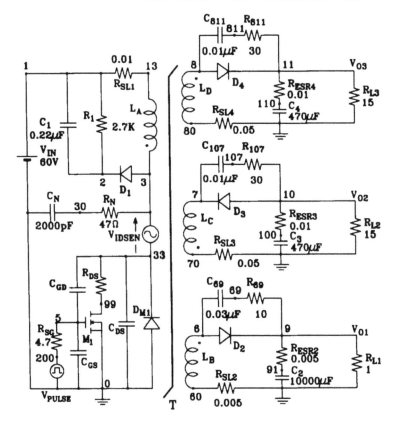

R_{DS} C_{DS} C_{GD} C_{GS} and T are non-linear elements
M1 is a near-ideal MOS transistor

Fig. A6.2.1   SPICE model of Flyback converter.

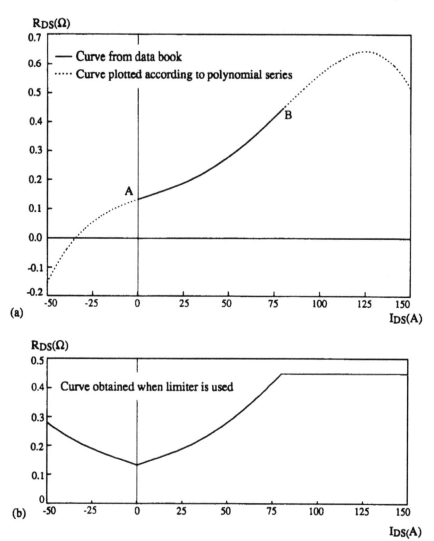

Fig. A6.2.2 Non-linear characteristics of on-state resistance R_{DS}.

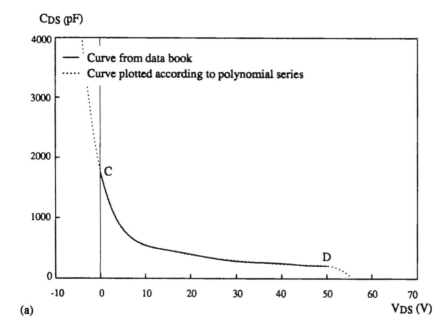

(a)

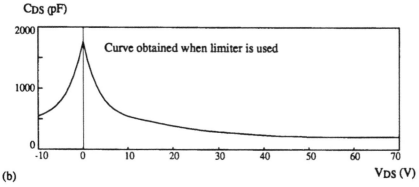

(b)

Fig. A6.2.3 Non-linear characteristics of capacitance C_{DS}.

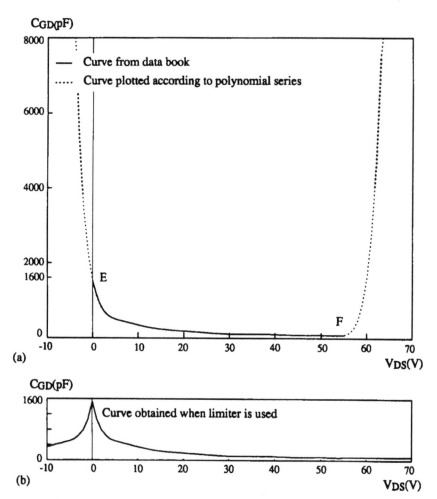

(a)

(b)

Fig. A6.2.4   Non-linear characteristics of capacitance C_{GD}.

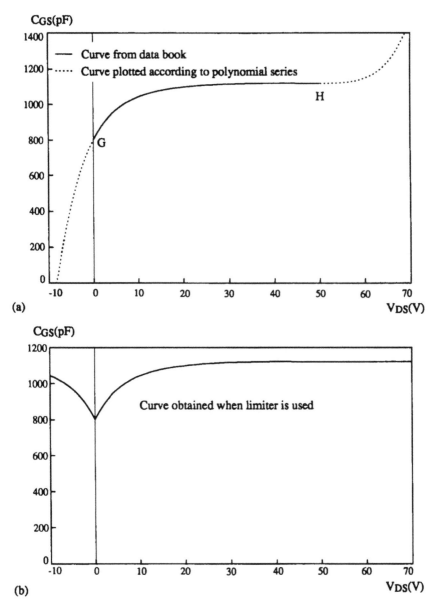

Fig. A6.2.5 Non-linear characteristics of capacitance C_{GS}.

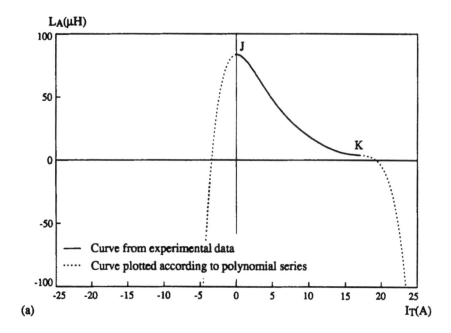

(a)

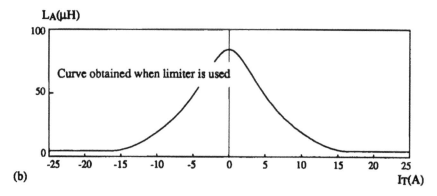

(b)

Fig. A6.2.6 Non-linear characteristics of inductance L_A.

A6.2 <u>CYCLE-BY-CYCLE SIMULATION OF FLYBACK CONVERTER</u>
<u>WITH NON-LINEAR RCL</u>

```
CYCLE-BY-CYCLE SIMULATION OF FLYBACK CONVERTER
*WITH NON-LINEAR RCL
*
*
*
.SUBCKT LIMITER 11 12 13 14
*V(11,12) = INPUT V
*V(13) = CLAMPING V
*V(14) = OUTPUT V
RI1 11 0 1MEG
RI2 12 0 1MEG
E1 17 0 POLY(1) 11 12    0 1
E2 16 0 POLY(1) 11 12    0 -1
D1 17 15 DMOD
D2 16 15 DMOD
*D1,D2 = FULL-WAVE RECTIFIER
R1 15 0 1K
R2 15 14 1MEG
D3 14 13 DMOD
*D3 = CLAMPING DIODE
.ENDS LIMITER
*
*
*
.SUBCKT RDS 1 4
VSENR 1 5
*VSENR IS A DUMMY VOLTAGE SOURCE USED TO SENSE THE
*CURRENT IN RDS.
ER 5 4 POLY(2) 9 0 6 0    0 0 0 0 1
*ER = V(6,0)*V(9,0) = S*(I*RO),
*WHICH IS USED TO MODEL THE VOLTAGE DROP OF RDS.
XLIMR 11 12 13 14 LIMITER
*THIS IS A SUBCIRCUIT CALL FOR A LIMITER TO BE USED
*TO GENERATE A VOLTAGE V2 REPRESENTING THE RANGE-
*LIMITED PARAMETER |I|.
*V(11,12) = LIMITER INPUT V
*V(13) = LIMITING V
*V(14) = LIMITER OUTPUT V
HR 11 12 POLY(1) VSENR   0 1
*HR IS THE INPUT TO THE LIMITER,
*REPRESENTING THE CURRENT I.
VLIM 13 0 DC 80
*THE LIMIT OF THE CONTROL CURRENT IDS IS 80A.
ES 6 0 POLY(1) 14 0   1 1.61082543625219E-2
+ -1.14345814997192E-4 6.58606268297644E-6
+ -3.79110038073517E-8
*ES = SCALING FACTOR S FOUND BY CURVE-FITTING
*V(6) = S
RS 6 0 1MEG
FI 0 9 POLY(1) VSENR   0 1
*FI = IDS
RO 9 0 0.13275
*V(9) = VOLTAGE DROP OF RO = I*RO
.ENDS RDS
*
*
*
```

```
.SUBCKT CGD 2 1 11 12
FC 2 1 POLY(2) VSENC VSENS   0 0 0 0 1
*FC = S*ICO = S*CGDO(dVGD/dt),
*WHICH IS USED TO MODEL THE CURRENT IN CGD.
*2,1 = TERMINALS OF CGD ( NOTE THAT NODES 1 AND 2 ARE
*INTERNAL NODES OF THE SUBCIRCUIT ), WHICH WILL BE
*CONNECTED TO THE EXTERNAL NODES 5 AND 33
*THROUGH THE STATEMENT
*XCGD 5 33 33 0 CGD.
*V(11,12) = CONTROL INPUT = VDS
XLIMCGD 11 12 13 14 LIMITER
*V(11,12) = LIMITER INPUT V
*V(13) = LIMITING V
*V(14) = LIMITER OUTPUT V
VLIM 13 0 DC 55
*THE LIMIT OF THE CONTROL VOLTAGE VDS IS 55V.
GS 0 6 POLY(1) 14 0   1 -3.846227360028892E-1
+ 9.604895369552782E-2 -1.379155595673776E-2
+ 1.191815341105608E-3 -6.486352085758587E-5
+ 2.273768905057493E-6 -5.119848935620762E-8
+ 7.148546703618407E-10 -5.629752656905672E-12
+ 1.910678387667590E-14
*GS = S
VSENS 6 0
*I IN VSENS = S
EGEN 7 0 POLY(1) 2 1   0 1
*EGEN = VGD
VSENC 7 8
*I IN VSENC = ICO = CGDO(dVGD/dt)
CGDO 8 0 1519.96P
.ENDS CGD
*
*
*
.SUBCKT CGS 2 1 11 12
FC 2 1 POLY(2) VSENC VSENS   0 0 0 0 1
*FC = S*ICO = S*CGSO(dVGS/dt),
*WHICH IS USED TO MODEL THE CURRENT IN CGS.
*2,1 = TERMINALS OF CGS, WHICH WILL BE CONNECTED TO
*EXTERNAL NODES 5 AND 0 THROUGH THE STATEMENT
*XCGS 5 0 33 0 CGS.
*V(11,12) = CONTROL VOLTAGE = VDS
XLIMCGS 11 12 13 14 LIMITER
*V(11,12) = LIMITER INPUT V
*V(13) = LIMITING V
*V(14) = LIMITER OUTPUT V
VLIM 13 0 DC 50
*THE LIMIT OF THE CONTROL VOLTAGE VDS IS 50V.
GS 0 6 POLY(1) 14 0   1 6.019353310102442E-2
+ -4.747819921395985E-3 2.331334542208174E-4
+ -7.168951092767750E-6 1.332441854698176E-7
+ -1.367686857859239E-9 5.953058763852476E-12
*GS = S
VSENS 6 0
```

```
*I IN VSENS = S
EGEN 7 0 POLY(1) 2 1  0 1
*EGEN = VGS
VSENC 7 8
*I IN VSENC = ICO = CGSO(dVGS/dt),
CGSO 8 0 801.9P
.ENDS CGS
*
*
*
.SUBCKT CDS 2 1 11 12
FC 2 1 POLY(2) VSENC VSENS  0 0 0 0 1
*FC = S*ICO = S*CDSO(dVDS/dt),
*WHICH IS USED TO MODEL THE CURRENT IN CDS.
*2,1 = TERMINALS OF CDS, WHICH WILL BE CONNECTED TO
*THE EXTERNAL NODES 33 AND 0, THROUGH THE STATEMENT
*XCDS 33 0 33 0 CDS.
*V(11,12) = CONTROL VOLTAGE = VDS
XLIMCDS 11 12 13 14 LIMITER
*V(11,12) = LIMITER INPUT V
*V(13) = LIMITING V
*V(14) = LIMITER OUTPUT V
VLIM 13 0 DC 50
*THE LIMIT OF THE CONTROL VOLTAGE VDS IS 50V.
GS 0 6 POLY(1) 14 0  1 -1.91146611015335E-1
+ 2.26383044718857E-2 -1.50024393298653E-3
+ 5.71003875678653E-5 -1.24306936501284E-6
+ 1.43891857514898E-8 -6.86572429196239E-11
*GS = S
VSENS 6 0
*I IN VSENS = S
EGEN 7 0 POLY(1) 2 1  0 1
*EGEN = VDS
VSENC 7 8
*I IN VSENC = ICO = CDSO(dVDS/dt)
CDSO 8 0 1760.5P
.ENDS CDS
*
*
*
*INPUT CIRCUIT
VIN 1 0 DC 60
RSL1 1 13 0.01
C1 2 1 0.22U  IC=63.9
R1 1 2 2.7K
D1 3 2 MUR480
VIDSEN 3 33
CN 30 0 2000P
RN 3 30 47
M1 99 5 0 0 MODM W=2 L=2U
.MODEL MODM NMOS (VTO=3.8)
DM1 0 33 DM
RSG 200 5 4.7
.MODEL DM D
*
```

```
XRDS 33 99 RDS
XCGD 5 33 33 0 CGD
XCGS 5 0 33 0 CGS
XCDS 33 0 33 0 CDS
VPULSE 200 0 PULSE(0 15 0 0.1U 0.1U 2.4U 6.3U)
*SWITCHING FREQUENCY = 158.7KHz
*DUTY CYCLE = 0.4
*
*
*
*MODELING OF TRANSFORMER
*TRANSFORMER EQUIVALENT CIRCUIT
VSENA 3 313
ELA 313 13 POLY(2) 121 0 120 0  0 0 0 0 1
*3,13 = TERMINALS OF LA
VSENB 6 660
ELB 660 60 POLY(2) 123 0 120 0  0 0 0 0 1
*6,60 = TERMINALS OF LB
VSENC 70 707
ELC 707 7 POLY(2) 125 0 120 0  0 0 0 0 1
*70,7 = TERMINALS OF LC
VSEND 8 880
ELD 880 80 POLY(2) 127 0 120 0  0 0 0 0 1
*8,80 = TERMINALS OF LD
*
*AMPLITUDE LIMITER FOR SCALING FACTOR CIRCUIT
XLIML 111 112 113 114 LIMITER
*XLIML IS THE AMPLITUDE LIMITER FOR THE CONTROL INPUT
*OF THE SCALING FACTOR CIRCUIT.
*V(111,112) = LIMITER INPUT V
*V(113) = LIMITING V
*V(114) = LIMITER OUTPUT V
FLA 0 90 POLY(1) VSENA  0 1
*FLA = ILA
FLB 0 90 POLY(1) VSENB  0 0.1429
*FLB = ILB*NB/NA
FLC 0 90 POLY(1) VSENC  0 0.4286
*FLC = ILC*NC/NA
FLD 0 90 POLY(1) VSEND  0 0.4286
*FLD = ILD*ND/NA
VSENT 90 0
*VSENT IS USED TO SENSE THE TOTAL MAGNETIZING CURRENT.
HL 111 112 POLY(1) VSENT  0 1
*HL IS A VOLTAGE REPRESENTING THE SUM OF MAGNETIZING
*CURRENTS (ILA + ILB*NB/NA + ILC*NC/NA + ILD*ND/NA).
*
VLIMIT 113 0 DC 16
*THE LIMIT OF THE CONTROL CURRENT IS 16A.
ES 120 0 POLY(1)  114 0  1 7.00320852353470E-3
+ -4.59114508905011E-2 8.28908977913860E-3
+ -6.94105259044459E-4 2.86182877067600E-5
+ -4.63590079013760E-7
*V(120) = ES = SCALING FACTOR S
RS 120 0 1MEG
```

```
FSLA 0 121 POLY(1) VSENA  0 1
*FSLA = ILA
LAO 121 0 84U IC=-1.76
*
FSLB 0 123 POLY(1) VSENB  0 1
*FSLB = ILB
LBO 123 0 1.7143U
*
FSLC 0 125 POLY(1) VSENC  0 1
*FSLC = ILC
LCO 125 0 15.43U
*
FSLD 0 127 POLY(1) VSEND  0 1
*FSLD = ILD
LDO 127 0 15.43U
*
K1 LAO LBO 0.97
K2 LAO LCO 0.97
K3 LAO LDO 0.97
K4 LBO LCO 0.97
K5 LBO LDO 0.97
K6 LCO LDO 0.97
*K1,K2,K3,K4,K5,K6 ARE THE COUPLING COEFFICIENTS
*AMONG LA,LB,LC, AND LD.
*END OF TRANSFORMER MODEL
*
*
*
*VO1 OUTPUT CIRCUIT
RSL2 60 0 0.005
D2 6 9 MBR1045
C69 6 69 0.03U
R69 69 9 10
C2 91 0 10000U IC=4.845
RESR2 9 91 0.005
RL1 9 0 1
*
*
*VO2 OUTPUT CIRCUIT
RSL3 70 0 0.05
D3 10 7 MUR460
C107 7 107 0.01U
R107 107 10 30
RESR3 10 100 0.01
C3 100 0 470U IC=-15.68
RL2 10 0 15
*
*
*VO3 OUTPUT CIRCUIT
RSL4 80 0 0.05
D4 8 11 MUR460
C811 8 811 0.01U
R811 811 11 30
RESR4 11 110 0.01
```

```
C4 110 0 470U IC=15.68
RL3 11 0 15
*
*
*THE FOLLOWING ARRANGEMENT IS FOR MEASUREMENT OF
*TRANSISTOR POWER DISSIPATION ONLY.
FAV 0 501 POLY(1) VIDSEN 0 1
RFAV 501 0 1
*V(501) = DRAIN CURRENT OF M1
GAV 0 500 POLY(2) 33 0 501 0  0 0 0 0 1
*V(33,0) = DRAIN VOLTAGE OF M1
*GAV = INSTANTANEOUS POWER DISSIPATION OF M1
CAV 500 0 1
*V(500) = ENERGY
RAV 500 0 1G
*AVERAGED POWER DISSIPATION OF M1 = V(500)/TIME
*
*
*
.MODEL DMOD D (N=10M)
*THE USE OF N=10M REDUCES THE FORWARD VOLTAGE DROP OF
*THE DIODE TO NEAR ZERO.
.MODEL MUR460 D(IS=289.1F RS=33.23M N=1 XTI=5 EG=1.11
+               CJO=70.72P M=.5733 VJ=.75 FC=.5 TT=123.3N)
.MODEL MUR480 D(IS=92.91F RS=58.12M N=1 XTI=5 EG=1.11
+               CJO=66.41P M=.5022 VJ=.75 FC=.5 TT=185N)
.MODEL MBR1045 D(IS=168.4N RS=8.013M N=1 XTI=0 EG=1.11
+               CJO=888.9P M=.4639 VJ=.75 FC=.5)
.OPTIONS ITL4=100 ITL5=0 LIMPTS=1000
*ITL4=100     RESETS THE TRANSIENT ANALYSIS TIMEPOINT
*             ITERATION LIMIT TO 100. THE DEFAULT
*             VALUE IS 10.
*ITL5=0       OMITS THE TRANSIENT ANALYSIS TOTAL
*             ITERATION LIMIT. THE DEFAULT VALUE IS 5000.
*LIMPTS=1000  RESETS THE TOTAL NUMBER OF POINTS THAT
*             CAN BE PRINTED OR PLOTTED. THE DEFAULT
*             VALUE IS 201.
.TRAN 0.1U 50U UIC
.PRINT TRAN I(VIDSEN) V(33) V(9)
.PLOT TRAN I(VIDSEN) V(33) V(9)
*MORE VARIABLES CAN BE ADDED TO THE .PRINT AND
*.PLOT LIST AT THE DISCRETION OF THE DESIGNER.
.PROBE
*THE .PROBE STATEMENT IS FOR PSPICE ONLY.
*DELETE .PROBE STATEMENT IF RUN BY SPICE.
.END
```

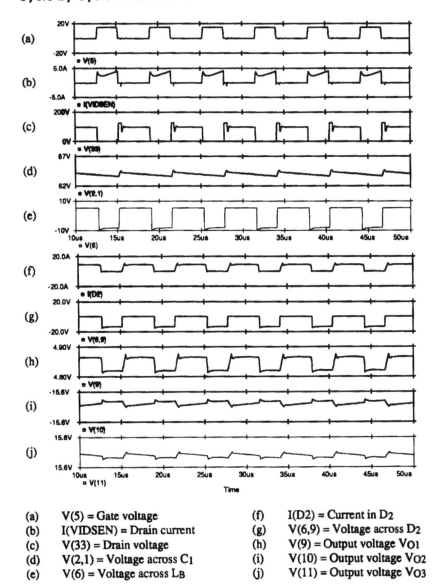

(a)	V(5) = Gate voltage	(f)	I(D2) = Current in D_2
(b)	I(VIDSEN) = Drain current	(g)	V(6,9) = Voltage across D_2
(c)	V(33) = Drain voltage	(h)	V(9) = Output voltage V_{O1}
(d)	V(2,1) = Voltage across C_1	(i)	V(10) = Output voltage V_{O2}
(e)	V(6) = Voltage across L_B	(j)	V(11) = Output voltage V_{O3}

Fig. A6.2.7   Simulated waveforms of Flyback converter with non-linear
              RCL.

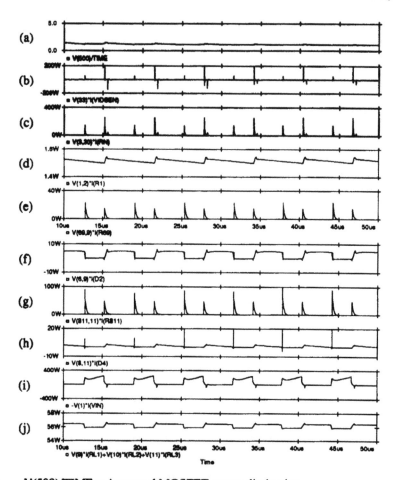

(a)   V(500)/TIME = Averaged MOSFET power dissipation
(b)   V(33)*I(VIDSEN) = Instantaneous MOSFET power dissipation
(c)   V(3,30)*I(RN) = Instantaneous power loss in R_N
(d)   V(1,2)*I(R1) = Instantaneous power loss in R_1
(e)   V(69,9)*I(R69) = Instantaneous power loss in R_{69}
(f)   V(6,9)*I(D2) = Instantaneous power loss in D_2
(g)   V(811,11)*I(R811) = Instantaneous power loss in R_{811}
(h)   V(8,11)*I(D4) = Instantaneous power loss in D4
(i)   -V(1)*I(VIN) = Instantaneous power input
(j)   V(9)*I(RL1)+V(10)*I(RL2)+V(11)*I(RL3) = Total instantaneous power output

Fig. A6.2.8   Power dissipations of Flyback converter with non-linear RCL.

7
Simulation of Low-Frequency Behaviors of Converters

The cycle-by-cycle simulations discussed in Chapter 6 are basically time-domain simulations (which are referred to as transient simulations in SPICE) based on the actual circuits. Such simulations are very useful for the prediction of detailed switching waveforms. However, they have the following deficiencies:

1. When applied to a complete regulator, any meaningful cycle-by-cycle simulation (which may consist of many cycles) may take a very long time to complete.
2. The frequency-domain simulations (referred to as ac simulations in SPICE) of the $\delta V_o(s)/\delta D(s)$ and $\delta V_o(s)/\delta V_i(s)$ transfer functions, which are essential for the design of the feedback and compensation circuit in the regulator, cannot be carried out directly based on the actual circuit.

To solve these problems, therefore, it is also necessary to perform simulations based on the low-frequency behavior model of the converter or regulator.

However, recalling the mathematical expressions used in Chapters 2, 3, and 5 to describe the low-frequency behavior models of converters, we realize that there is a need to simulate these expressions if we wish to use such models to predict the low-frequency behaviors of converters.

In this chapter the methods used to simulate mathematical expressions in the SPICE program will be studied in Section 7.1. Simulation examples and further justifications for simulations based on the low-frequency behavior models will be discussed in Section 7.2. Some new techniques of developing a combined model for both continuous and discontinuous modes of operation will be described in Sections 7.3 and 7.4. These techniques will be further extended to current-controlled converters in Section 7.5 and to converters with multiple outputs in Section 7.6. It will be found that the combined model is extremely useful for analysis of converters under large-signal conditions and for the design of regulators.

Typical simulation examples will be given in the form of attachments to illustrate the power of computer simulation. These examples may also be used by circuit designers as starting points for developing their own converter models or simulation files for specific applications.

7.1 SPICE SIMULATION OF MATHEMATICAL EXPRESSIONS

A careful examination of the mathematical expressions used in Chapters 2, 3, 5, 8, and 9 to describe the low-frequency behaviors of converters reveals that the functions required to simulate these expressions are basically addition, subtraction, division, square, square root, sine, and cosine. In this section the methods used to obtain these functions will be studied.

Implementation of the addition and subtraction is straightforward if variables are represented by voltages/currents because these functions involve only a series/parallel connection of the sources. The simulation of multiplication, division, square, and square root, however, requires the use of controlled sources.

SPICE actually allows direct simulation of controlled voltages or currents that are expressible as polynomials of the controlling voltages or currents. The following two examples illustrates how this is done:

Example 1. For a one-dimensional voltage-controlled voltage source E_o, which is expressible as a polynomial function of a control voltage V_C, so that

$$E_o = V_{dc} + P_1 V_C + P_2 V_C^2 + P_3 V_C^3 + P_4 V_C^4 + \cdots \qquad (7.1)$$

the required SPICE statement format to simulate it is

$$E0 \ N+ \ N- \ POLY(1) \ NC+ \ NC- \ V_{dc} \ P_1 \ P_2 \ P_3 \ P_4 \ \ldots$$

where E0 is the name, and $N+$, $N-$ are the positive and negative nodes, of the controlled voltage source. $NC+$ and $NC-$ are the controlling nodes, the voltage at $NC+$ with respect to $NC-$ being the controlling voltage V_C. V_{dc}, P_1, P_2, P_3, P_4 . . . are the polynomial coefficients.

Example 2. For a two-dimensional voltage-controlled voltage source such as

$$E_o = V_{dc} + P_1 V_{C1} + P_2 V_{C2} + P_3 V_{C1}^2 + P_4 V_{C1} V_{C2} + P_5 V_{C2}^2$$
$$+ P_6 V_{C1}^3 + P_7 V_{C1}^2 V_{C2} + P_8 V_{C1} V_{C2}^2 + P_9 V_{C2}^3 + \cdots \qquad (7.2)$$

the required SPICE statement is

```
E0 N+ N- POLY(2) NC1+ NC1- NC2+ NC2- Vdc P1 P2 P3 P4
+ P5 P6 P7 P8 P9 ...
```

where E0, N+ and N− have meanings similar to those in Example 1. Here NC1+ and NC1− are the controlling nodes that define the first control voltage V_{C1}. NC2+ and NC2− are the controlling nodes that define the second control voltage V_{C2}. V_{dc}, P_1, P_2, P_3, P_4, P_5, P_6, P_7, P_8, and P_9 are the polynomial coefficients, as shown in Eq. (7.2).

More about the simulation of nonlinear controlled sources can be found in Sections 3.9–3.12 of Appendix B at the end of this book.

Making use of the one-dimensional voltage-controlled voltage source given in Example 1, it is straightforward to implement the squaring function. Figure 7.1 gives an example showing how an output voltage $E_o = 10$

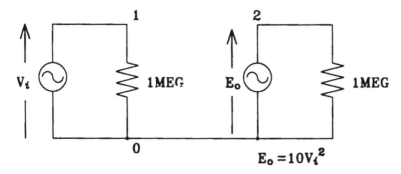

```
VI 1 0
*SPECIFY THE INPUT VOLTAGE IN THE VI STATEMENT
R1 1 0 1MEG
EO 2 0 POLY(1)  1 0  0 0 10
*VOLTAGE AT NODE 2 IS THE OUTPUT VOLTAGE
R2 2 0 1MEG
```

FIGURE 7.1 Simulation of squaring.

V_i^2 is obtained. All we need is a simple statement like

```
EO  2  0  POLY(1)  1  0  0  0  10
```

It should be noted that the two 1 MEGΩ resistances shown in **Fig. 7.1** are actually dummy resistances just to satisfy the SPICE requirement that at least two circuit elements are connected to each node.

Making use of the two-dimensional voltage-controlled voltage source given in Example 2, we can implement the function of multiplication. Figure 7.2 shows how an output voltage $E_o = 100\,V_{i1}\,V_{i2}$ can be obtained. Again, all we need is a simple statement:

```
EO  3  0  POLY(2)  1  0  2  0  0  0  0  0  100
```

Simulations of square root and division are more complicated and require the use of feedback components. Examples of these circuits, developed in the form of subcircuits, will be discussed separately in the following two subsections.

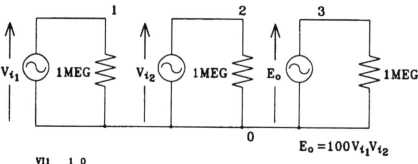

```
VI1   1 0
*SPECIFY INPUT 1 IN THE VI1 STATEMENT
R1    1 0 1MEG
VI2   2 0
*SPECIFY INPUT 2 IN THE VI2 STATEMENT
R2    2 0 1MEG
EO    3 0 POLY(2)  1 0  2 0  0 0 0 0 100
*VOLTAGE AT NODE 3 IS THE OUTPUT VOLTAGE
R3    3 0 1MEG
```

FIGURE 7.2 Simulation of multiplication.

7.1.1 Simulation of Square Root

A subcircuit to generate the square-root function is shown in Fig. 7.3. It is basically an operational amplifier with voltage feedback $E_f = E_o^2$. Applying the virtual ground concept, it is found that

$$V_d \approx 0 \tag{7.3}$$
$$E_f = E_o^2 = V_i \tag{7.4}$$

and

$$E_o = \sqrt{V_i} \tag{7.5}$$

Therefore, the output E_o is the square root of the input V_i.

The square-root subcircuit listing is also given in Fig. 7.3. It is interesting to see that if the feedback voltage E_f in the square-root subcircuit is modified to

$$E_f = K_2 E_o^2 + K_1 E_o \tag{7.6}$$

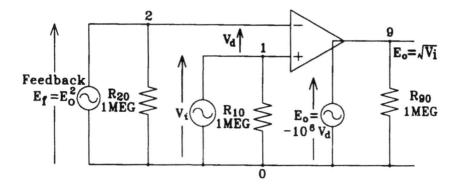

```
.SUBCKT SQR 1 9
• NODE 1 IS THE INPUT NODE
• NODE 9 IS THE OUTPUT NODE
R10  1 0  1MEG
EF   2 0  POLY(1) 9 0 0 0 1
R20  2 0  1MEG
EO   9 0  POLY(1) 2 1 0 −1MEG
R90  9 0  1MEG
.ENDS SQR
```

FIGURE 7.3   Square-root subcircuit.

the new output voltage E_o, if the simulation converges, will become a solution to the equation $E_f = V_i$, or

$$K_2 E_o^2 + K_1 E_o - V_i = 0 \qquad\qquad (7.7)$$

Similarly, if we arrange to have

$$E_f = E_o^N \qquad\qquad (7.8)$$

the output E_o, if it converges to a solution, will be an Nth root of V_i.

7.1.2 Simulation of Division

A subcircuit to simulate a divider is shown in Fig. 7.4. It is basically an operational amplifier with a negative voltage feedback $E_f = V_{i2}E_o$. Ap-

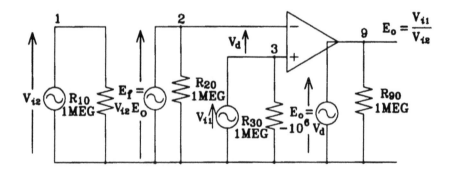

```
.SUBCKT DIV 3 1 9
* NODE 3 IS THE INPUT NODE FOR THE DIVIDEND
* NODE 1 IS THE INPUT NODE FOR THE DIVISOR
* NODE 9 IS THE OUTPUT NODE, V(9) = V(3)/V(1)
R30 3 0 1MEG
R10 1 0 1MEG
E20 2 0 POLY(2)  1 0  9 0  0 0 0 0 1
*E20 = Ef
R20 2 0 1MEG
EO  9 0 POLY(1) 2 3 0 -1MEG
R90 9 0 1MEG
.ENDS DIV
```

FIGURE 7.4   Divider subcircuit.

plying the virtual ground concept again, it is found that

$$V_d \approx 0 \tag{7.9}$$

$$E_f = V_{i2}E_o = V_{i1} \tag{7.10}$$

$$\therefore E_o = \frac{V_{i1}}{V_{i2}} \tag{7.11}$$

The divider subcircuit listing is also given in Fig. 7.4.

It should be noted that the nonconvergence problem may occur if V_{i2} is very small becuase E_o will then tend to be infinity.

7.1.3 Simulation of Sine and Cosine Functions

In the simulation of sine or cosine function, it is necessary only to expand the sine or cosine function into a series with ascending powers and place the coefficients appropriately in a one-dimensional controlled-source statement. For example, to generate an output voltage E_o so that

$$E_o = \sin V_i = 0 + V_i + 0 - \frac{V_i^3}{3!} + 0 + \frac{V_i^5}{5!} + 0 - \frac{V_i^7}{7!} + \cdots$$

$$\tag{7.12}$$

the statement as given below will be appropriate:

```
E0 9 0 POLY(1) 1 0 0 1 0 -.16666667 0
+ 8.3333333E-3 0 -1.984127E-4 0 2.7557319E-6 0
+ -2.5052108E-8 0 1.6059043E-10 0
+ -7.6471637E-13 0 2.8114572E-15 0 -8.2206352E-18
+ 0 1.9572941E-20 0 -3.8681701E-23 0
+ 6.4469502E-26 0 -9.1836898E-29 0 1.1309962E-31
+ 0 -1.2161250E-34 0
```

The preceding statement means that if an input voltage V_i is applied to node 1 (with respect to node 0), the output voltage E_o from node 9 (also with respect to node 0) will be sin V_i.

Similarly, for a cosine function, since we have

$$E_o = \cos V_i = 1 + 0 - \frac{V_i^2}{2!} + 0 + \frac{V_i^4}{4!} + 0 - \frac{V_i^6}{6!} + \cdots \tag{7.13}$$

the required SPICE statement to generate cos V_i will be:

```
E0 9 0 POLY(1) 1 0 1 0 -0.5 0 4.1666667E-2 0
+ -1.3888889E-3 0 2.4801587E-5 0 -2.7557319E-7
+ 0 2.0876756E-9 0 -1.1470745E-11 0 4.7794773E-14
+ 0 -1.5619206E-16 0 4.1103176E-19 0
+ -8.8967913E-22 0 1.6117375E-24 0
+ -2.4795962E-27 0 3.2798892E-30 0
+ -3.7699876E-33 0
```

7.2 SIMULATING CONVERTERS BASED ON LOW-FREQUENCY BEHAVIOR MODELS

When the simulation techniques described in Section 7.1 are applied, all the converter behavior models developed in Chapter 2 can be entered into SPICE. SPICE simulations based on these behavior models can thus be carried out to predict their frequency-response and transient-response characteristics.

As an example, let us consider the electronic transformer shown in Fig. 7.5. The low-frequency behavior model of the electronic transformer, as derived in Chapter 2, is given in Fig. 7.6. A circuit model entered into SPICE is shown in Fig. 7.7.

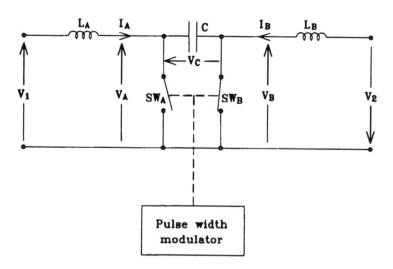

FIGURE 7.5   Basic circuit of electronic transformer.

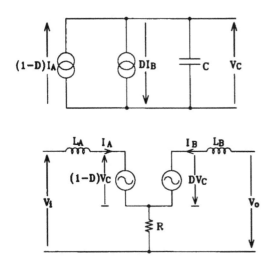

FIGURE 7.6 Low-frequency behavior model of electronic transformer.

The operation of the SPICE model given in Fig. 7.7 can be explained as follows:

1. The duty cycle D of the circuit is modeled as a voltage source having a value between 0 V and 1 V. In this example, $D = 0.5$.
2. The incremental changes δD and δV_i are assumed to be independent voltage sources.
3. The current source $[1 - D - \delta D]I_A$ and $[D + \delta D]I_B$ are modeled as nonlinear voltage-controlled current sources.
4. The voltage sources $[1 - D - \delta D]V_C$ and $[D + \delta D]V_C$ are modeled as nonlinear voltage-controlled voltage sources.
5. The two 1-GΩ resistors are dummy resistances added only to satisfy the requirements of SPICE.
6. R_1, R_2, and R_3 are the series resistances of L_A, L_B, and C_L, respectively. These resistances were not included in the analysis given in Chapter 3, so as to keep the algebraic manipulations at a manageable level of complexity. But they can be easily included in the SPICE program to improve the accuracy of simulation.

The complete simulation file is given as Attachment 7.1 (at the end of this chapter). The simulated $\delta V_o(s)/\delta D(s)$ and $\delta V_o(s)/\delta V_i(s)$ transfer functions for the given component values and duty cycle are shown in Figs. A7.1.2 and A7.1.3, respectively. It is found that, by using this simulation method,

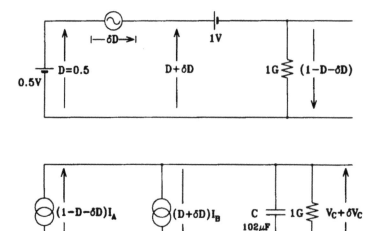

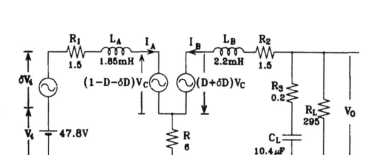

FIGURE 7.7 SPICE model of electronic transformer.

the transfer functions for extreme-case combinations of component values and duty cycle can be easily and quickly generated for analysis and design purposes.

The result of a transient simulation, assuming sudden changes in the duty cycle between 0.5 and 0.7, is shown in Fig. A7.1.4 of Attachment 7.1.

It may appear that the simulated results in Attachment 7.1 can actually be obtained by analytical methods. Why do we bother to do computer simulations?

A comparison between the analytical method and the computer simulation should reveal the following differences between the two approaches:

1. While the analytical method can provide good insight into the operation and characteristics of converters, the work involved can be too complicated. When additional components, such as the effective series resistance of the output filtering capacitor or an extra filter at the input of the converter, are added to the model, the analysis can become very difficult. Such difficulties, however, do not exist in computer simulations.
2. While theoretical transient analyses are difficult to perform for nonlinear circuits, transient responses are easy to simulate.

In addition to the reasons described earlier, we have one more very important motivation to perform computer simulations: whereas, in theoretical analysis, the continuous and discontinuous modes of operation have to be analyzed separately, it is possible for a combined SPICE model, which is valid for both continuous and discontinuous modes of operation, to be developed for simulations covering both modes of operation. It will be found that such a combined model is extremely useful for the computer-aided design of regulators.

Based on the simulation techniques described in Section 7.1, the development of a combined converter behavior model for both continuous and discontinuous modes of operation will be discussed in the next section.

7.3 DEVELOPING A COMBINED MODEL FOR BOTH CONTINUOUS AND DISCONTINUOUS MODES OF OPERATION

The philosophy we use to develop a common SPICE model for both continuous and discontinuous modes of operation can be outlined as follows [83]:

1. We first develop an inductor equivalent circuit to simulate the averaged current in the energy-storage inductor of the converter. In the simulation of the inductor current, we make use of a resistive current generator to model the discontinuous-mode current in the inductor. An additional inductive current component will be added to the discontinuous-mode current to simulate the inductor current under continuous-mode condition. Since the inductive current component will drop to zero automatically in the discontinuous-mode

operation, there is no need for a separate discontinuous-mode model.

2. The simulated inductor current is then weighted according to the duty cycle of the power switch and redistributed appropriately to related branches of the circuit to simulate the averaged currents in the input and output circuits of the converter.

Based on the concept mentioned above, the SPICE model for the buck-boost converter for both continuous and discontinuous modes of operation will be developed in the following subsection.

7.3.1   Combined Model of Buck-Boost Converter

Figure 7.8 is an example of a buck-boost converter to illustrate the actual modeling method. From the discontinuous-mode current waveform shown in Fig. 7.8(b), it is found that if the averaged value of the inductor current is defined as I_L, the averaged input current supplied by the input voltage V_i will be equal to

$$I_i = \frac{D}{D + D_2} I_L \tag{7.14}$$

where D_2 is the duty cycle of the flywheel diode D_f. The output current delivered to the output circuit, consisting of C_L and R_L connected in parallel, is

$$I_o = \frac{D_2}{D + D_2} I_L \tag{7.15}$$

Based on Eqs. (7.14) and (7.15), the SPICE model of the converter can be found as shown in Fig. 7.8(c).

It will be shown that, although the model shown in Fig. 7.8(c) is developed based on the discontinuous-mode operation, the same model can be extended to both modes of operation by adding elements to the inductor equivalent circuit to simulate the inductive component of the inductor current.

The problem we now have to solve is: How do we actually model the inductor equivalent circuit shown in Fig. 7.8(c), which is supposed to generate outputs representing the inductor current I_L and the value of D_2?

Based on the waveforms given in Fig. 7.9 for the three different operating conditions (discontinuous mode, critical condition, and continuous

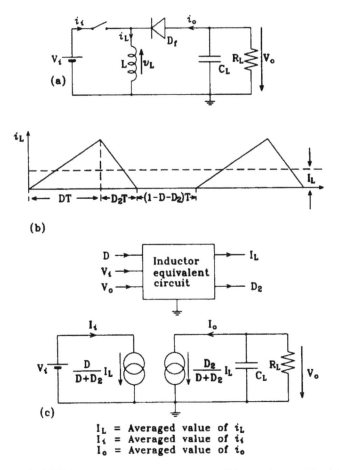

FIGURE 7.8 (a) Buck-boost converter, (b) waveform of i_L, (c) SPICE model.

mode), the development of the inductor equivalent circuit can be explained as follows:

1.   For discontinuous-mode operation, the voltage across the inductor, as shown in Fig. 7.9(a), is given by

$$v_L = V_i \qquad \text{for } 0 < t < DT \tag{7.16}$$

$$v_L = -V_o \qquad \text{for } DT < t < (D + D_2)T \tag{7.17}$$

$$v_L = 0 \qquad \text{for } (D + D_2)T < t < T \tag{7.18}$$

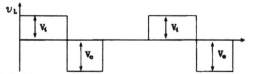

(a) Discontinuous—mode voltage waveform.

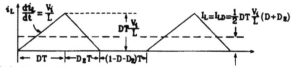

(b) Discontinuous—mode current waveform.

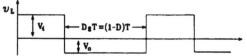

(c) Critical condition voltage waveform.

(d) Critical condition current waveform.

(e) Continuous—mode voltage waveform.

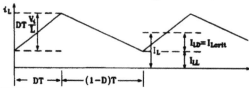

(f) Continuous—mode current waveform.

FIGURE 7.9 Inductor voltage and current waveforms of buck-boost converter.

The averaged inductor voltage is

$$V_L = DV_i - D_2V_o \qquad (7.19)$$

where (for discontinuous-mode operation)

$$D_2V_o = DV_i \qquad (7.20)$$

$$D_2 = D \frac{V_i}{V_o} \tag{7.21}$$

$$V_L = 0 \tag{7.22}$$

Although $V_L = 0$, the actual averaged inductor current for discontinuous-mode operation, which is denoted as I_{LD}, is not zero. The value of I_{LD} can be found from Fig. 7.9(b) as

$$I_{LD} = \frac{1}{2} DT \frac{V_i}{L} (D + D_2) \tag{7.23}$$

The current I_{LD} is simulated by the discontinuous-mode current generator in the combined model shown in Fig. 7.10(a). However, since Eq. (7.23) is valid only for frequencies lower than one-half of the converter switching frequency f_s, an RC circuit with a cutoff frequency of $(1/2)f_s$ (a 1-Ω resistance in parallel with a $1/(\pi f_s)$-F capacitance) is used in Fig. 7.10(a) to remove the high-frequency component of the current. Only the current in V_{SEND} is used to represent I_{LD}. (V_{SEND} is a dummy voltage source used to sense the current I_{LD}.)

2. For the critical condition between the continuous and discontinuous modes of operation, the voltage and current waveforms are shown in Fig. 7.9(c) and 7.9(d). We then have

$$D_2 = (1 - D) \tag{7.24}$$

Substitution of Eq. (7.24) into Eq. (7.23) gives the corresponding critical value of I_{LD}:

$$I_{LD} = I_{Lcrit} = \frac{1}{2} DT \frac{V_i}{L} \tag{7.25}$$

In order to simulate Eq. (7.25), all that is required is to ensure that the parameter D_2 is maintained at the value of $(1 - D)$ for the critical condition.

3. For continuous-mode operation, the waveforms are as shown in Figs. 7.9(e) and 7.9(f), where

$$D_2 = 1 - D \tag{7.26}$$

$$V_L = DV_i - (1 - D)V_o = DV_i - D_2V_o \tag{7.27}$$

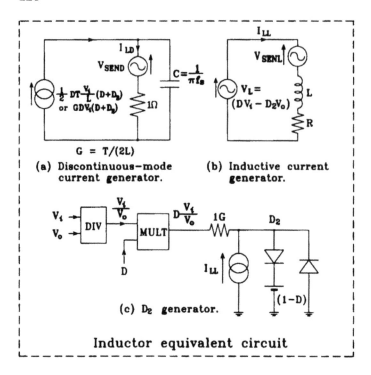

(a) Discontinuous—mode current generator.

(b) Inductive current generator.

(c) D_2 generator.

Inductor equivalent circuit

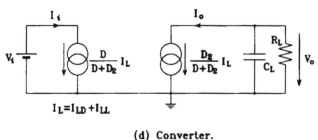

$I_L = I_{LD} + I_{LL}$

(d) Converter.

FIGURE 7.10 Combined model of buck-boost converter for both continuous and discontinuous modes of operation.

$$I_L = I_{LD} + I_{LL} = I_{Lcrit} + I_{LL} \qquad (7.28)$$

$$I_{LL} = \frac{1}{L} \int_0^t V_L \, dt = \frac{1}{L} \int_0^t (DV_i - D_2V_0) \, dt \qquad (7.29)$$

The current I_{LL} is now modeled by the inductive current generator shown in Fig. 7.10(b). Note that a small resistance R is added in series with the inductance L to satisfy the SPICE rule that a voltage source cannot be connected to an inductor with zero resistance (because it implies an indeterminable dc current).

4.   The D_2 generator shown in Fig. 7.10(c) is used to generate a parameter D_2, such that

$$D_2 = D \frac{V_i}{V_o} \qquad \text{for discontinuous-mode operation}$$

$$D_2 = (1 - D) \qquad \text{for critical condition and} \\ \text{continuous-mode operation}$$

Special attention should be paid to the D_2 generator circuit shown in Fig. 7.10(c). There, by making use of a divider and a multiplier, an algebraic expression DV_i/V_o (in the form of a voltage) is generated to represent the variable D_2 for discontinuous-mode operation to satisfy Eq. (7.21).

In continuous-mode operation, however, we have either

$$I_{LL} > 0 \tag{7.30}$$

or

$$D \frac{V_i}{V_o} > (1 - D) \tag{7.31}$$

In either case, the diode clamping circuit shown in Fig. 7.10(c), consisting of two diodes and the voltage source $(1 - D)$, clamps the value of D_2 to $(1 - D)$ to satisfy Eq. (7.26).

Figure 7.11 shows the actual SPICE arrangement to simulate the buck-boost converter. The corresponding input file and two simulated $\delta V_o(s)/\delta D(s)$ characteristics (one for continuous mode, one for discontinuous mode) are given in Attachment 7.2 (at the end of this chapter). In this file the following should be noted:

1.   The following parameters are assumed in the listing given in A7.2 (refer to Fig. 7.8(a) for the circuit):

dc input voltage $V_i = 20$ V
Steady-state duty cycle $D = 0.5$
Switching frequency $f_s = 100$ kHz

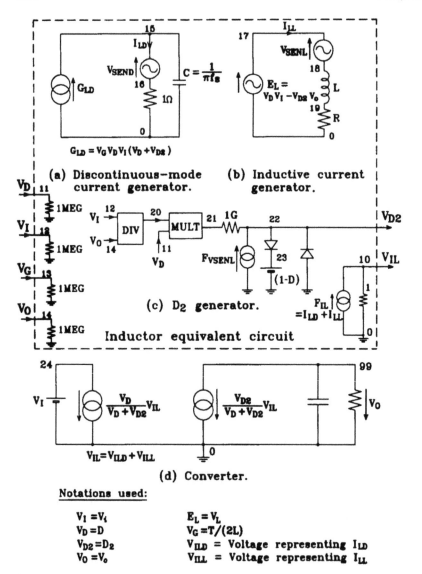

$G_{LD} = V_G V_D V_I (V_D + V_{D2})$

(a) Discontinuous-mode   (b) Inductive current
    current generator.   generator.

(c) D$_2$ generator.

Inductor equivalent circuit

$V_{IL} = V_{ILD} + V_{ILL}$

(d) Converter.

Notations used:

$V_I = V_i$ $E_L = V_L$
$V_D = D$ $V_G = T/(2L)$
$V_{D2} = D_2$ V_{ILD} = Voltage representing I_{LD}
$V_O = V_o$ V_{ILL} = Voltage representing I_{LL}

FIGURE 7.11   Actual SPICE arrangement of combined model for buck-boost converter.

Inductance $L = 5$ μH
Effective series resistance of $L = 5$ mΩ
Filtering capacitance $C_L = 1000$ μF
Load resistance $R_L = 0.5$ Ω (for continuous-mode operation)

2. As will be explained in the next subsection, the buck-boost model in this file can be easily modified to a buck or boost model.
3. The statements for the inductor equivalent circuit (between the "*START OF INDUCTOR EQUIVALENT CIRCUIT" and the "*END OF INDUCTOR EQUIVALENT CIRCUIT" statements) are extremely useful because similar statements are used in many other converter behavior models as well.

7.3.2 Models for Buck and Boost Converters

Applying the principle described above, the buck-boost model in Fig. 7.10 can be easily modified to form a buck or boost model. The necessary modifications required to change it to a buck model are listed in Table 7.1. The modifications required to change it to a boost model are listed in Table

Table 7.1 Changes Required to Modify a Buck-Boost Model to a Buck Model[a]

Component to be changed	Changed from (buck-boost)	To (buck)	Remarks
The V_i input to the inductor equivalent circuit	V_i	$(V_i - V_o)$	Meaning: 1. I_{LD} changed from $\frac{1}{2} DT \frac{V_i}{L} (D + D_2)$ to $\frac{1}{2} DT \frac{(V_i - V_o)}{L} (D + D_2)$ 2. D_2 changed from DV_i/V_o to $D(V_i - V_o)/V_o$ 3. V_L changed from $DV_i - D_2 V_o$ to $D(V_i - V_o) - D_2 V_o$
The output current I_o of converter	$\dfrac{D_2}{D + D_2} I_L$	$-I_L$	Negative sign implies change of direction only, as shown in Fig. 7.12(d).

[a]Refer to Fig. 7.12 for the detailed arrangements of the buck model and compare it with the buck-boost model in Fig. 7.10.

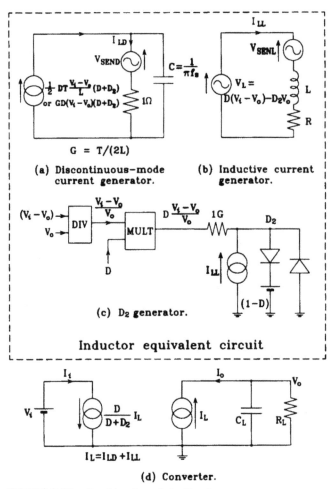

(a) Discontinuous–mode current generator.

G = T/(2L)

(b) Inductive current generator.

(c) D₂ generator.

Inductor equivalent circuit

$I_L = I_{LD} + I_{LL}$

(d) Converter.

FIGURE 7.12 Combined model of buck converter.

7.2. The remarks given in the tables explain the implications of the modifications.

Note that the low-frequency behavior model given in Attachment 7.2 also contains comments describing how it can be modified into a buck or boost model.

7.4 COMBINED MODEL FOR THE ĆUK CONVERTER

Applying the techniques introduced in Section 7.3, we will study the development of a combined model (for both continuous and discontinuous

Table 7.2 Changes Required to Modify a Buck-Boost Model to a Boost Model[a]

Component to be changed	Changed from (Buck-Boost)	To (Boost)	Remarks
The V_o input to the inductor equivalent circuit	V_o	$(V_o - V_i)$	Meaning: 1. D_2 changed from DV_i/V_o to $DV_i/(V_o - V_i)$ 2. V_L changed from $DV_i - D_2V_o$ to $DV_i - D_2(V_o - V_i)$
The output current I_o of converter	$\dfrac{D_2}{D + D_2} I_L$	$-\dfrac{D_2}{D + D_2} I_L$	Negative sign indicates change of direction only, as shown in Fig. 7.13(d).
The input current I_i of converter	$\dfrac{D}{D + D_2} I_L$	I_L	

[a]Refer to Fig. 7.13 for the detailed arrangements of the boost model and compare it with the buck-boost model in Fig. 7.10.

modes) for the Ćuk converter. It will, however, be found that the steps involved are more complex for the following reasons:

1. There are two inductors in a Ćuk converter.
2. The definition of "discontinuous-mode" operation of the Ćuk converter is different from that of the buck, buck-boost, or boost converter. In a buck, buck-boost, or boost converter, the discontinuous mode of operation is defined as the operating condition in which the inductor current falls to zero during the time $DT < t < T$. However, in a Ćuk converter, it should be defined as the operating condition in which the flywheel diode current, shown as i_{Df} in Fig. 7.14, falls to zero within the time $DT < t < T$.
3. The two inductors in a Ćuk converter may be magnetically coupled together to reduce the current ripples. This further complicates the low-frequency behavior model.

In the following subsection the development of a low-frequency behavior model for a Ćuk converter with uncoupled inductors will be described.

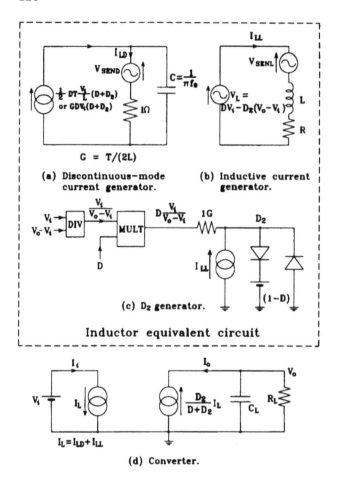

(a) Discontinuous-mode current generator.

(b) Inductive current generator.

(c) D_2 generator.

Inductor equivalent circuit

(d) Converter.

FIGURE 7.13 Combined model of boost converter.

7.4.1 Ćuk Converter with Uncoupled Inductors

Let us consider the Ćuk converter circuit shown in Fig. 7.14 and the waveforms shown in Fig. 7.15 for discontinuous-mode operation.

It may appear that, since a Ćuk converter consists effectively of a boost converter followed by a buck converter, the inductor currents in L_A and L_B should look like those shown as i_{LA} and i_{LB} in Figs. 7.15(b) and 7.15(e). The actual current waveforms, however, resemble those shown as (i_{LA} + I_{LO}) and (i_{LB} − I_{LO}) in Figs. 7.15(c) and 7.15(f), meaning that there is an additional dc component I_{LO} flowing in the inductors [38]. The principle

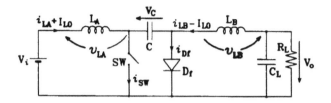

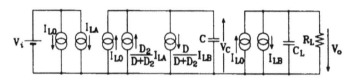

FIGURE 7.14 Development of combined model for Ćuk converter.

of conservation of energy requires that the circuit should, in the steady state, satisfy the power relationship

$$\text{Output power} = \eta \times \text{input power} \tag{7.32}$$

where η is the conversion efficiency of the converter. We therefore have, for discontinuous-mode operation,

$$V_o(I_{LBD} - I_{LO}) = \eta V_i(I_{LAD} + I_{LO}) \tag{7.33}$$

$$I_{LO} = \frac{V_o I_{LBD} - \eta V_i I_{LAD}}{V_o + \eta V_i} \tag{7.34}$$

By applying the principle described in Section 7.3, a model of the Ćuk converter (for both continuous and discontinuous modes) that takes into account the effect of I_{LO} can be developed as shown in Fig. 7.16. As we study this model, our attention should be focused on the following:

1. Functionally, circuit blocks (a) and (b) in Fig. 7.16 are similar to blocks (a) and (b) in Fig. 7.13, which are for a boost converter.

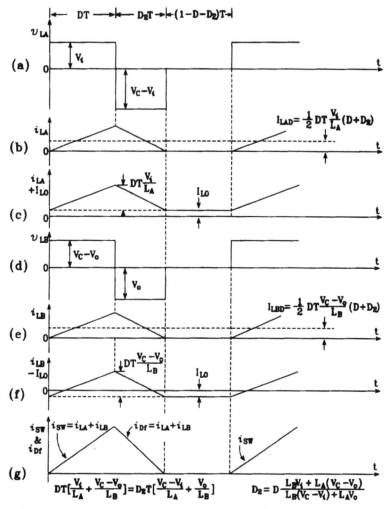

FIGURE 7.15 Voltage and current waveforms of Ćuk converter for discontinuous-mode operation.

2. Circuit blocks (c) and (d) in Fig. 7.16 are similar to blocks (a) and (b) in Fig. 7.12, which are for a buck converter.
3. Circuit block (e) in Fig. 7.16, containing V_{LO}, L_A, L_B, and $(R_A + R_B)$ in series, is used to model the current I_{LO} in Eq. (7.34). It should be noted that it is only in the steady state that I_{LO} is given by Eq. (7.34). Under transient conditions, the rate of change of I_{LO} is lim-

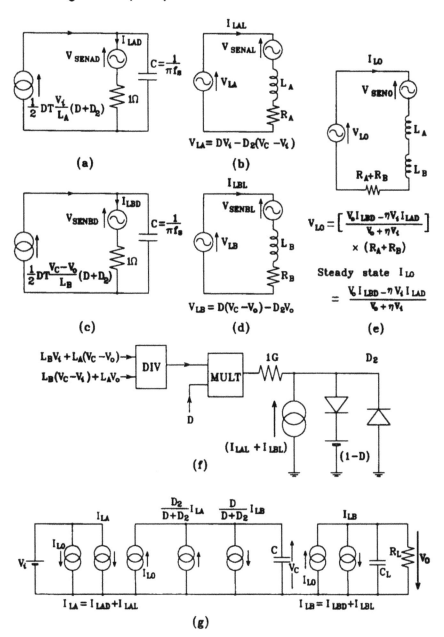

FIGURE 7.16   Combined model of Ćuk converter.

ited by the effective inductance in the circuit, as implied by circuit block (e) in Fig. 7.16.

4. I_{LO} forms an additional current component in the input circuit, the output circuit, and the circuit containing the energy storage capacitor C, as shown in Fig. 7.16(g).

5. Circuit block (f) is used to generate the function D_2 (the duty cycle of flywheel diode D_f), as explained below.

In order to find D_2, we have to refer back to the waveforms of the currents in the transistor switch SW and the flywheel diode D_f, shown as i_{sw} and i_{Df}, respectively, in Fig. 7.15(g). Since the rise in i_{sw} during $0 < t < DT$ should be equal to the fall in i_{Df} during $DT < t < (D + D_2)T$, we have (refer to Fig. 7.14 for the circuit):

$$\text{Rise in } i_{sw} = \text{fall in } i_{Df} \tag{7.35}$$

$$\text{Rise in } (i_{LA} + i_{LB}) = \text{fall in } (i_{LA} + i_{LB}) \tag{7.36}$$

$$DT\left[\frac{V_i}{L_A} + \frac{V_C - V_o}{L_B}\right] = D_2T\left[\frac{V_C - V_i}{L_A} + \frac{V_o}{L_B}\right] \tag{7.37}$$

$$\therefore D_2 = D\,\frac{L_B V_i + L_A(V_C - V_o)}{L_B(V_C - V_i) + L_A V_o} \tag{7.38}$$

The variable D_2 is obtained by applying voltages $[L_B V_i + L_A(V_C - V_o)]$ and $[L_B(V_C - V_i) + L_A V_o]$ to the D_2 generator shown in circuit block (f) of Fig. 7.16. All the components in Fig. 7.16 together form a SPICE model of the Ćuk converter (with uncoupled inductors).

7.4.2 Ćuk Converter with Coupled Inductors

The Ćuk converter model described in the last subsection is based on the assumption that the two inductors L_A and L_B are not coupled. When L_A and L_B are coupled together, the SPICE model will also need the following modifications in order to simulate the effects of magnetic coupling (refer to Fig. 7.17):

1. The inductors L_A and L_B need to be linked by a SPICE statement such as

    K LA LB VALUE

    where VALUE is the coupling coefficient. The polarity of coupling is shown in Figs. 7.17(b), 7.17(d), and 7.17(e).

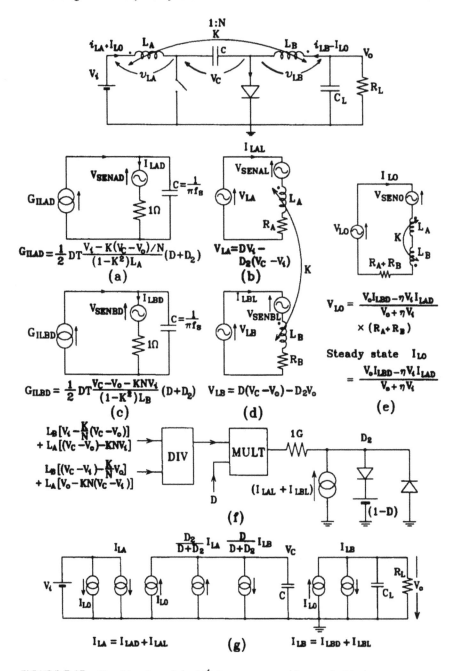

FIGURE 7.17 Combined model of Ćuk converter with coupled inductors.

2. New expressions for I_{LAD} (discontinuous-mode I_{LA}), I_{LBD} (discon-
    tinuous-mode I_{LB}), and D_2 have to be found, that take into account
    the effects of coupling.

It is found that, although it is a simple matter to incorporate the coupling
coefficient between L_A and L_B into the model, it requires a further examination of the current waveforms in order to find the new expressions
for I_{LAD}, I_{LBD}, and D_2.

Figure 7.18 shows the new current waveforms of the Ćuk converter with
coupled inductors. (Note that, although the shapes of the current waveforms in Fig. 7.18 are similar to those shown in Fig. 7.15, the scales are
different.)

In order to determine the new current components I_{LAD} and I_{LBD}, we
need to find the new di_{LA}/dt and di_{LB}/dt first. Consider the equivalent
circuit of two coupled inductors, as shown in Fig. 7.19. We have

$$v_{LA} = L_A \frac{di_{LA}}{dt} + M \frac{di_{LB}}{dt} \tag{7.39}$$

$$v_{LB} = M \frac{di_{LA}}{dt} + L_B \frac{di_B}{dt} \tag{7.40}$$

Solving for the unknown di_{LA}/dt, we have

$$\frac{di_{LA}}{dt} = \frac{\begin{vmatrix} v_{LA} & M \\ v_{LB} & L_B \end{vmatrix}}{\begin{vmatrix} L_A & M \\ M & L_B \end{vmatrix}} \tag{7.41}$$

$$= v_{LA} \frac{L_B}{L_A L_B - M^2} - v_{LB} \frac{M}{L_A L_B - M^2} \tag{7.42}$$

$$= \frac{v_{LA}}{L_A} \frac{1}{1 - M^2/(L_A L_B)} - \frac{v_{LB}}{L_A} \frac{M/L_B}{1 - M^2/(L_A L_B)} \tag{7.43}$$

where

$$\frac{M^2}{L_A L_B} = \frac{K^2 L_A L_B}{L_A L_B} = K^2,$$

$$\frac{M}{L_B} = \frac{K[L_A L_B]^{1/2}}{L_B} = K \left[\frac{L_A}{L_B} \right]^{1/2} = \frac{K}{N} \tag{7.44}$$

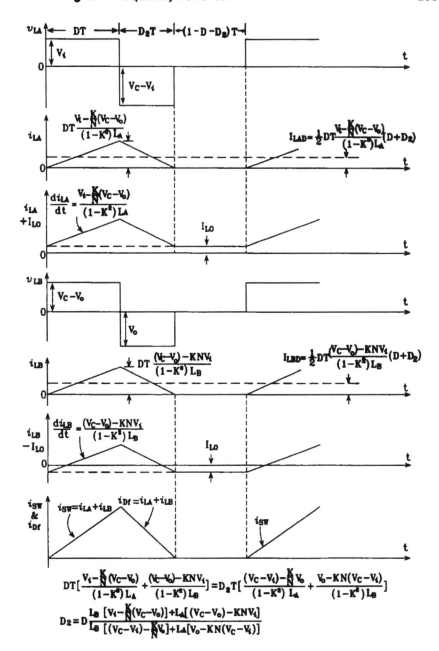

FIGURE 7.18 New current waveforms for Ćuk converter with coupled inductors.

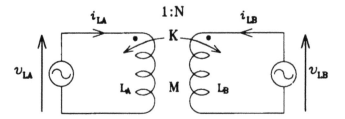

FIGURE 7.19 Equivalent circuit of two coupled inductors.

$$\therefore \frac{di_{LA}}{dt} = \frac{v_{LA}}{L_A} \frac{1}{1 - K^2} - \frac{v_{LB}}{L_A} \frac{K/N}{1 - K^2} \qquad (7.45)$$

$$= \frac{v_{LA} - (K/N)\, v_{LB}}{(1 - K^2)L_A} \qquad (7.46)$$

Equation (7.46) implies that, as far as the primary loop containing L_A is concerned, it has an induced emf from the secondary equal to $(-K/N)\, v_{LB}$ and a total effective inductance equal to $(1 - K^2)L_A$.

The value of di_{LA}/dt for $0 < t < DT$ may now be found by substituting $v_{LA} = V_i$ and $v_{LB} = (V_C - V_o)$ into Eq. (7.46):

$$\frac{di_{LA}}{dt} = \frac{V_i - (K/N)(V_C - V_o)}{(1 - K^2)L_A} \qquad (7.47)$$

The current I_{LAD} can therefore be determined as

$$I_{LAD} = \frac{1}{2} DT \frac{V_i - (K/N)(V_C - V_o)}{(1 - K^2)L_A} (D + D_2) \qquad (7.48)$$

Based on a similar analysis starting with Eqs. (7.39) and (7.40), the values of di_{LB}/dt for $0 < t < DT$ and I_{LBD} can be determined as

$$\frac{di_{LB}}{dt} = \frac{(V_C - V_o) - KNV_i}{(1 - K^2)L_B} \qquad (7.49)$$

$$I_{LBD} = \frac{1}{2} DT \frac{(V_C - V_o) - KNV_i}{(1 - K^2)L_B} (D + D_2) \qquad (7.50)$$

Equations (7.48) and (7.50) are simulated by circuit blocks (a) and (c), respectively, in Fig. 7.17.

Now, let us proceed to determine the new value of D_2. Applying the same technique of analysis, we find the values of di_{LA}/dt and di_{LB}/dt during the time $DT < t < (D + D_2)T$ as

$$\frac{di_{LA}}{dt} = -\frac{(V_C - V_i) - (K/N)V_o}{(1 - K)^2 L_A} \tag{7.51}$$

$$\frac{di_{LB}}{dt} = -\frac{V_o - KN(V_C - V_i)}{(1 - K^2)L_B} \tag{7.52}$$

Since the total rise of $(i_{LA} + i_{LB})$ during $0 < t < DT$ must be equal to the total fall of $(i_{LA} + i_{LB})$ during $DT < t < (D + D_2)T$, we have

$$DT \left[\frac{di_{LA}}{dt} + \frac{di_{LB}}{dt}\right]\Bigg|_{0<t<DT} = D_2T \left[\frac{-di_{LA}}{dt} + \frac{-di_{LB}}{dt}\right]\Bigg|_{DT<t<(D+D_2)T} \tag{7.53}$$

Substitution of Eqs. (7.47), (7.49), (7.51), and (7.52) into Eq. (7.53) gives

$$DT \left[\frac{V_i - (K/N)(V_C - V_o)}{(1 - K^2)L_A} + \frac{(V_C - V_o) - KNV_i}{(1 - K^2)L_B}\right]$$
$$= D_2T \left[\frac{(V_C - V_i) - (K/N)V_o}{(1 - K^2)L_A} + \frac{V_o - KN(V_C - V_i)}{(1 - K^2)L_B}\right] \tag{7.54}$$

We therefore have D_2 equal to

$$D_2 = D \frac{L_B[V_i - (K/N)(V_C - V_o)] + L_A[(V_C - V_o) - KNV_i]}{L_B[(V_C - V_i) - (K/N)V_o] + L_A[V_o - KN(V_C - V_i)]} \tag{7.55}$$

In summary, it can be stated that the equations given below, together with the circuit shown in Fig. 7.17, completely model the low-frequency behavior of the coupled-inductor Ćuk converter in either the continuous or discontinuous mode of operation:

$$I_{LAD} = \tfrac{1}{2}DT \frac{V_i - (K/N)(V_C - V_o)}{(1 - K^2)L_A} (D + D_2) \qquad \text{[From Eq. (7.48)]} \tag{7.56}$$

$$V_{LA} = DV_i - D_2(V_C - V_i) \tag{7.57}$$

$$I_{LBD} = \tfrac{1}{2}DT\frac{(V_C - V_o) - KNV_i}{(1 - K^2)L_B}(D + D_2) \qquad \text{[from Eq. (7.50)]}$$

$$\tag{7.58}$$

$$V_{LB} = D(V_C - V_o) - D_2V_o \tag{7.59}$$

$$D_2 = D\frac{L_B[V_i - (K/N)(V_C - V_o)] + L_A[V_C - V_o) - KNV_i]}{L_B[(V_C - V_i) - (K/N)V_o] + L_A[V_o - KN(V_C - V_i)]}$$

$$\tag{7.60}$$

$$I_{LO} = \frac{V_o I_{LBD} - \eta V_i I_{LAD}}{V_o + \eta V_i} \qquad \text{[from Eq. (7.34)]} \tag{7.61}$$

(It is necessary to make an estimate for η if it is not known.)

$$I_{LA} = I_{LAD} + I_{LAL} \tag{7.62}$$

$$I_{LB} = I_{LBD} + I_{LBL} \tag{7.63}$$

For the purpose of illustration, the circuit shown in Fig. 7.20 is modeled and simulated. The SPICE input file and some simulated results are given in Attachment 7.3 at the end of this chapter.

In this simulation file, the following should be noted:

1. Figure A7.3.2 is the result of a transient simulation in which a slow change of loading current from 0.15 to 5 A is assumed. The transition of operation from discontinuous to continuous mode is seen to occur at a loading current of 0.3 A.
2. Figure A7.3.3 is a simulated $\delta V_o(s)/\delta D(s)$ transfer function for continuous-mode operation, obtained by specifying a loading resistance $R_L = 8\ \Omega$ and a duty cycle $D = 0.4$.

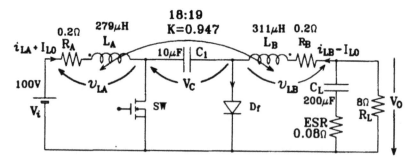

FIGURE 7.20 Ćuk converter with coupled inductors.

3. Figure A7.3.4 is a simulated $\delta V_o(s)/\delta D(s)$ transfer function for discontinuous-mode operation, obtained by specifying a loading resistance $R_L = 300 \ \Omega$ and a duty cycle $D = 0.26$.

7.5 MODELING CURRENT-CONTROLLED CONVERTERS

The modeling of a current-controlled converter is similar to that of an ordinary converter except that the duty cycle D of the current-controlled converter is not directly a known input but must be calculated from the given current-control input and the then existing operating conditions. Based on the example of a buck converter, the techniques of developing the low-frequency behavior models of current-controlled converters will be discussed in this section.

Refer back to the constant-frequency current-controlled buck converter in Fig. 5.7(a) of Chapter 5, which is redrawn here as Fig. 7.21(a). The current waveform of the converter for discontinuous-mode operation is shown in Fig. 7.21(b), and that for continuous-mode operation in Fig. 7.21(c). In these waveform diagrams, it is assumed that, because of the propagation delay in the amplifiers (A_1 and A_2) and the S–R flip-flop, the switching of SW is delayed by λT. (Note that this delay can significantly affect the duty cycle D.)

Figure 7.22 shows the modeling arrangement of the current-controlled buck converter. Compared with the ordinary buck converter model given in Fig. 7.12, it can be found that there is an additional "D generator" in Fig. 7.22(d) to generate the duty cycle D for the current-controlled converter. The function of the D generator is to calculate the value of D based on the control input (I_c) and the then existing operation condition. The current waveforms shown in Fig. 7.21 can help us to find D as a function of I_c, m_1, m_c, λ, T, and I_{LL}.

From the discontinuous-mode current waveform shown in Fig. 7.21(b), we have

$$I_c = (DT - \lambda T)(m_1 + m_c) \tag{7.64}$$

$$D = \frac{I_c}{(m_1 + m_c)T} + \lambda \tag{7.65}$$

where I_c represents the control input, and $I_c = V_c/(KR_f)$.

From the continuous-mode current waveform shown in Fig. 7.21(c), we have

$$I_c - I_{LL} = (DT - \lambda T)(m_1 + m_c) \tag{7.66}$$

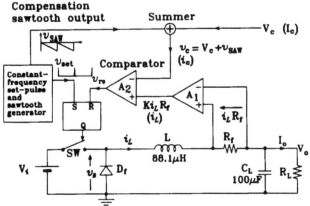

(a) Converter circuit with compensation slope.

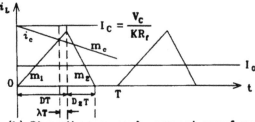

(b) Discontinuous-mode current waveform.

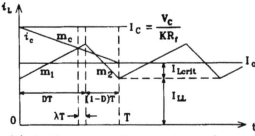

(c) Continuous-mode current waveform.

FIGURE 7.21 Current-controlled buck converter with compensation slope and its current waveforms.

$$D = \frac{I_c - I_{LL}}{(m_1 + m_c)T} + \lambda \qquad (7.67)$$

It is obvious that Eq. (7.67) alone can be used as the expression for the duty cycle D for both the continuous and discontinuous modes of operation because, in the discontinuous-mode operation, I_{LL} automatically drops to

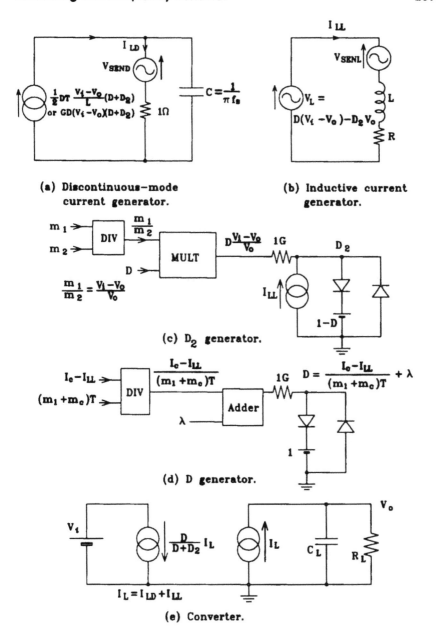

(a) Discontinuous−mode current generator.

(b) Inductive current generator.

(c) D_2 generator.

(d) D generator.

(e) Converter.

FIGURE 7.22   Schematic representation of combined model of current-controlled buck converter.

zero to enable Eq. (7.67) to reduce to Eq. (7.65). Based on Eq. (7.67), the D generator is constructed as shown in Fig. 7.22(d).

For the purpose of illustration, the SPICE listing and some simulation results for the current-controlled converter shown in Fig. 7.21(a) are given as Attachment 7.4 at the end of this chapter. In this simulation file, the following circuit parameters are assumed:

dc input voltage = 100 V
Control input I_c = 10 A
Compensation slope m_c = 0.6 × 10^6 A/s
Switching frequency f_s = 50 kHz
Delay factor λ = 0.05
Load resistance R_L = 9.8 Ω
Sampling resistance R_f is negligibly small

Figure A7.4.2 is a simulated $\delta V_o(s)/\delta V_i(s)$ characteristic of the converter that shows its inherent ability to reject the ripple in the dc input voltage. Figure A7.4.3 is a simulated $\delta V_o(s)/\delta I_c(s)$ transfer function. Both figures show the single-pole characteristic of the converter, even under continuous-mode operating condition.

7.6 MODELING MULTIPLE-OUTPUT AND PUSH-PULL CONVERTERS

Many converters are designed to provide multiple outputs. The usual method is to use a high-frequency power transformer with multiple secondary windings, each winding providing one output. In this section the low-frequency behavior models of high-frequency transformers, and converters with multiple outputs, will be studied. The effects of push-pull operation on the low-frequency behavior model will also be considered.

7.6.1 Behavior Model of Transformer

As far as the low-frequency behavior is concerned, the effect of the high-frequency power transformer in a converter is similar to that of an idealized transformer that can operate down to zero frequency (because it is capable of stepping the dc output voltage of the converter up or down).

Consider as an example the circuit shown in Fig. 7.23(a), which is a buck-boost converter. Figure 7.23(b) represents the low-frequency behavior model of the buck-boost converter, the details of which can be found in Fig. 7.10. The circuit shown in Fig. 7.23(c) is a flyback converter, which is effectively a buck-boost converter with a high-frequency transformer T.

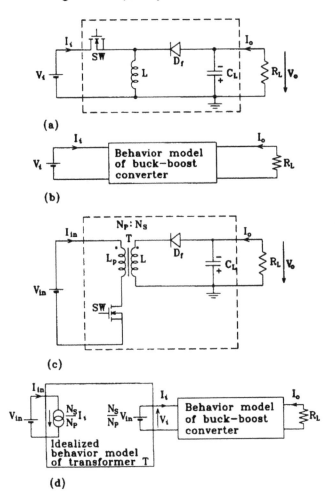

FIGURE 7.23 Low-frequency behavior model of flyback converter. (a) Buck-boost converter. (b) Behavior model of buck-boost converter. (c) Flyback converter. (d) Behavior model of flyback converter.

It should be easy to understand that if the transformer T in the flyback converter shown in Fig. 7.23(c) is a one-to-one transformer without leakage inductance (but with a primary inductance $L_P = L$), the low-frequency behavior model of the flyback converter will be the same as that of the buck-boost converter. If, however, the transformer T has a secondary to primary turns ratio of N_S/N_P, we shall have a step-up or step-down ratio of N_S/N_P in the output dc voltage, too. In order to simulate this effect, the idealized transformer as shown in Fig. 7.23(d) may be used as a low-frequency behavior model of the high-frequency transformer.

7.6.2   Modeling Converters with Multiple Outputs

If the converter in Fig. 7.23(c) has multiple outputs, the behavior model in Fig. 7.23(d) will have to be modified to include the multiple-output circuits. Figure 7.24 shows how it is done. There, the multiple-output flyback converter shown in Fig. 7.24(a) is modeled as the circuit shown in Fig. 7.24(b). Each of the two output behavior models in Fig. 7.24(b) is similar to the buck-boost model in Fig. 7.10. (Note that a coupling coefficient K needs to be specified for the inductances L_1 and L_2.)

7.6.3   Modeling Push-Pull Converters

The low-frequency behavior models of transformer-coupled push-pull, half-bridge, and full-bridge converters (all will be referred to as push-pull

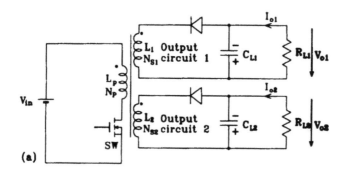

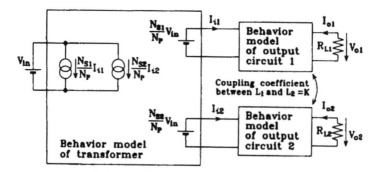

FIGURE 7.24   (a) Flyback converter with multiple outputs and (b) its low-frequency behavior model.

converters) are actually similar to that of the buck converter, which is shown in Fig. 7.12. The following modifications, however, are required for the push-pull model:

1. The parameter D in the push-pull model should be interpreted as twice the duty cycle of each switching transistor.
2. The parameter T in the discontinuous-mode current generator of the push-pull model should be interpreted as one-half of the switching period of each switching transistor.
3. The parameter f_s in the capacitance expression $C = 1/(\pi f_s)$ of the push-pull model should be interpreted as twice the switching frequency of each switching transistor.

7.6.4 Multiple-Output Simulation Example

In this subsection, we shall apply the modeling methodology described above to model and simulate a dual-output transformer-coupled push-pull converter. Both the frequency-response (ac) and transient-response simulations will be performed. In these simulations, one output of the converter is purposely arranged to operate in the discontinuous-mode operation, while the other operates in the continuous-mode operation.

The circuit to be simulated is shown in Fig. 7.25. Depending on the dc input voltage (V_{in}) and the output loading current (I_{o1} or I_{o2}), each of the

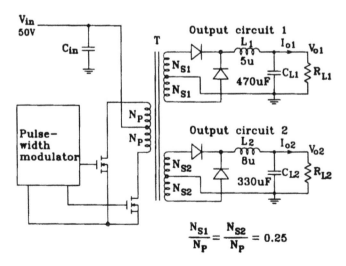

FIGURE 7.25 Transformer-coupled push-pull converter with two outputs. Switching frequency of each transistor = 50 kHz.

two output circuits may operate independently either in continuous mode or discontinuous mode.

The complete modeling and simulation file are given in Attachment 7.5 (at the end of this chapter). In the SPICE equivalent circuit shown in Fig. A7.5.1, the transformer T is modeled as a subcircuit. Each of the two converter output circuits is modeled effectively as a buck converter subcircuit. (Note that a transformer-coupled push-pull converter is functionally a push-pull version of the buck converter.) The effective series resistances of various components, such as the dc supply (R_{IN}), transformer windings (R_T, R_{48}, R_{59}), output rectifiers (D_{o1}, D_{o2}), and filtering capacitors (R_{C1}, R_{C2}), are also included in the model. The detailed arrangement of the converter subcircuit is given in Fig. A7.5.2.

In Attachment 7.5, the following simulations are performed:

1. A frequency-response (ac) simulation $\delta V_o(s)/\delta D(s)$ for $D = 0.5$ is carried out, with output 1 operating in discontinuous mode ($R_{L1} = 10\,\Omega$) and output 2 operating in continuous mode ($R_{L2} = 0.25\,\Omega$).
2. A transient simulation for $D = 0.5$ and with step changes in I_{o2} (output current of output 2) is also carried out. In this simulation, the loading resistance of output 1 is kept constant at $10\,\Omega$ to maintain discontinuous-mode operation, while the current I_{o2} is switched between $(V_{o2}/0.25)$A and $[(V_{o2}/0.25) + 4]$A. (For both values of I_{o2}, continuous-mode operation is maintained for output 2.)

The simulations shown in Figs. A7.5.3–A7.5.5 illustrate clearly the first-order characteristics of the discontinuous-mode-operated output circuit 1 and the second-order characteristics of the continuous-mode-operated output circuit 2.

7.7 SUMMARY AND FURTHER REMARKS

In this chapter the simulation of the low-frequency behaviors of converters has been discussed. A method to develop a combined model for both continuous and discontinuous modes of operation has been introduced and applied to various types of converters. Such a combined model is essential to the simulation of converters under large-signal conditions because practically all converters may make transitions between the two modes of operation. Simulation examples of various types of converters have also been given in the form of attachments. These examples can be used as starting points for circuit designers to develop their own converter models and simulation files for specific applications.

Simulations based on the low-frequency behavior models are particularly useful for the analysis of feedback compensation circuits in regulators and for the prediction of regulator performance.

Details about how the cycle-by-cycle simulation and the low-frequency behavior simulation can be used to help design converters and regulators will be discussed in the next chapter.

EXERCISES

1. Based on the simulation techniques introduced in Section 7.1, design a SPICE model to generate an output current, such that

    $$I_{o1} = 10^{-12}[e^{40V_i} - 1]$$
    $$I_{o2} = \ln V_i$$

    where V_i is a given input voltage.
2. In what ways is the combined converter model for both continuous and discontinuous modes of operation more useful than the separate models derived in Chapter 2? Is the combined model more useful than the separate models as far as manual analyses are concerned?
3. Compare the differences between the cycle-by-cycle simulation described in Chapter 6 and the low-frequency behavior simulation described in this chapter. Consider aspects of
    a. methods of modeling and simulation
    b. applications.
4. Starting from the buck-boost converter model file given in Attachment 7.2, develop a low-frequency behavior model for a boost converter with the same set of circuit parameters (same V_i, L, R_L, C_L, T, etc.).
5. Assume that a third output circuit is added to the dual-output, transformer-coupled push-pull converter shown in Fig. 7.25. The additional circuit is identical to that of output circuit 1. Develop a low-frequency behavior model of the new converter. (You may use the circuit file given in Attachment 7.5 as a starting point.)
6. Assume that the MOS transistor SW and flywheel diode D_f in the Ćuk converter circuit shown in Fig. 7.20 are replaced by two active and bidirectional switches, so that the converter becomes an electronic transformer. Develop a low-frequency behavior model for this electronic transformer.

ATTACHMENT 7.1: BEHAVIOR SIMULATION OF ELECTRONIC TRANSFORMER

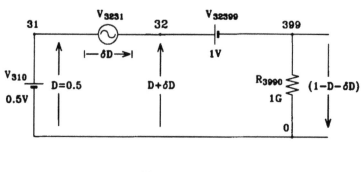

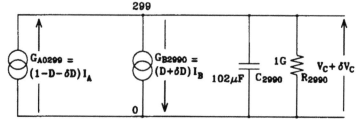

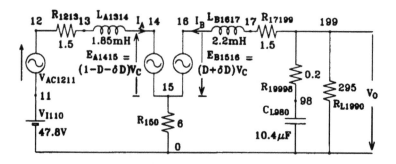

Fig. A7.1.1 SPICE model of electronic transformer.

A7.1.1   INPUT LISTING FOR SIMULATION OF dVo(S)/dD(S)
          TRANSFER FUNCTION

```
ELECTRONIC TRANSFORMER
*DUTY CYCLE TO OUTPUT VOLTAGE TRANSFER FUNCTION
*(REFER TO FIG.A7.1.1 FOR THE CIRCUIT.)
*
*
*SIMULATION OF DUTY CYCLE
*V(32)=DUTY CYCLE D
*V(0,399)=1-D
V310 31 0 DC 0.5
*THIS SETS THE STEADY DUTY CYCLE TO 0.5
V3231 32 31 AC 1
*V3231 SIMULATES THE FUNCTION OF dD
V32399 32 399 DC 1
R3990 399 0 1G
*
*
*SIMULATION OF DUTY-CYCLE CONTROLLED CURRENT SOURCES
GA0299 0 299 POLY(2) 0 399  12 13  0 0 0 0 0.66666667
*GA0299=(1-D)*IA
*IA IS MODELED AS V(12,13)/R1213
*V(299)=VOLTAGE ACROSS C
GB2990 299 0 POLY(2) 32 0  199 17  0 0 0 0 0.66666667
*GB2990=D*IB
*IB IS MODELED AS V(199,17)/R17199
C2990 299 0 102UF
R2990 299 0 1G
*
*
*INPUT CIRCUIT
VI110 11 0 DC 47.8
VAC1211 12 11 AC 0
*THE ZERO VOLTAGE OF VAC1211 ACTUALLY INACTIVATES VAC1211,
*TO ALLOW FOR A dVO(S)/dD(S) SIMULATION.
R1213 12 13 1.5
LA1314 13 14 1.85MH
EA1415 14 15 POLY(2) 0 399  299 0  0 0 0 0 1
*EA1415=(1-D)*VC
R150 15 0 6
*
*
*OUTPUT CIRCUIT
EB1516 15 16 POLY(2) 32 0  299 0  0 0 0 0 1
*EB1516=D*VC
LB1617 16 17 2.2MH
R17199 17 199 1.5
R19998 199 98 0.2
CL980 98 0 10.4UF
RL1990 199 0 295
*
*
.AC DEC 13 1 10K

.PRINT AC VDB(0,199) VP(0,199)
.PLOT AC VDB(0,199) VP(0,199)
.PROBE
*THE .PROBE STATEMENT IS FOR PSPICE ONLY.
*DELETE .PROBE STATEMENT IF RUN BY SPICE.
.END
```

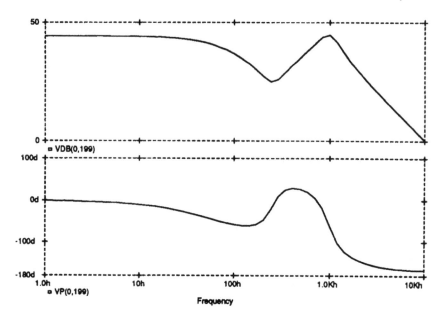

VDB(0,199) = Amplitude of $\delta V_0(s)/\delta D(s)$
VP(0,199) = Phase of $\delta V_0(s)/\delta D(s)$

Fig. A7.1.2   Simulated $\delta V_0(s)/\delta D(s)$ transfer function of electronic transformer.

A7.1.2   INPUT LISTING FOR SIMULATION OF dVo(S)/dVi(S)
          TRANSFER FUNCTION

```
ELECTRONIC TRANSFORMER
*SUPPLY INPUT VOLTAGE TO OUTPUT VOLTAGE TRANSFER FUNCTION
*(REFER TO FIG. A7.1.1 FOR THE CIRCUIT.)
*
*
*SIMULATION OF DUTY CYCLE
*V(32)=DUTY CYCLE D
*V(0,399)=1-D
V310 31 0 DC 0.5
*THIS SETS THE STEADY DUTY CYCLE TO 0.5
V3231 32 31 AC 0
*THE ZERO VOLTAGE OF V3231 ACTUALLY INACTIVATES V3231,
*TO ALLOW FOR A dVO(S)/dVI(S) SIMULATION.
V32399 32 399 DC 1
R3990 399 0 1G
*
*
*SIMULATION OF DUTY-CYCLE CONTROLLED CURRENT SOURCES
```

```
GA0299 0 299 POLY(2) 0 399  12 13  0 0 0 0 0.66666667
*GA0299=(1-D)*IA
*IA IS MODELED AS V(12,13)/R1213
*V(299)=VOLTAGE ACROSS C
GB2990 299 0 POLY(2) 32 0  199 17  0 0 0 0 0.66666667
*GB2990=D*IB
*IB IS MODELED AS V(199,17)/R17199
C2990 299 0 102UF
R2990 299 0 1G
*
*
*INPUT CIRCUIT
VI110 11 0 DC 47.8
VAC1211 12 11 AC 1
*VAC1211 SIMULATES THE FUNCTION OF dVI.
R1213 12 13 1.5
LA1314 13 14 1.85MH
EA1415 14 15 POLY(2) 0 399  299 0  0 0 0 0 1
*EA1415=(1-D)*VC
R150 15 0 6
*
*
*OUTPUT CIRCUIT
EB1516 15 16 POLY(2) 32 0  299 0  0 0 0 0 1
*EB1516=D*VC
LB1617 16 17 2.2MH
R17199 17 199 1.5
R19998 199 98 0.2
CL980 98 0 10.4UF
RL1990 199 0 295
*
*
.AC DEC 13 1 10K
.PRINT AC VDB(0,199) VP(0,199)
.PLOT AC VDB(0,199) VP(0,199)
.PROBE
*THE .PROBE STATEMENT IS FOR PSPICE ONLY.
*DELETE .PROBE STATEMENT IF RUN BY SPICE.
.END
```

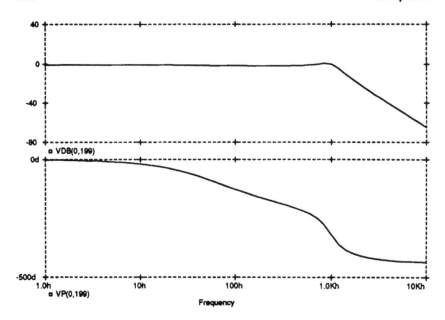

VDB(0,199) = Amplitude of $\delta V_o(s)/\delta V_i(s)$
VP(0,199) = Phase of $\delta V_o(s)/\delta V_i(s)$

Fig. A7.1.3 Simulated $\delta V_o(s)/\delta V_i(s)$ transfer function of
electronic transformer.

A7.1.3   INPUT LISTING FOR TRANSIENT SIMULATION

```
ELECTRONIC TRANSFORMER
*TRANSIENT SIMULATION ASSUMING A CHANGE IN DUTY CYCLE
*(REFER TO FIG. A7.1.1 FOR THE CIRCUIT.)
*
*
*SIMULATION OF DUTY CYCLE
*V(32)=DUTY CYCLE D
*V(0,399)=1-D
V310 31 0 DC 0.5
*THIS SETS THE STEADY DUTY CYCLE TO 0.5
V3231 32 31 PWL (0 0   2M 0   2.1M 0.2   35M 0.2   35.1M 0
+ 1 0)
*THE ABOVE STATEMENT SPECIFIES A PULSE IN THE DUTY CYCLE
V32399 32 399 DC 1
R3990 399 0 1G
*
*
*SIMULATION OF DUTY-CYCLE CONTROLLED CURRENT SOURCES
GA0299 0 299 POLY(2) 0 399   12 13   0 0 0 0 .66666667
*GA0299=(1-D)*IA
```

```
*IA IS MODELED AS V(12,13)/R1213
*V(299)=VOLTAGE ACROSS C
GB2990 299 0 POLY(2) 32 0  199 17  0 0 0 0 .66666667
*GB2990=D*IB
*IB IS MODELED AS V(199,17)/R17199
C2990 299 0 102UF
R2990 299 0 1G
*
*
*INPUT CIRCUIT
VI110 11 0 DC 47.8
VAC1211 12 11 AC 0
*THE ZERO VOLTAGE OF VAC1211 ACTUALLY INACTIVATES VAC1211,
*TO ALLOW FOR TRANSIENT SIMULATION.
R1213 12 13 1.5
LA1314 13 14 1.85MH
EA1415 14 15 POLY(2) 0 399  299 0  0 0 0 0 1
*EA1415=(1-D)*VC
R150 15 0 6
*
*
*OUTPUT CIRCUIT
EB1516 15 16 POLY(2) 32 0  299 0  0 0 0 0 1
*EB1516=D*VC
LB1617 16 17 2.2MH
R17199 17 199 1.5
R19998 199 98 0.2
CL980 98 0 10.4UF
RL1990 199 0 295
*
*
.TRAN .4M 44M
.PRINT TRAN V(32,0) V(0,199)
.PLOT TRAN V(32,0) V(0,199)
.PROBE
*THE .PROBE STATEMENT IS FOR PSPICE ONLY.
*DELETE .PROBE STATEMENT IF RUN BY SPICE.
.END
```

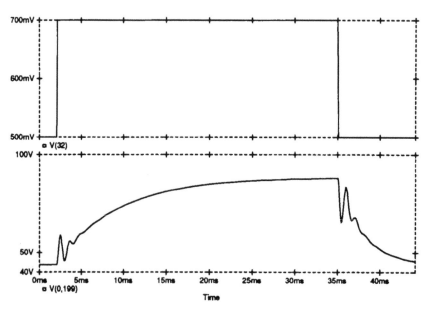

V(32)=Duty cycle input (1000mV=1)
V(0,199)=Output voltage V_0

Fig. A7.1.4 Simulated transient response of electronic transformer.

ATTACHMENT 7.2: BEHAVIOR SIMULATION OF BUCK-BOOST CONVERTER

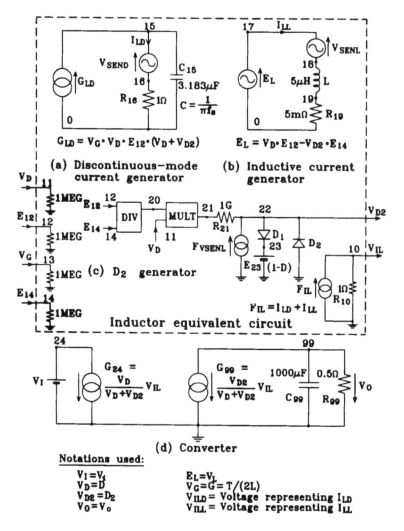

Fig. A7.2.1 Combined behavior model of Buck-Boost converter.

A7.2   SPICE INPUT LISTING OF BUCK-BOOST CONVERTER

```
BUCK-BOOST CONVERTER (ALL MODES)
*DUTY CYCLE TO OUTPUT VOLTAGE TRANSFER FUNCTION
*
*
*DIVIDER SUBCIRCUIT
*V(9)=V(3)/V(1)
.SUBCKT DIV 3 1 9
R30 3 0 1MEG
R10 1 0 1MEG
E20 2 0 POLY(2) 1 0 9 0   0 0 0 0 1
R20 2 0 1MEG
E90 9 0 POLY(1) 2 3   0 -1MEG
R90 9 0 1MEG
.ENDS DIV
*
*
*
*START OF INDUCTOR EQUIVALENT CIRCUIT
*
*V(11) = DUTY CYCLE INPUT
R11 11 0 1MEG
*
*V(12) = VI FOR BUCK-BOOST OR BOOST
*V(12) = VI-VO FOR BUCK
R12 12 0 1MEG
*
*V(13) = VG = T/(2L)
R13 13 0 1MEG
*
*V(14) = VO FOR BUCK-BOOST OR BUCK
*V(14) = VO-VI FOR BOOST
R14 14 0 1MEG
*
*
*DISCONTINUOUS-MODE CURRENT GENERATOR
GLD 0 15 POLY(1) 100 0   0 1
*GLD = V(100) =. VG*VD*V(12)*(VD+VD2)
*NOTE THAT V(12) = VI FOR BUCK-BOOST OR BOOST
*V(12) = VI-VO FOR BUCK
*V(15) = VILD
VSEND 15 16
R16 16 0 1
C15 15 0 3.183U
*C15 = 1/(3.1416*FS) = 3.183U
*TO GET VD*V(12) :
E103 103 0 POLY(2) 11 0 12 0   0 0 0 0 1
*V(11) = VD
*V(12) = VI FOR BUCK-BOOST OR BOOST
*V(12) = (VI-VO) FOR BUCK
*V(103) = VD*V(12)
R103 103 0 1MEG
```

```
*TO GET VD+VD2 :
E101 101 0 POLY(2) 22 0 11 0   0 1 1
*V(22) = VD2, V(11) = VD
*V(101) = VD+VD2
R101 101 0 1MEG
*TO GET VG*VD*V(12)*(VD+VD2) :
E100 100 0 POLY(3) 13 0 103 0 101 0
+0 0 0 0 0 0 0 0 0 0 0 0 0 1
*V(100) = V(13)*V(103)*V(101) = VG*VD*V(12)*(VD+VD2)
R100 100 0 1MEG
*
*
*INDUCTIVE CURRENT GENERATOR
EL 17 0 POLY(1) 110 0   0 1
*V(17) = V(110) = VD*V(12)-VD2*V(14)
*FOR BUCK-BOOST : V(12) = VI, V(14) = VO
*FOR BUCK : V(12) = VI-VO, V(14) = VO
*FOR BOOST : V(12) = VI, V(14) = VO-VI
VSENL 17 18
L 18 19 5U
R19 19 0 5M
*TO GET EL = VD*V(12)-VD2*V(14) :
E104 104 0 POLY(2) 22 0 14 0   0 0 0 0 1
*V(22) = VD2
*V(14) = VO FOR BUCK-BOOST OR BUCK
*V(14) = VO-VI FOR BOOST
*V(104) = VD2*V(14)
R104 104 0 1MEG
E110 110 0 POLY(2) 103 0 104 0   0 1 -1
*V(110) = EL = VD*V(12)-VD2*V(14)
R110 110 0 1MEG
*
*
*TOTAL INDUCTOR CURRENT GENERATOR
FIL 0 10 POLY(2) VSEND VSENL   0 1 1
R10 10 0 1
*V(10) = VILD+VILL = VIL(REPRESENTING IL)
*
*
*D2 GENERATOR
X1 12 14 20 DIV
*V(12) = VI FOR BUCK-BOOST OR BOOST
*V(12) = VI-VO FOR BUCK
*V(14) = VO FOR BUCK-BOOST OR BUCK
*V(14) = VO-VI FOR BOOST
*V(20) = V(12)/V(14)
E21 21 0 POLY(2) 20 0 11 0   0 0 0 0 1
*V(21) = VD*V(12)/V(14)
R21 21 22 1G
FVSENL 0 22 POLY(1) VSENL   0 1
*FVSENL = ILL
D1 22 23 DIDEAL
*DIDEAL IS A DIODE MODEL HAVING
```

```
*ALMOST ZERO FORWARD VOLTAGE DROP
*(N=1M, AS INDICATED IN THE .MODEL DIDEAL
*STATEMENT TOWARDS THE END OF THIS LISTING)
E23 23 0 POLY(1) 11 0  1 -1
*V(11) = VD
*V(23) = 1-VD
D2 0 22 DIDEAL
*V(22) = VD2
*
*END OF INDUCTOR EQUIVALENT CIRCUIT
*
*
*
*BUCK-BOOST CONVERTER
*INPUT CIRCUIT
VI 24 0 DC 20
*VI = DC SUPPLY VOLTAGE
X2 11 101 300 DIV
*V(11) = VD, V(101) = VD+VD2
*V(300) = VD/(VD+VD2)
*
G24 24 0 POLY(2) 10 0 300 0  0 0 0 0 1
*V(10) = VIL
*G24 = VIL*VD/(VD+VD2)
*THE ABOVE G24 STATEMENT IS FOR BUCK-BOOST
*AND BUCK ONLY.
*FOR BOOST, CHANGE IT TO :
*G24 24 0 POLY(1) 10 0  0 1
*
*OUTPUT CIRCUIT
X3 22 101 400 DIV
*V(22) = VD2, V(101) = VD+VD2
*V(400) = VD2/(VD+VD2)
*
G99 99 0 POLY(2) 10 0 400 0  0 0 0 0 1
*V(10) = VIL
*G99 = VIL*VD2/(VD+VD2)
*THE ABOVE G99 STATEMENT IS FOR BUCK-BOOST ONLY.
*FOR BOOST, CHANGE IT TO :
*G99 0 99 POLY(2) 10 0 400 0  0 0 0 0 1
*FOR BUCK, CHANGE IT TO :
*G99 0 99 POLY(1) 10 0  0 1
CL 99 0 1000U
RL 99 0 .5
*
*
*DUTY CYCLE INPUT
*THIS PART OF THE CIRCUIT IS NOT SHOWN IN FIG.A7.2.1.
VAC 1 0 AC 1
VD 11 1 DC 0.5
*DUTY CYCLE = 0.5
*
*
```

```
*V(12) INPUT
E12 12 0 POLY(1) 24 0  0 1
*THE ABOVE E12 STATEMENT IS FOR
*BUCK-BOOST AND BOOST ONLY
*FOR BUCK, CHANGE IT TO :
*E12 12 0 POLY(2) 24 0 99 0  0 1 -1
*
*
*VG INPUT
VG 13 0 DC 1
*VG = T/(2L) = 1
*
*
*V(14) INPUT
E14 14 0 POLY(1) 0 99  0 1
*THE ABOVE E14 STATEMENT IS FOR BUCK-BOOST ONLY.
*FOR BUCK, CHANGE IT TO :
*E14 14 0 POLY(1) 99 0  0 1
*FOR BOOST, CHANGE IT TO :
*E14 14 0 POLY(2) 99 0 24 0  0 1 -1
*
*
.MODEL DIDEAL D (N=1M)
.AC DEC 10 1 50K
.PRINT AC VDB(0,99) VP(0,99)
.PLOT AC VDB(0,99) VP(0,99)
*THE .PRINT AND .PLOT OUTPUTS
*FOR BUCK AND BOOST CONVERTERS SHOULD BE CHANGED
*TO VDB(99,0) AND VP(99,0).
.OPTIONS ITL1 = 50
.PROBE
*THE .PROBE STATEMENT IS FOR PSPICE ONLY.
*DELETE .PROBE STATEMENT IF RUN BY SPICE.
.END
```

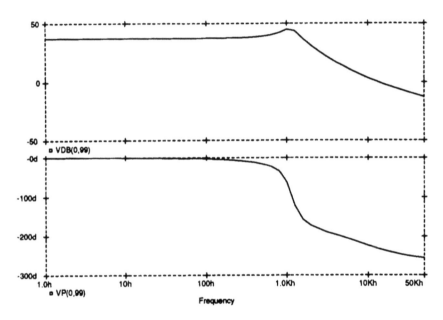

VDB(0,99) = Amplitude of $\delta V_0(s)/\delta D(s)$
VP(0,99) = Phase of $\delta V_0(s)/\delta D(s)$
$R_L = 0.5\Omega$

Fig. A7.2.2   Simulated dVo(s)/dD(s) transfer function of Buck-Boost
           converter for continuous-mode operation.

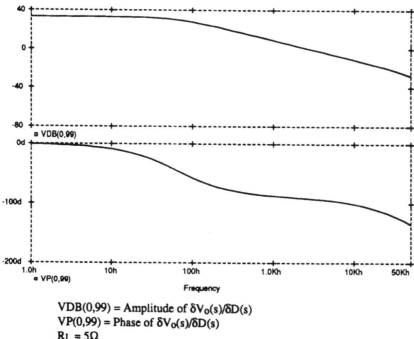

VDB(0,99) = Amplitude of $\delta V_o(s)/\delta D(s)$
VP(0,99) = Phase of $\delta V_o(s)/\delta D(s)$
$R_L = 5\Omega$

Fig. A7.2.3 Simulated dVo(s)/dD(s) transfer function of Buck-Boost
converter for discontinuous-mode operation.

ATTACHMENT 7.3: BEHAVIOR SIMULATION OF ĆUK CONVERTER

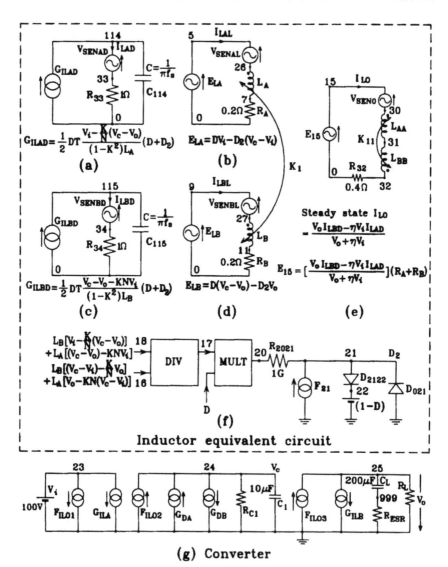

Fig. A7.3.1 Combined behavior model of Ćuk converter.

A7.3.1   <u>INPUT LISTING FOR SIMULATION OF CUK CONVERTER</u>
         <u>(OUTPUT VOLTAGE AGAINST LOADING CURRENT)</u>

```
CUK CONVERTER
*OUTPUT VOLTAGE vs OUTPUT CURRENT CHARACTERISTIC
*
*
*
*DIVIDER SUBCIRCUIT
*V(9)=V(3)/V(1)
.SUBCKT DIV 3 1 9
R30 3 0 1MEG
R10 1 0 1MEG
E20 2 0 POLY(2) 1 0 9 0   0 0 0 0 1
R20 2 0 1MEG
E90 9 0 POLY(1) 2 3   0 -1MEG
R90 9 0 1MEG
.ENDS DIV
*
*
*
*START OF INDUCTOR EQUIVALENT CIRCUIT
*K=0.947, K*K=0.897, (1-K*K)=0.103
*N=19/18=1.056
*KN=1, K/N=0.897
*LA=279U, LB=311U
*
VD 3 0 DC 0.26
*V(3)=DUTY CYCLE INPUT D
R3 3 0 1MEG
*
E1 1 0 POLY(1) 23 0   0 1
*V(1)=INPUT VOLTAGE VI
R1 1 0 1MEG
*
VT 2 0 DC 10U
*V(2)=SWITCHING PERIOD INPUT T
R2 2 0 1MEG
*
*
*DISCONTINUOUS-MODE CURRENT GENERATOR A
GILAD 0 114 POLY(3) 109 0 103 0 128 0   0 0 0 0 0 0 0 0
+ 0 0 0 0 0 0 0.5
*GILAD=0.5*V(109)*V(103)*V(128)
C114 114 0 3.183U
VSENAD 114 33
R33 33 0 1
*
E109 109 0 POLY(2) 2 0 3 0   0 0 0 0 1
*V(109)=T*D
R109 109 0 1MEG
*
E103 103 0 POLY(2) 3 0 21 0   0 1 1
```

```
*V(103)=D+D2
R103 103 0 1MEG
*
X4 105 124 128 DIV
*V(128)=(VI-(VC-VO)*K/N)/(LA*(1-K*K))
*
E105 105 0 POLY(2) 1 0 101 0  0 1 -0.897
*V(105)=VI-(VC-VO)*K/N
R105 105 0 1MEG
*
E101 101 0 POLY(2) 24 0 25 0  0 1 1
*V(101)=VC-VO
R101 101 0 1MEG
*
V124 124 0 DC 28.74U
*V(124)=LA*(1-K*K)=279U*0.103=28.74U
R124 124 0 1MEG
*
*
*DISCONTINUOUS-MODE CURRENT GENERATOR B
GILBD 0 115 POLY(3) 109 0 103 0 129 0  0 0 0 0 0 0 0 0 0
+ 0 0 0 0 0 0 0.5
*GILBD=0.5*V(109)*V(103)*V(129)
C115 115 0 3.183U
VSENBD 115 34
R34 34 0 1
*
*V(109)=T*D
*V(103)=D+D2
*
X5 107 125 129 DIV
*V(129)=((VC-VO)-K*N*VI)/(LB*(1-K*K))
*
E107 107 0 POLY(2) 101 0 1 0  0 1 -1
*V(107)=(VC-VO)-K*N*VI
R107 107 0 1MEG
*V(101)=VC-VO
*V(1)=VI
*
V125 125 0 DC 32.03U
*V(125)=LB*(1-K*K)=311U*0.103=32.03U
R125 125 0 1MEG
*
*
*INDUCTIVE CURRENT GENERATOR A
K1 LA LB 0.947
*K1=K=0.947
ELA 5 0 POLY(2) 110 0 111 0  0 1 -1
*V(5)=D*VI-(D2*(VC-VI))
VSENAL 5 26
LA 26 7 279U
RA 7 0 0.2
*
```

```
E110 110 0 POLY(2) 3 0 1 0   0 0 0 0 1
*V(110)=D*VI
R110 110 0 1MEG
*
E111 111 0 POLY(2) 21 0 102 0   0 0 0 0 1
*V(21)=D2
*V(111)=D2*(VC-VI)
R111 111 0 1MEG
*
E102 102 0 POLY(2) 24 0 1 0   0 1 -1
*V(102)=VC-VI
R102 102 0 1MEG
*
*
*INDUCTIVE CURRENT GENERATOR B
ELB 9 0 POLY(2) 112 0 113 0   0 1 -1
*V(9)=(D*(VC-VO))-(D2*VO)
VSENBL 9 27
LB 27 11 311U
RB 11 0 0.2
*
E112 112 0 POLY(2) 3 0 101 0   0 0 0 0 1
*V(112)=D*(VC-VO)
R112 112 0 1MEG
*V(3)=D
*V(101)=VC-VO
*
E113 113 0 POLY(2) 21 0 0 25   0 0 0 0 1
*V(113)=D2*VO
R113 113 0 1MEG
*V(21)=D2
*V(0,25)=VO
*
*
*ILO GENERATOR
VSEN0 15 30
LAA 30 31 279U
LBB 32 31 311U
K11 LAA LBB 0.947
R32 32 0 0.4
*
E15 15 0 POLY(1) 14 0   0 0.4
*V(15)=V(14)*(RA+RB), (RA+RB)=0.4
*
X1 118 13 14 DIV
*V(14)=(ILBD*VO-n*VI*ILAD)/(n*VI+VO)
*
E13 13 0 POLY(2) 1 0 0 25   0 0.88 1
*V(13)=nVI+VO, n=CONVERSION EFFICIENCY=0.88
R13 13 0 1MEG
*
E118 118 0 POLY(2) 117 0 116 0   0 1 -1
*V(118)=ILBD*VO-n*VI*ILAD
```

```
R118 118 0 1MEG
*
E116 116 0 POLY(2) 121 0 1 0   0 0 0 0 0.88
*V(116)=n*VI*ILAD, n=CONVERSION EFFICIENCY=0.88
R116 116 0 1MEG
*
H1 121 0 POLY(1) VSENAD  0 1
*V(121)=ILAD
R121 121 0 1MEG
*
E117 117 0 POLY(2) 122 0 0 25  0 0 0 0 1
*V(117)=ILBD*VO
R117 117 0 1MEG
*V(122)=ILBD
*V(0,25)=VO
*
H2 122 0 POLY(1) VSENBD 0 1
*V(122)=ILBD
R122 122 0 1MEG
*
*
*D2 GENERATOR
X2 18 16 17 DIV
*V(17)=V(18)/V(16)
*
E18 18 0 POLY(2) 105 0 107 0   0 311U 279U
*V(105)=VI-(VC-VO)*K/N
*V(107)=(VC-VO)-K*N*VI
*V(18)=LB*V(105)+LA*V(107), LA=279U, LB=311U
R18 18 0 1MEG
*
E16 16 0 POLY(2) 106 0 108 0   0 311U 279U
*V(16)=LB*V(106)+LA*V(108), LA=279U LB=311U
R16 16 0 1MEG
*
E106 106 0 POLY(2) 102 0 0 25   0 1 -0.897
*V(102)=VC-VI
*V(0,25)=VO
*V(106)=(VC-VI)-VO*K/N
R106 106 0 1MEG
*
E108 108 0 POLY(2) 0 25 102 0   0 1 -1
*V(108)=VO-K*N*(VC-VI)
R108 108 0 1MEG
*V(0,25)=VO
*V(102)=VC-VI
*
E20 20 0 POLY(2) 17 0 3 0    0 0 0 0 1
*V(20)=V(17)*D
R2021 20 21 1G
F21 0 21 POLY(2) VSENAL VSENBL  0 1 1
D2122 21 22 DMOD
E22 22 0 POLY(1) 3 0  1 -1
```

```
D021 0 21 DMOD
*V(21)=D2
*
*END OF INDUCTOR EQUIVALENT CIRCUIT
*
*
*
*CONVERTER
VI 23 0 DC 100V
*V(23)=VI
FILO1 23 0 POLY(1) VSEN0 0 1
GILA 23 0 POLY(1) 119 0   0 1
*V(119)=ILA
*
FILO2 0 24 POLY(1) VSEN0 0 1
GDA 0 24 POLY(2) 104 0 119 0   0 0 0 0 1
*V(104)=D2/(D+D2)
*V(119)=ILA
GDB 24 0 POLY(2) 154 0 120 0   0 0 0 0 1
*V(154)=D/(D+D2)
*V(120)=ILB
*
X3 21 103 104 DIV
*V(104)=D2/(D+D2)
*
H119 119 0 POLY(2) VSENAD VSENAL   0 1 1
*V(119)=ILAD+ILAL=ILA
R119 119 0 1MEG
*
X6 3 103 154 DIV
*V(154)=D/(D+D2)
*
H120 120 0 POLY(2) VSENBD VSENBL   0 1 1
*V(120)=ILBD+ILBL=ILB
R120 120 0 1MEG
*
C1 24 0 10U
RC1 24 0 1G
*V(24)=VC
*
FILO3 0 25 POLY(1) VSEN0 0 1
GILB 25 0 POLY(1) 120 0   0 1
*V(0,25)=VO
*V(120)=ILB
CL 25 999 200U
RESR 999 0 0.08
RL 25 0 10K
*
*ACTIVE LOAD
VSIRL 50 25
*VSIRL IS USED TO SENSE THE LOADING CURRENT.
ILOAD 0 50 PWL(0 0.15   0.15 0.15   5 5)
*ILOAD IS A CURRENT SOURCE SIMULATING AN ACTIVE LOAD.
```

```
*
*
*
.MODEL DMOD D (N=1M)
.NODESET V(21)=0.6
.TRAN 0.1 5 0
.PRINT TRAN V(0,25) I(VSIRL)
.PLOT TRAN V(0,25) I(VSIRL)
.OPTIONS ITL1=300 PIVTOL=0
.PROBE
*THE .PROBE STATEMENT IS FOR PSPICE ONLY.
*DELETE .PROBE STATEMENT IF RUN BY SPICE.
.END
```

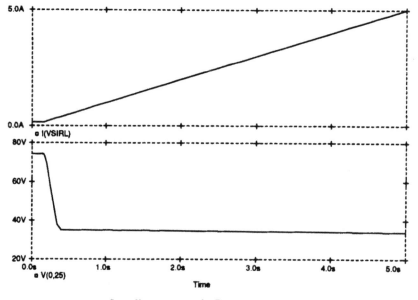

I(VSIRL) = Loading current in R_L

V(0,25) = V_0

Fig. A7.3.2   Simulated output voltage for different values of loading
                     current for Ćuk converter.

A7.3.2 <u>INPUT LISTING FOR SIMULATION OF CUK CONVERTER
(CONTINUOUS-MODE dVO/dD TRANSFER FUNCTION)</u>

```
CUK CONVERTER
*dVO/dD CHARACTERISTIC FOR RL=8 AND D=0.4 (RESULTS SHOWN
*IN FIG. A7.3.3)
*(THE RESULTS SHOWN IN FIG. A7.3.4 ARE OBTAINED BY
*SPECIFYING RL=300 AND D=0.26.)
*
*
*
*DIVIDER SUBCIRCUIT
*V(9)=V(3)/V(1)
.SUBCKT DIV 3 1 9
R30 3 0 1MEG
R10 1 0 1MEG
E20 2 0 POLY(2) 1 0 9 0  0 0 0 0 1
R20 2 0 1MEG
E90 9 0 POLY(1) 2 3  0 -1MEG
R90 9 0 1MEG
.ENDS DIV
*
*
*
*START OF INDUCTOR EQUIVALENT CIRCUIT
*K=0.947, K*K=0.897, (1-K*K)=0.103
*N=19/18=1.056
*KN=1, K/N=0.897
*LA=279U, LB=311U
*
VDAC 99 0 AC 1
VD 3 99 DC 0.4
*V(3)=DUTY CYCLE INPUT D
R3 3 0 1MEG
*
E1 1 0 POLY(1) 23 0  0 1
*V(1)=INPUT VOLTAGE VI
R1 1 0 1MEG
*
VT 2 0 DC 10U
*V(2)=SWITCHING PERIOD INPUT T
R2 2 0 1MEG
*
*
*DISCONTINUOUS-MODE CURRENT GENERATOR A
GILAD 0 114 POLY(3) 109 0 103 0 128 0  0 0 0 0 0 0 0 0
+ 0 0 0 0 0 0 0.5
*GILAD=0.5*V(109)*V(103)*V(128)
C114 114 0 3.183U
VSENAD 114 33
R33 33 0 1
*
E109 109 0 POLY(2) 2 0 3 0  0 0 0 0 1
*V(109)=T*D
```

```
R109 109 0 1MEG
*
E103 103 0 POLY(2)  3 0 21 0   0 1 1
*V(103)=D+D2
R103 103 0 1MEG
*
X4 105 124 128 DIV
*V(128)=(VI-(VC-VO)*K/N)/(LA*(1-K*K))
*
E105 105 0 POLY(2)  1 0 101 0   0 1 -0.897
*V(105)=VI-(VC-VO)*K/N
R105 105 0 1MEG
*
E101 101 0 POLY(2)  24 0 25 0   0 1 1
*V(101)=VC-VO
R101 101 0 1MEG
*
V124 124 0 DC 28.74U
*V(124)=LA*(1-K*K)=279U*0.103=28.74U
R124 124 0 1MEG
*
*
*DISCONTINUOUS-MODE CURRENT GENERATOR B
GILBD 0 115 POLY(3) 109 0 103 0 129 0  0 0 0 0 0 0 0 0
+ 0 0 0 0 0 0.5
*GILBD=0.5*V(109)*V(103)*V(129)
C115 115 0 3.183U
VSENBD 115 34
R34 34 0 1
*
*V(109)=T*D
*V(103)=D+D2
*
X5 107 125 129 DIV
*V(129)=((VC-VO)-K*N*VI)/(LB*(1-K*K))
*
E107 107 0 POLY(2)  101 0 1 0   0 1 -1
*V(107)=(VC-VO)-K*N*VI
R107 107 0 1MEG
*V(101)=VC-VO
*V(1)=VI
*
V125 125 0 DC 32.03U
*V(125)=LB*(1-K*K)=311U*0.103=32.03U
R125 125 0 1MEG
*
*
K1 LA LB 0.947
*K1=K
*
*INDUCTIVE CURRENT GENERATOR A
ELA 5 0 POLY(2) 110 0 111 0   0 1 -1
*V(5)=D*VI-(D2*(VC-VI))
```

```
VSENAL 5 26
LA 26 7 279U
RA 7 0 0.2
*
E110 110 0 POLY(2) 3 0 1 0  0 0 0 0 1
*V(110)=D*VI
R110 110 0 1MEG
*
E111 111 0 POLY(2) 21 0 102 0  0 0 0 0 1
*V(21)=D2
*V(111)=D2*(VC-VI)
R111 111 0 1MEG
*
E102 102 0 POLY(2) 24 0 1 0  0 1 -1
*V(102)=VC-VI
R102 102 0 1MEG
*
*
*INDUCTIVE CURRENT GENERATOR B
ELB 9 0 POLY(2) 112 0 113 0  0 1 -1
*V(9)=(D*(VC-VO))-(D2*VO)
VSENBL 9 27
LB 27 11 311U
RB 11 0 0.2
*
E112 112 0 POLY(2) 3 0 101 0  0 0 0 0 1
*V(112)=D*(VC-VO)
R112 112 0 1MEG
*V(3)=D
*V(101)=VC-VO
*
E113 113 0 POLY(2) 21 0 0 25  0 0 0 0 1
*V(113)=D2*VO
R113 113 0 1MEG
*V(21)=D2
*V(0,25)=VO
*
*
*ILO GENERATOR
VSENO 15 30
LAA 30 31 279U
LBB 32 31 311U
K11 LAA LBB 0.947
R32 32 0 0.4
*
E15 15 0 POLY(1) 14 0  0 0.4
*V(15)=V(14)*(RA+RB),  (RA+RB)=0.4
*
X1 118 13 14 DIV
*V(14)=(ILBD*VO-n*VI*ILAD)/(n*VI+VO)
*
E13 13 0 POLY(2) 1 0 0 25  0 0.88 1
*V(13)=nVI+VO, n=CONVERSION EFFICIENCY=0.88
```

```
R13 13 0 1MEG
*
E118 118 0 POLY(2) 117 0 116 0   0 1 -1
*V(118)=ILBD*VO-n*VI*ILAD
R118 118 0 1MEG
*
E116 116 0 POLY(2) 121 0 1 0   0 0 0 0 0.88
*V(116)=n*VI*ILAD, n=CONVERSION EFFICIENCY=0.88
R116 116 0 1MEG
*
H1 121 0 POLY(1) VSENAD  0 1
*V(121)=ILAD
R121 121 0 1MEG
*
E117 117 0 POLY(2) 122 0 0 25  0 0 0 0 1
*V(117)=ILBD*VO
R117 117 0 1MEG
*V(122)=ILBD
*V(0,25)=VO
*
H2 122 0 POLY(1) VSENBD 0 1
*V(122)=ILBD
R122 122 0 1MEG
*
*
*D2 GENERATOR
X2 18 16 17 DIV
*V(17)=V(18)/V(16)
*
E18 18 0 POLY(2) 105 0 107 0   0 311U 279U
*V(105)=VI-(VC-VO)*K/N
*V(107)=(VC-VO)-K*N*VI
*V(18)=LB*V(105)+LA*V(107), LA=279U, LB=311U
R18 18 0 1MEG
*
E16 16 0 POLY(2) 106 0 108 0   0 311U 279U
*V(16)=LB*V(106)+LA*V(108), LA=279U LB=311U
R16 16 0 1MEG
*
E106 106 0 POLY(2) 102 0 0 25   0 1 -0.897
*V(102)=VC-VI
*V(0,25)=VO
*V(106)=(VC-VI)-VO*K/N
R106 106 0 1MEG
*
E108 108 0 POLY(2) 0 25 102 0   0 1 -1
*V(108)=VO-K*N*(VC-VI)
R108 108 0 1MEG
*V(0,25)=VO
*V(102)=VC-VI
*
E20 20 0 POLY(2) 17 0 3 0   0 0 0 0 1
*V(20)=V(17)*D
```

```
R2021 20 21 1G
F21 0 21 POLY(2) VSENAL VSENBL  0 1 1
D2122 21 22 DMOD
E22 22 0 POLY(1) 3 0  1 -1
D021 0 21 DMOD
*V(21)=D2
*
*END OF INDUCTOR EQUIVALENT CIRCUIT
*
*
*
*CONVERTER
VI 23 0 DC 100V
*V(23)=VI
FILO1 23 0 POLY(1) VSEN0 0 1
GILA 23 0 POLY(1) 119 0  0 1
*V(119)=ILA
*
FILO2 0 24 POLY(1) VSEN0 0 1
GDA 0 24 POLY(2) 104 0 119 0  0 0 0 0 1
*V(104)=D2/(D+D2)
*V(119)=ILA
GDB 24 0 POLY(2) 154 0 120 0  0 0 0 0 1
*V(154)=D/(D+D2)
*V(120)=ILB
*
X3 21 103 104 DIV
*V(104)=D2/(D+D2)
*
H119 119 0 POLY(2) VSENAD VSENAL  0 1 1
*V(119)=ILAD+ILAL=ILA
R119 119 0 1MEG
*
X6 3 103 154 DIV
*V(154)=D/(D+D2)
*
H120 120 0 POLY(2) VSENBD VSENBL  0 1 1
*V(120)=ILBD+ILBL=ILB
R120 120 0 1MEG
*
C1 24 0 10U
RC1 24 0 1G
*V(24)=VC
*
FILO3 0 25 POLY(1) VSEN0 0 1
GILB 25 0 POLY(1) 120 0  0 1
*V(0,25)=VO
*V(120)=ILB
CL 25 999 200U
RESR 999 0 0.08
RL 25 0 8
*
*
```

```
*
*
.MODEL DMOD D (N=1M)
.NODESET V(21)=0.6
.AC DEC 10 1 50K
.PRINT AC VDB(0,25) VP(0,25)
.PLOT AC VDB(0,25) VP(0,25)
.OPTIONS ITL1=300 PIVTOL=0
.PROBE
*THE .PROBE STATEMENT IS FOR PSPICE ONLY.
*DELETE .PROBE STATEMENT IF RUN BY SPICE.
.END
```

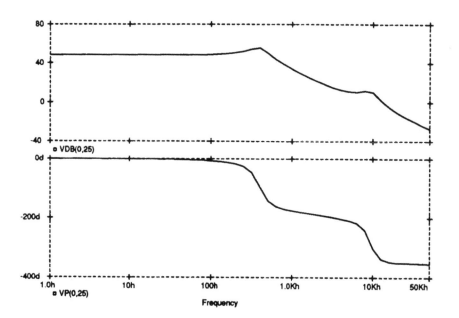

$VDB(0,25)$ = Amplitude of $\delta V_o(s)/\delta D(s)$
$VP(0,25)$ = Phase of $\delta V_o(s)/\delta D(s)$
$D=0.4$
$R_L=8\Omega$

Fig. A7.3.3   Simulated $\delta V_o(s)/\delta D(s)$ transfer function of Ćuk converter in continuous-mode operation.

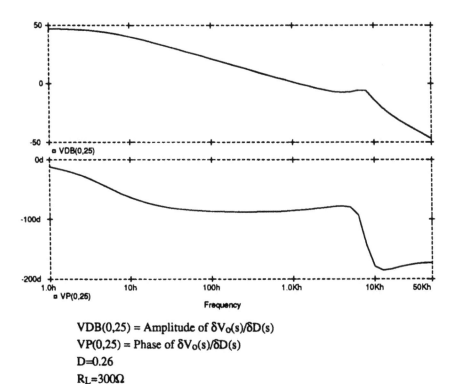

VDB(0,25) = Amplitude of $\delta V_o(s)/\delta D(s)$
VP(0,25) = Phase of $\delta V_o(s)/\delta D(s)$
D=0.26
R_L=300Ω

Fig. A7.3.4   Simulated dVo(s)/dD(s) transfer function of Čuk
converter in discontinuous-mode operation.

ATTACHMENT 7.4:   BEHAVIOR SIMULATION OF CURRENT-CONTROLLED BUCK CONVERTER

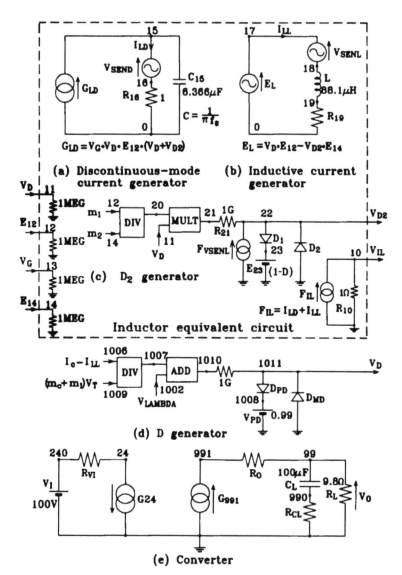

Fig. A7.4.1   SPICE model of current-controlled Buck converter.

A7.4.1 <u>INPUT LISTING FOR SIMULATION OF CURRENT-CONTROLLED</u>
<u>BUCK CONVERTER (CONTINUOUS-MODE dVO/dVI)</u>

```
CURRENT-CONTROLLED BUCK CONVERTER (ALL MODES)
*dVO/dVI, CONTINUOUS-MODE OPERATION
*
*DIVIDER SUBCIRCUIT
*V(9)=V(3)/V(1)
.SUBCKT DIV 3 1 9
R30 3 0 1MEG
R10 1 0 1MEG
E20 2 0 POLY(2) 1 0 9 0  0 0 0 0 1
R20 2 0 1MEG
E90 9 0 POLY(1) 2 3  0 -1MEG
R90 9 0 1MEG
.ENDS DIV
*
*
*
*START OF INDUCTOR EQUIVALENT CIRCUIT
*
*V(11) = DUTY CYCLE INPUT
R11 11 0 1MEG
*
*V(12) = VI-VO FOR BUCK
R12 12 0 1MEG
*
*V(13) = VG = T/(2L)
R13 13 0 1MEG
*
*V(14) = VO FOR BUCK
R14 14 0 1MEG
*
*
*DISCONTINUOUS-MODE CURRENT GENERATOR
GLD 0 15 POLY(1) 100 0  0 1
*GLD = V(100) = VG*VD*V(12)*(VD+VD2)
*V(12) = VI-VO FOR BUCK
*V(15) = VILD
VSEND 15 16
R16 16 0 1
C15 15 0 6.366U
*TO GET VD*V(12) :
E103 103 0 POLY(2) 11 0 12 0  0 0 0 0 1
*V(11) = VD
*V(12) = (VI-VO) FOR BUCK
*V(103) = VD*V(12)
R103 103 0 1MEG
*TO GET VD+VD2 :
E101 101 0 POLY(2) 22 0 11 0  0 1 1
*V(22) = VD2, V(11) = VD
*V(101) = VD+VD2
R101 101 0 1MEG
```

```
*TO GET VG*VD*V(12)*(VD+VD2) :
E100 100 0 POLY(3) 13 0 103 0 101 0
+0 0 0 0 0 0 0 0 0 0 0 0 0 0 1
*V(100) = VG*VD*V(12)*(VD+VD2)
R100 100 0 1MEG
*
*
*INDUCTIVE CURRENT GENERATOR
EL 17 0 POLY(1) 110 0  0 1
*V(17) = V(110) = VD*V(12)-VD2*V(14)
*FOR BUCK : V(12) = VI-VO, V(14) = VO
VSENL 17 18
L 18 19 88.1U
R19 19 0 10M
*TO GET EL = VD*V(12)-VD2*V(14) :
E104 104 0 POLY(2) 22 0 14 0  0 0 0 0 1
*V(22) = VD2
*V(14) = VO FOR BUCK
*V(104) = VD2*V(14)
R104 104 0 1MEG
E110 110 0 POLY(2) 103 0 104 0  0 1 -1
*V(110) = EL = VD*V(12)-VD2*V(14)
R110 110 0 1MEG
*
*
*TOTAL INDUCTOR CURRENT GENERATOR
FIL 0 10 POLY(2) VSEND VSENL  0 1 1
R10 10 0 1
*V(10) = VILD+VILL = VIL(REPRESENTING IL)
*
*
*D2 GENERATOR
X1 12 14 20 DIV
*V(12) = VI-VO FOR BUCK
*V(14) = VO FOR BUCK
*V(20) = V(12)/V(14)
E21 21 0 POLY(2) 20 0 11 0  0 0 0 0 1
*V(21) = VD*V(12)/V(14)
R21 21 22 1G
FVSENL 0 22 POLY(1) VSENL  0 1
*FVSENL = ILL
D1 22 23 DIDEAL
*DIDEAL IS A DIODE MODEL HAVING ALMOST
*ZERO FORWARD VOLTAGE DROP (N=1M,
*AS INDICATED IN THE .MODEL DIDEAL
*STATEMENT TOWARDS THE END OF THIS
*LISTING)
E23 23 0 POLY(1) 11 0  1 -1
*V(11) = VD
*V(23) = 1-VD
D2 0 22 DIDEAL
*V(22) = VD2
*
```

```
*END OF INDUCTOR EQUIVALENT CIRCUIT
*
*
*
*BUCK CONVERTER
*INPUT CIRCUIT
VI 240 0 DC 100 AC 1
*VI CONSISTS OF TWO COMPONENTS: DC 100 AND AC 1
RVI 240 24 0.5
*VI = DC SUPPLY VOLTAGE
X2 11 101 300 DIV
*V(11) = VD, V(101) = VD+VD2
*V(300) = VD/(VD+VD2)
*
G24 24 0 POLY(2) 10 0 300 0  0 0 0 0 1
*V(10) = VIL
*G24 = VIL*VD/(VD+VD2)
*
*OUTPUT CIRCUIT
G991 0 991 POLY(1) 10 0   0 1
RO 991 99 0.5
CL 99 990 100U
RCL 990 0 215M
RL 99 0 9.8
*
*
*D GENERATOR
X4 1006 1009 1007 DIV
*V(1007) = V(1006)/V(1009) = (IC-ILL)/((MC+M1)*T)
*V(1006) = IC-ILL
*V(1009) = (MC+M1)*T
*
E1006 1006 0 POLY(1) 114 1001   0 1
*V(1006) = IC-ILL
*V(114) = IC
*V(1001) = ILL
R1006 1006 0 1MEG
*
HVSENL 1001 0 POLY(1) VSENL   0 1
*V(1001) = ILL
R1001 1001 0 1MEG
*
E1009 1009 0 POLY(2) 1003 0 130 0  0 0 0 0 1
*V(1009) = (MC+M1)*T
*V(1003) = MC+M1
*V(130)=T
R1009 1009 0 1MEG
*
E1003 1003 0 POLY(1) 12 0   600K 11350.73779
*V(1003) = MC+M1
*MC = 600K
*V(12) = VI-VO
*M1 = (VI-VO)/L = 11350.73779*(VI-VO)
```

```
R1003 1003 0 1MEG
*
E1010 1010 0 POLY(2) 1007 0 1002 0  0 1 1
*V(1010)=(IC-ILL)/((MC+M1)*T)+LAMBDA
*V(1007) = (IC-ILL)/((MC+M1)*T)
*V(1002) = LAMBDA
R1010 1010 1011 1G
DPD 1011 1008 DIDEAL
VPD 1008 0 0.99
DMD 0 1011 DIDEAL
*V(1011) = D = DUTY CYCLE
*
V1002 1002 0  0.05
*V(1002) = LAMBDA = 0.05
R1002 1002 0 1MEG
*
VIC 114 0 DC 10
*VIC = IC = CONTROL CURRENT INPUT = 10
R114 114 0 1MEG
*
*
*DUTY CYCLE INPUT
ED 11 0 POLY(1) 1011 0  0 1
*V(1011) = D
*
*
*V(12) INPUT
E12 12 0 POLY(2) 24 0 99 0  0 1 -1
*V(12) = VI-VO
*
*
*VG INPUT
VG 13 0 DC 0.113507377
*VG = T/(2L) = 20U/(2*88.1U) = 0.113507377
*
*
*VT INPUT
VT 130 0 DC 20U
*VT = SWITCHING PERIOD = 20U
R130 130 0 1MEG
*
*
*V(14) INPUT
E14 14 0 POLY(1) 99 0  0 1
*
*
.MODEL DIDEAL D (N=1M)
.AC DEC 10 1 10K
.PRINT AC VDB(99) VP(99)
.PLOT AC VDB(99) VP(99)
.OPTIONS ITL1=200
*SETTING ITL1=200 HELPS DC ANALYSIS TO CONVERGE
*(DEFAULT VALUE=100).
.PROBE
*THE .PROBE STATEMENT IS FOR PSPICE ONLY.
*DELETE .PROBE STATEMENT IF RUN BY SPICE.
.END
```

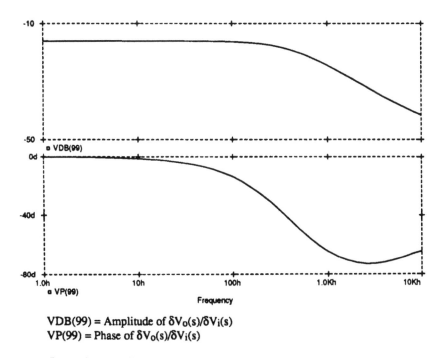

VDB(99) = Amplitude of $\delta V_o(s)/\delta V_i(s)$
VP(99) = Phase of $\delta V_o(s)/\delta V_i(s)$

Operating conditions :
$V_i = 100V$
$I_c = 10A$
$R_L = 9.8\Omega$

Fig. A7.4.2   Simulated $\delta V_o(s)/\delta V_i(s)$ transfer function of current-controlled
              Buck converter in continuous-mode operation.

A7.4.2 INPUT LISTING FOR SIMULATION OF CURRENT-CONTROLLED
        BUCK CONVERTER (CONTINUOUS-MODE dVO/dIC)

```
CURRENT-CONTROLLED BUCK CONVERTER (ALL MODES)
*dVO/dIC FOR CONTINUOUS-MODE OPERATION
*
*DIVIDER SUBCIRCUIT
*V(9)=V(3)/V(1)
.SUBCKT DIV 3 1 9
R30 3 0 1MEG
R10 1 0 1MEG
E20 2 0 POLY(2) 1 0 9 0  0 0 0 0 1
R20 2 0 1MEG
E90 9 0 POLY(1) 2 3  0 -1MEG
R90 9 0 1MEG
.ENDS DIV
*
*
*
*START OF INDUCTOR EQUIVALENT CIRCUIT
*
*V(11) = DUTY CYCLE INPUT
R11 11 0 1MEG
*
*V(12) = VI-VO FOR BUCK
R12 12 0 1MEG
*
*V(13) = VG = T/(2L)
R13 13 0 1MEG
*
*V(14) = VO FOR BUCK
R14 14 0 1MEG
*
*
*DISCONTINUOUS-MODE CURRENT GENERATOR
GLD 0 15 POLY(1) 100 0  0 1
*GLD = V(100) = VG*VD*V(12)*(VD+VD2)
*V(12) = VI-VO FOR BUCK
*V(15) = VILD
VSEND 15 16
R16 16 0 1
C15 15 0 6.366U
*TO GET VD*V(12) :
E103 103 0 POLY(2) 11 0 12 0  0 0 0 0 1
*V(11) = VD
*V(12) = (VI-VO) FOR BUCK
*V(103) = VD*V(12)
R103 103 0 1MEG
*TO GET VD+VD2 :
E101 101 0 POLY(2) 22 0 11 0  0 1 1
*V(22) = VD2, V(11) = VD
*V(101) = VD+VD2
R101 101 0 1MEG
```

```
*TO GET VG*VD*V(12)*(VD+VD2) :
E100 100 0 POLY(3) 13 0 103 0 101 0
+0 0 0 0 0 0 0 0 0 0 0 0 1
*V(100) = VG*VD*V(12)*(VD+VD2)
R100 100 0 1MEG
*
*
*INDUCTIVE CURRENT GENERATOR
EL 17 0 POLY(1) 110 0  0 1
*V(17) = V(110) = VD*V(12)-VD2*V(14)
*FOR BUCK : V(12) = VI-VO, V(14) = VO
VSENL 17 18
L 18 19 88.1U
R19 19 0 10M
*TO GET EL = VD*V(12)-VD2*V(14) :
E104 104 0 POLY(2) 22 0 14 0  0 0 0 0 1
*V(22) = VD2
*V(14) = VO FOR BUCK
*V(104) = VD2*V(14)
R104 104 0 1MEG
E110 110 0 POLY(2) 103 0 104 0  0 1 -1
*V(110) = EL = VD*V(12)-VD2*V(14)
R110 110 0 1MEG
*
*
*TOTAL INDUCTOR CURRENT GENERATOR
FIL 0 10 POLY(2) VSEND VSENL  0 1 1
R10 10 0 1
*V(10) = VILD+VILL = VIL(REPRESENTING IL)
*
*
*D2 GENERATOR
X1 12 14 20 DIV
*V(12) = VI-VO FOR BUCK
*V(14) = VO FOR BUCK
*V(20) = V(12)/V(14)
E21 21 0 POLY(2) 20 0 11 0  0 0 0 0 1
*V(21) = VD*V(12)/V(14)
R21 21 22 1G
FVSENL 0 22 POLY(1) VSENL  0 1
*FVSENL = ILL
D1 22 23 DIDEAL
*DIDEAL IS A DIODE MODEL HAVING ALMOST
*ZERO FORWARD VOLTAGE DROP (N=1M,
*AS INDICATED IN THE .MODEL DIDEAL
*STATEMENT TOWARDS THE END OF THIS
*LISTING)
E23 23 0 POLY(1) 11 0  1 -1
*V(11) = VD
*V(23) = 1-VD
D2 0 22 DIDEAL
*V(22) = VD2
*
```

```
*END OF INDUCTOR EQUIVALENT CIRCUIT
*
*
*
*BUCK CONVERTER
*INPUT CIRCUIT
VI 240 0 DC 100V
RVI 240 24 0.5
*VI = DC SUPPLY VOLTAGE
X2 11 101 300 DIV
*V(11) = VD, V(101) = VD+VD2
*V(300) = VD/(VD+VD2)
*
G24 24 0 POLY(2) 10 0 300 0  0 0 0 0 1
*V(10) = VIL
*G24 = VIL*VD/(VD+VD2)
*
*OUTPUT CIRCUIT
G991 0 991 POLY(1) 10 0  0 1
RO 991 99 0.5
CL 99 990 100U
RCL 990 0 215M
RL 99 0 9.8
*
*
*D GENERATOR
X4 1006 1009 1007 DIV
*V(1007) = V(1006)/V(1009) = (IC-ILL)/((MC+M1)*T)
*V(1006) = IC-ILL
*V(1009) = (MC+M1)*T
*
E1006 1006 0 POLY(1) 114 1001  0 1
*V(1006) = IC-ILL
*V(114) = IC
*V(1001) = ILL
R1006 1006 0 1MEG
*
HVSENL 1001 0 POLY(1) VSENL  0 1
*V(1001) = ILL
R1001 1001 0 1MEG
*
E1009 1009 0 POLY(2) 1003 0 130 0  0 0 0 0 1
*V(1009) = (MC+M1)*T
*V(1003) = MC+M1
*V(130)=T
R1009 1009 0 1MEG
*
E1003 1003 0 POLY(1) 12 0  600K 11350.73779
*V(1003) = MC+M1
*MC = 600K
*V(12) = VI-VO
*M1 = (VI-VO)/L = 11350.73779*(VI-VO)
R1003 1003 0 1MEG
```

```
*
E1010 1010 0 POLY(2) 1007 0 1002 0  0 1 1
*V(1010) = (IC-ILL)/((MC+M1)*T)+LAMBDA
*V(1007) = (IC-ILL)/((MC+M1)*T)
*V(1002) = LAMBDA
R1010 1010 1011 1G
DPD 1011 1008 DIDEAL
VPD 1008 0 0.99
DMD 0 1011 DIDEAL
*V(1011) = D = DUTY CYCLE INPUT
*
V1002 1002 0  0.05
*V(1002) = LAMBDA = 0.05
R1002 1002 0 1MEG
*
VIC 114 0 DC 10 AC 1
*VIC = IC = CONTROL CURRENT INPUT
*VIC CONSISTS OF TWO COMPONENTS: DC 10 AND AC 1
R114 114 0 1MEG
*
*
*DUTY CYCLE INPUT
ED 11 0 POLY(1) 1011 0  0 1
*V(1011) = D
*
*
*V(12) INPUT
E12 12 0 POLY(2) 24 0 99 0  0 1 -1
*V(12) = VI-VO
*
*
*VG INPUT
VG 13 0 DC 0.113507377
*VG = T/(2L) = 20U/(2*88.1U) = 0.113507377
*
*
*VT INPUT
VT 130 0 DC 20U
*VT = SWITCHING PERIOD = 20U
R130 130 0 1MEG
*
*
*V(14) INPUT
E14 14 0 POLY(1) 99 0  0 1
*
*
.MODEL DIDEAL D (N=1M)
.AC DEC 10 1 10K
.PRINT AC VDB(99) VP(99)
.PLOT AC VDB(99) VP(99)
.OPTIONS ITL1=200
*SETTING ITL1=200 HELPS DC ANALYSIS TO CONVERGE
*(DEFAULT VALUE=100).
.PROBE
*THE .PROBE STATEMENT IS FOR PSPICE ONLY.
*DELETE .PROBE STATEMENT IF RUN BY SPICE.
.END
```

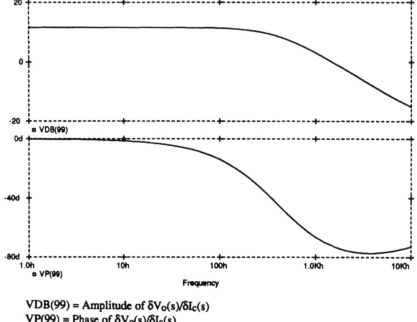

VDB(99) = Amplitude of $\delta V_o(s)/\delta I_c(s)$
VP(99) = Phase of $\delta V_o(s)/\delta I_c(s)$

Operating conditions :
$V_i = 100V$
$I_c = 10A$
$R_L = 9.8\Omega$

Fig. A7.4.3   Simulated $\delta V_o(s)/\delta I_c(s)$ transfer function of current-controlled
           Buck converter in continuous-mode operation .

ATTACHMENT 7.5: BEHAVIOR SIMULATION OF DUAL-OUTPUT TRANSFORMER-COUPLED PUSH-PULL CONVERTER

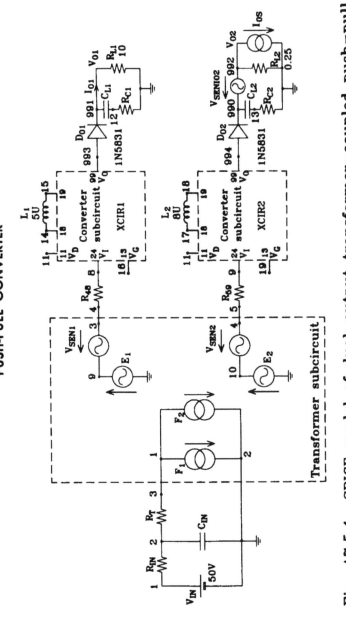

Fig. A7.5.1 SPICE model of dual-output transformer-coupled push-pull converter.

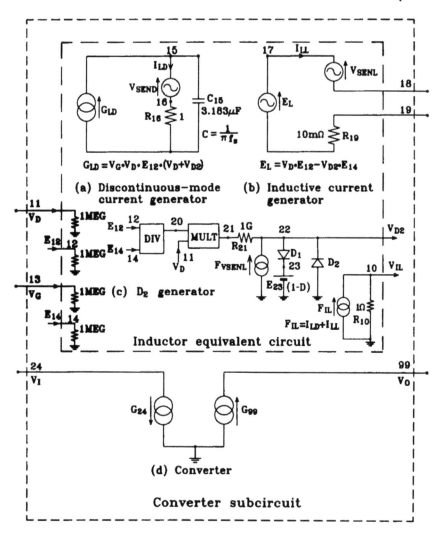

Fig. A7.5.2   SPICE model of converter subcircuit.

A7.5 <u>SPICE INPUT LISTING OF DUAL-OUTPUT CONVERTER</u>

```
DUAL-OUTPUT CONVERTER (ALL MODES)
*dVO/dD AND TRANSIENT SIMULATIONS
*
*
*DIVIDER SUBCIRCUIT
*V(9) = V(3)/V(1)
.SUBCKT DIV 3 1 9
R30 3 0 1MEG
R10 1 0 1MEG
E20 2 0 POLY(2) 1 0 9 0  0 0 0 0 1
R20 2 0 1MEG
E90 9 0 POLY(1) 2 3  0 -1MEG
R90 9 0 1MEG
.ENDS DIV
*
*
*
*TRANSFORMER SUBCIRCUIT
*V(1,2)=INPUT, V(3)=OUTPUT 1, V(4)=OUTPUT 2
.SUBCKT TRANSFORMER 1 2 3 4
F1 1 2 POLY(1) VSEN1  0 0.25
*NS1:NP = 0.25
F2 1 2 POLY(1) VSEN2  0 0.25
*NS2:NP = 0.25
E1 9 0 POLY(1) 1 2  0 0.25
*NS1:NP = 0.25
E2 10 0 POLY(1) 1 2  0 0.25
*NS2:NP = 0.25
VSEN1 9 3
VSEN2 10 4
.ENDS TRANSFORMER
*
*
*
.SUBCKT CONVERTER 11 13 18 19 24 99
*V(11) = DUTY CYCLE INPUT
*V(13) = VG INPUT
*V(24) = INPUT VOLTAGE
*V(99) = OUTPUT VOLTAGE
*18,19 = INDUCTOR TERMINALS
*
*
*START OF INDUCTOR EQUIVALENT CIRCUIT
*V(11) = DUTY CYCLE INPUT
R11 11 0 1MEG
*
*V(12) = VI-VO
R12 12 0 1MEG
*
*V(13) = VG = T/(2L), T = 10U
R13 13 0 1MEG
```

```
*
*V(14) = VO
R14 14 0 1MEG
*
*
*DISCONTINUOUS-MODE CURRENT GENERATOR
GLD 0 15 POLY(1) 100 0  0 1
*GLD = VG*VD*(VI-VO)*(VD+VD2)
*V(15) = VILD
VSEND 15 16
R16 16 0 1
C15 15 0 3.183U
*
E103 103 0 POLY(2) 11 0 12 0  0 0 0 0 1
*V(11) = VD
*V(12) = (VI-VO)
*V(103) = VD*(VI-VO)
R103 103 0 1MEG
*
E101 101 0 POLY(2) 22 0 11 0  0 1 1
*V(22) = VD2, V(11) = VD
*V(101) = VD+VD2
R101 101 0 1MEG
*
E100 100 0 POLY(3) 13 0 103 0 101 0
+0 0 0 0 0 0 0 0 0 0 0 0 0 1
*V(100) = VG*VD*(VI-VO)*(VD+VD2)
R100 100 0 1MEG
*
*
*INDUCTIVE CURRENT GENERATOR
EL 17 0 POLY(1) 110 0  0 1
*V(17) = V(110) = VD*(VI-VO)-VD2*VO
*
VSENL 17 18
R19 19 0 10M
*
E104 104 0 POLY(2) 22 0 14 0  0 0 0 0 1
*V(22) = VD2
*V(14) = VO
*V(104) = VD2*VO
R104 104 0 1MEG
*
E110 110 0 POLY(2) 103 0 104 0  0 1 -1
*V(110) = EL = VD*(VI-VO)-VD2*VO
R110 110 0 1MEG
*
*
*TOTAL INDUCTOR CURRENT GENERATOR
FIL 0 10 POLY(2) VSEND VSENL  0 1 1
R10 10 0 1
*V(10) = VILD+VILL = VIL(REPRESENTING IL)
*
```

```
*
*D2 GENERATOR
X1 12 14 20 DIV
*V(12) = VI-VO
*V(14) = VO
*V(20) = (VI-VO)/VO
E21 21 0 POLY(2) 20 0 11 0   0 0 0 0 1
*V(21) = VD*(VI-VO)/VO
R21 21 22 1G
FVSENL 0 22 POLY(1) VSENL  0 1
*FVSENL = ILL
D1 22 23 DIDEAL
E23 23 0 POLY(1) 11 0  1 -1
*V(11) = VD                      (
*V(23) = 1-VD
D2 0 22 DIDEAL
*V(22) = VD2
*END OF INDUCTOR EQUIVALENT CIRCUIT
*
*
*CONVERTER
*INPUT CIRCUIT
X2 11 101 300 DIV
*V(11) = VD, V(101) = VD+VD2
*V(300) = VD/(VD+VD2)
*
G24 24 0 POLY(2) 10 0 300 0  0 0 0 0 1
*V(10) = VIL
*G24 = VIL*VD/(VD+VD2)
*
*OUTPUT CIRCUIT
G99 0 99 POLY(1) 10 0   0 1
*V(10) = VIL
*
*
*V(12) INPUT
E12 12 0 POLY(2) 24 0 99 0   0 1 -1
*
*
*V(14) INPUT
E14 14 0 POLY(1) 99 0   0 1
*
*
.ENDS CONVERTER
*
*
*
*MAIN CIRCUIT
VIN 1 0 DC 50V
*V(1) = PRIMARY INPUT VOLTAGE = 50V
RIN 1 2 0.1
CIN 2 0 1000U
RT 2 3 0.1
```

```
*RIN AND RT ARE NOT SHOWN IN FIG. 7.25.
X1 3 0 4 5 TRANSFORMER
*V(3,0) = TRANSFORMER PRIMARY INPUT
*V(4), V(5) = TRANSFORMER SECONDARY OUTPUTS
R48 4 8 0.025
R59 5 9 0.025
*
*CONVERTER CIRCUIT 1
XCIR1 11 16 14 15 8 993 CONVERTER
*V(11) = DUTY CYCLE INPUT
*V(16) = VG = T/(2L) INPUT
*14, 15 = INDUCTOR TERMINALS
*V(8) = CONVERTER INPUT VOLTAGE
*V(993) = CONVERTER OUTPUT VOLTAGE
L1 14 15 5U
VG1 16 0 1
*VG1 = T/(2L1) = 10U/(2*5U) = 1
DO1 993 991 D1N5831
CL1 991 12 470U
RC1 12 0 20M
RL1 991 0 10
*V(991) = OUTPUT 1 = VO1
*
*CONVERTER CIRCUIT 2
XCIR2 11 19 17 18 9 994 CONVERTER
*V(11) = DUTY CYCLE INPUT
*V(19) = VG = T/(2L) INPUT
*17, 18 = INDUCTOR TERMINALS
*V(9) = CONVERTER INPUT VOLTAGE
*V(994) = CONVERTER OUTPUT VOLTAGE
L2 17 18 8U
VG2 19 0 0.625
*VG2 = T/(2L2) = 10U/(2*8U) = 0.625
DO2 994 990 D1N5831
CL2 990 13 330U
RC2 13 0 20M
VSENIO2 990 992
RL2 992 0 0.25
*V(992) = OUTPUT 2 = VO2
IOS 992 0 PULSE (0 4 20M 0.1M 0.1M 20M 40M)
*IOS SIMULATES AN ACTIVE LOAD SWITCHING BETWEEN 0A AND 4A.
*DURING AC SIMULATION, IOS WILL BE DISABLED.
*
*DUTY CYCLE INPUT
VAC 110 0 AC 1
*DURING TRANSIENT SIMULATION, VAC WILL BE DISABLED.
VD 11 110 DC 0.5
*DUTY CYCLE = 0.5
*
*
.MODEL DIDEAL D (N=1M)
.MODEL D1N5831 D (IS=89.43U RS=3.991M N=1 XTI=0 EG=1.11
+ CJO=3.488N M=.5044 VJ=.75 FC=.5)
```

```
.AC DEC 10 1 10K
.PRINT AC VDB(991) VP(991) VDB(992) VP(992)
.PLOT AC VDB(991) VP(991) VDB(992) VP(992)
.TRAN 0.1M 100M
.PRINT TRAN I(VSENIO2) V(991) V(992)
.PLOT TRAN I(VSENIO2) V(991) V(992)
.OPTIONS LIMPTS=2000
.PROBE
*THE .PROBE STATEMENT IS FOR PSPICE ONLY.
*DELETE .PROBE STATEMENT IF RUN BY SPICE.
.END
```

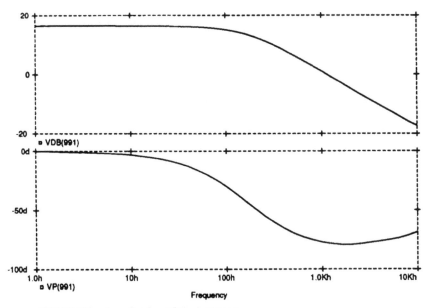

VDB(991) = Amplitude of $\delta V_{o1}(s)/\delta D(s)$
VP(991) = Phase of $\delta V_{o1}(s)/\delta D(s)$
$R_{L1} = 10\Omega$

Fig. A7.5.3   Simulated $\delta V_{o1}(s)/\delta D(s)$ transfer function of output 1 (discontinuous mode).

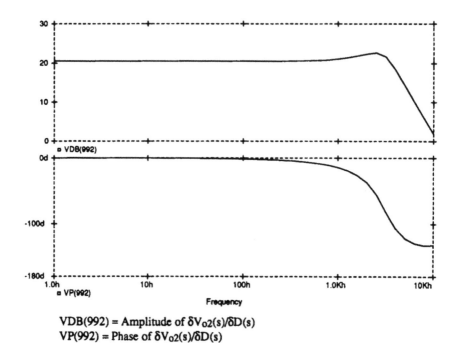

VDB(992) = Amplitude of $\delta V_{o2}(s)/\delta D(s)$
VP(992) = Phase of $\delta V_{o2}(s)/\delta D(s)$
$R_{L2} = 0.25\Omega$

Fig. A7.5.4   Simulated $\delta V_{o2}(s)/\delta D(s)$ transfer function of output 2
              (continuous mode).

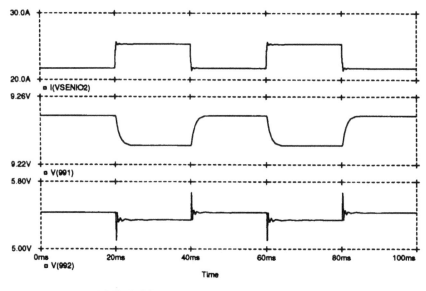

I(VSENIO2) = I_{O2}
V(991) = V_{O1} (discontinuous-mode output)
V(992) = V_{O2} (continuous-mode output)

Fig. A7.5.5 Simulated transient responses.

8
Computer-Aided Design of Converters and Regulators

Because of the nonlinear nature of converter and regulator circuits, CAD (computer-aided design) tools are highly desirable for the design of such circuits. Based on SPICE, the techniques of using circuit simulation programs to help design converters and regulators, the commonly encountered problems, and solutions to these problems will be studied in this chapter.

Since SPICE is basically a numerical simulation program, circuit designers should be aware that all ideas should come from themselves and that SPICE is simply a tool for trying out and verifying their ideas. In the design process, many repeated trials and iterations will be required. The essential prerequisites for a successful CAD exercise are:

1. Circuit designers must have a good and intuitive understanding, at least qualitatively, of the circuit operation, so that they can lead the iteration toward a converged design.
2. Designers must be able to develop an appropriate model for the circuit to be designed.
3. Designers must be familiar with the CAD tool used.

A major objective of the earlier chapters and the four appendixes at the end of this book is, in fact, to enable the reader to acquire these prerequisites.

In any simulation and design exercise, the designer must, however, also be aware of the limitations of the model and the CAD tool used. Because of these limitations, the simulation results and subsequent design can sometimes be completely meaningless.

In Sections 8.1–8.3, the philosophy and methods of using SPICE to help design converter and regulator circuits will be discussed. Some commonly encountered problems that are due to the limitations of the model and the CAD tool, and the methods to solve these problems, will also be examined there. In Section 8.4, a detailed example of converter and regulator design, complete with simulation files, will be given. Finally, in Section 8.5 the practical implementation of the design will be briefly outlined.

8.1 INTERACTION BETWEEN THE DESIGNER AND CAD TOOLS

Figure 8.1 shows a typical CAD process, starting with a preliminary design of the circuit. The circuit is first simulated by the simulation program. The results of simulation are examined and digested by the circuit designer.

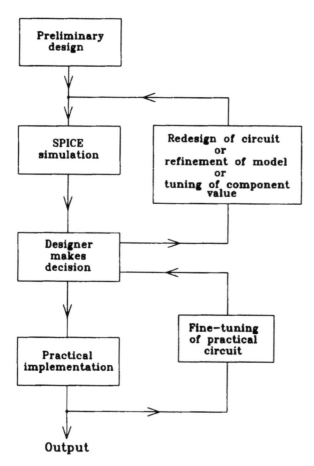

FIGURE 8.1 The CAD design process.

Based on the information obtained, the designer then decides what to do next. It may be a redesign of the circuit, a refinement of the model, or a tuning of component values. Usually, many repeated simulations and iterations are required in order to obtain a final design.

Note that the so-called preliminary design shown in Fig. 8.1 may indeed be a rudimentary design, which can be conceived with minimum effort. If, however, a reasonably fast simulation machine (with a mathematics coprocessor) is available, a design with well-defined characteristics can still be obtained within a reasonably short time.

When a design has been finalized by using CAD tools, the simulated performance can be used as objectives to achieve in the practical implementation of the design. A fine-tuning of the practical circuit, possibly involving further simulations and iterations, may also be required to optimize the performance, as indicated in Fig. 8.1.

8.2 COMPUTER-AIDED DESIGN OF CONVERTER CIRCUITS

A computer loaded with a good simulation program and a library of circuit and component models is virtually equivalent to a workbench that is well equipped with components and programmable instruments for testing, measuring, and recording purposes. Exactly how such a workbench should be used to help design circuits is really left to the discretion of the designer. A major advantage of computer simulation compared with experimental work is the well-controlled environment in which designers can try out, verify, and optimize their designs without various practical limitations and potential hazards. Combining simulation programs with other programs to achieve design automation is also possible.

When applied to the design of converters, cycle-by-cycle simulation is particularly useful for the following applications:

1. To analyze the operation of existing or new converter topologies: A designer can start with practically any circuit, or any set of guessed component values, and then carry out simulations and iterations to verify or optimize a design. If such a wild-guess trial-and-error process were performed experimentally, it could be costly and potentially dangerous.

2. To predict the voltage and current stresses of circuit components for typical and worst-case conditions: Such information is necessary for the determination of voltage and current ratings of components. Cycle-by-cycle simulations are particularly useful for the design and adjustment of the snubber circuit to reduce the voltage stress while keeping the loss low.

3. To determine the switching losses of semiconductor devices under rigidly controlled testing conditions (such as precise rise/fall time of driving voltage and well-defined component parameters). Theoretical analysis or practical measurements of switching losses are often difficult.

4. To help examine the effects of stray capacitances, leakage inductances, and variable coupling coefficients of transformers: Although such information is very useful to the circuit designer, it is not so easy to predict theoretically or to obtain experimentally.

5. To analyze the switching trajectory of power switches for various conditions such as start-up, steady state, or power shutdown.

In the cycle-by-cycle simulation of converters, particular attention should, however, be paid to the following:

1. In order to reduce the time required for the switching waveforms of a converter to reach a steady state, an initial condition should be assigned to the output filtering capacitor and inductor (and possibly other large capacitors and inductors as well) of the converter. If the assigned initial condition is different from the actual final value, the simulated output voltage across the output filtering capacitor will show an exponential change or damped oscillation before settling down to a steady value. Based on the shape of the simulated output voltage waveform, the designer should be able to estimate a closer initial condition for the next simulation.

2. In the simulation of circuits that have large-amplitude and high-frequency ringing, a long simulation time may be needed because of the large number of calculations required for the rapidly changing voltages and currents. In the worst case, nonconvergence problems may also appear. Such phenomena often occur in converters in which transistors or diodes do not have snubber circuits connected across them. The long simulation time may discourage designers from using such simulations to help them design the circuit. However, the problem itself may also be interpreted as an indicator, signaling the designer to do something to reduce the ringing.

3. When more 5000 points of transient analysis are required, a .OPTION ITL5 = 0 statement should be included in the input file to remove the 5000-point default limit.

4. In the simulation of a converter circuit, the designer needs to decide how much detail of the gate/base driving circuitry should be included in the simulation. In theory, a more complete model will give more accurate results. However, a more complex model will

also increase the simulation time and the possibility of nonconvergence.
5. In deciding whether the nonlinear properties of a component should
    be included in the model, the designer should be fully aware that
    whereas inclusion would improve the accuracy of simulation (when
    the signal swing is large), it would also increase the simulation time
    and the possibility of nonconvergence.
6. In most cases, it is unnecessary and undesirable to include the pulse-
    width modulator in the cycle-by-cycle simulation of a converter.
    Even in the low-frequency behavior simulation of a complete reg-
    ulator circuit, the pulse-width modulator is normally modeled only
    as a simple $\delta D / \delta V_c$ transfer function (where V_c is the control input
    and D the duty cycle of the output pulse).

For the purpose of illustration, a detailed design example of a half-bridge
converter will be given in Section 8.4.

8.3 COMPUTER-AIDED DESIGN OF FEEDBACK AND COMPENSATION CIRCUITS

The design of feedback and compensation circuits for switching regulators
requires a wide range of knowledge, including converter characteristics and
feedback theory, and a good deal of practical experience. However, because of the nonlinearity of the converter circuit and the feedback amplifier
(under large-signal conditions), even expert designers would find it difficult
to predict the regulator performance precisely. As a result, a lot of trial-
and-error effort is still required in the actual design and implementation
of the feedback compensation circuit. The purpose of this section is to
present a new design approach using CAD tools to help the designer
achieve the desired regulator specifications more efficiently.

More specifically, the objectives of this section are:

1. To introduce the design philosophy
2. To introduce the techniques of simulating the low-frequency be-
    haviors of feedback loops in switching regulators for both small and
    large signals
3. To suggest methods of using CAD tools to design feedback and
    compensation circuits to meet specifications
4. To identify potential problems and suggest methods of solution

8.3.1 Design Philosophy

In the design of a feedback compensation circuit, a common method is to
assume a Bode plot (straight-line approximation of the frequency-response

characteristic) of the converter characteristic and then, based on the locations of poles and zeros, to design the feedback circuit. The disadvantages
of such an approach are:

1. Because of the limitations of manual analysis, many crude assump-
    tions and approximations are made, so that the design may not be
    realistic.
2. Because of the nonlinear nature of the regulator circuit, it is prac-
    tically impossible to predict manually whether a design would meet
    the specifications under large-signal conditions.

The CAD design approach we use here is different from the conventional
manual one. Instead of assuming a straight-line approximation (Bode plot)
of the converter characteristic, we start with a complete and accurate low-
frequency behavior model of the regulator as shown in Fig. 8.2. This
regulator model also takes into account the nonlinear characteristic of the
converter and is valid for both small-signal and large-signal conditions. In
designing the feedback compensation circuit, our objective here is to ensure
that the regulator meets:

1. The steady-state dc line and load regulation specifications
2. The transient line and load regulation specifications, for both small-
    and large-signal conditions

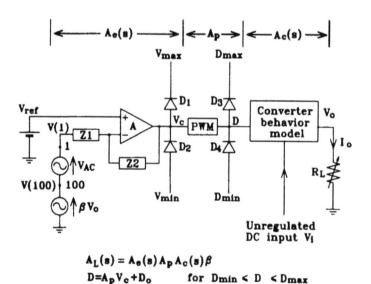

$$A_L(s) = A_e(s) A_p A_c(s) \beta$$
$$D = A_p V_c + D_o \qquad \text{for } D_{min} < D < D_{max}$$

FIGURE 8.2   Regulator model for simulation and design purposes.

Ideally, the required closed-loop regulator specifications should first be transformed into a frequency-domain objective loop-gain characteristic, and then the designer should design a feedback and compensation circuit to achieve the objective frequency characteristic. It will be found that, based on the model given in Fig. 8.2, it is easy to find a dc loop gain for the objective frequency characteristic. If the required unity-gain bandwidth and phase margin of the loop-gain characteristics are given, the model (Fig. 8.2) will also help the designer to determine accurately the frequency compensation characteristic required.

In the design process, many iterations may be necessary. However, for each iteration, simulations based on the model shown in Fig. 8.2 can be easily performed to ensure that the design meets all the required specifications (including large-signal transient specifications).

8.3.2   Modeling the Switching Regulator

The arrangement shown in Fig. 8.2 is a model of the regulator for simulation and design purposes. The essential features of the model can be explained as follows:

1. The behavior model for the converter is developed using the method introduced in Chapter 7. Converter models thus developed are valid for both continuous and discontinuous modes of operation and for both small and large signals. Such a model is absolutely necessary if meaningful large-signal transient simulations are to be carried out.

2. The pulse-width modulator (PWM) is modeled as a simple transfer function relating the output duty cycle D to the control input voltage V_c:

$$D = A_p V_c + D_o \qquad \text{for } D_{min} \leq D \leq D_{max} \qquad (8.1)$$

where A_p and D_o are constants that can be determined from PWM data sheets, D_{min} is the minimum value of D, and D_{max} is the maximum value of D. The clamping diodes D_3 and D_4 are used to limit the value of the duty cycle.

3. It is assumed that the frequency-response characteristic of the error amplifier (including frequency compensation) is determined predominantly by the impedance ratio Z_2/Z_1 and that the amplifier A can be modeled as a simple voltage-controlled voltage source in series with a small output resistance. It should, however, be noted that this assumption is not always valid because some error ampli-

fiers are designed to have very high output impedance (e.g., 10 MEGΩ). If necessary, the poles and zeros of the amplifier A should be included in the model.

4. The clamping diodes D_1 and D_2 are used to model the saturation/clipping effects of the error amplifier. This is necessary because the amplifier normally has a high gain so that saturation/clipping may occur easily when there is a large change in the loading current of the regulator.

5. The artificial SPICE signal source V_{AC}, added in series with the feedback circuit, is only for the convenience of determining the loop-gain characteristic of the feedback system under closed-loop conditions. In the steady state, we have the loop-gain $A_L(s)$ given by

$$A_L(s) = A_e(s)A_p A_c(s)\beta = \frac{-V(100)}{V(1)} \tag{8.2}$$

where $A_e(s)$ is the gain of the error amplifier (including frequency compensation), A_p the effective gain of the pulse-width modulator, $A_c(s)$ the $\delta V_o(s)/\delta D(s)$ transfer function of the converter, β the voltage feedback ratio, $V(100)$ the SPICE notation for the voltage at node 100, and $V(1)$ the SPICE notation for the voltage at node 1. The ratio $-V(100)/V(1)$ can be easily plotted as the output of an ac simulation in PSpice.

Based on the regulator model shown in Fig. 8.2, the design of the feedback compensation circuit, Z_1 and Z_2, will be discussed in the next subsection.

8.3.3 Design Procedures

The procedure we use here for the design of the feedback and compensation circuit is actually an interative process to achieve a set of objective specifications. As a starting point, we assume that all components in the SPICE model shown in Fig. 8.2, except the feedback impedances Z_1 and Z_2, are known. Our task is to find suitable components for Z_1 and Z_2 (each of Z_1 and Z_2 may contain a number of passive components) to enable the regulator to meet the line and load regulation specifications. The following are the design procedures:

1. We begin by assuming that Z_1 and Z_2 are resistive and

$$A_e(s) = \frac{Z_2}{Z_1} = \frac{R_2}{R_1} = 1 \tag{8.3}$$

An ac simulation is then performed to determine the loop-gain characteristic of the feedback system. Assuming V_{AC} as the input, the simulated plot of the ac voltage gain $-V(100)/V(1)$ is exactly the loop-gain characteristic we want:

$$A_L(s) = \frac{-V(100)}{V(1)} = \frac{R_2}{R_1} A_p A_c(s)\beta = A_p A_c(s)\beta \qquad (8.4)$$

2. If the ac simulation in procedure 1 indicates that the closed-loop system, with $R_2/R_1 = 1$, is stable, we can proceed to the dc simulation described in procedure 3. If, however, the ac simulation indicates a possible instability, the gain ratio R_2/R_1 should be suitably reduced before proceeding to procedure 3.
3. Direct current simulations are now performed to find:
    a. The dc output voltage V_o versus dc input voltage V_i characteristic (line regulation) for both the minimum and maximum loading current I_o, as shown in Fig. 8.3(a).
    b. The dc output voltage V_o versus dc output current I_o characteristic (load regulation) for both the maximum and minimum input voltage V_i, as shown in Fig. 8.3(b).
    Based on these characteristics, the worst-case steady-state line and load regulation factors for $R_2/R_1 = 1$ can be found.
4. From negative feedback theory, it is known that if, under the linear and open-loop condition, the steady-state output voltage changes by ΔV_{oo}, the change under the closed-loop condition ΔV_{oc} will be reduced to

$$\Delta V_{oc} = \frac{\Delta V_{oo}}{1 + A_L(0)} \qquad (8.5)$$

$$= \frac{\Delta V_{oo}}{1 + (R_2/R_1)A_p A_c(0)\beta} \qquad (8.6)$$

From Eq. (8.6), we have

$$\Delta V_{oo} = \Delta V_{oc}[1 + (R_2/R_1)A_p A_c(0)\beta] \qquad (8.7)$$

From Eq. (8.7) and the simulated characteristics shown in Fig. 8.3, we should be able to find the open-loop, steady-state line and load

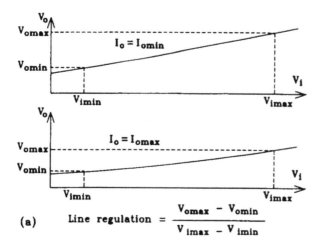

(a)   Line regulation $= \dfrac{V_{omax} - V_{omin}}{V_{imax} - V_{imin}}$

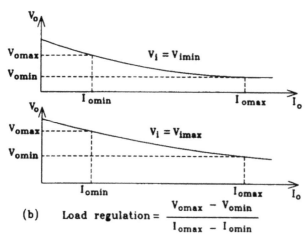

(b)   Load regulation $= \dfrac{V_{omax} - V_{omin}}{I_{omax} - I_{omin}}$

FIGURE 8.3   Determination of (a) line and (b) load regulations.

regulation characteristics of the converter. The line and load regulation factors for the open-loop system are simply equal to those
obtained in Fig. 8.3 multiplied by $[1 + (R_2/R_1)A_pA_c(0)\beta]$, where
$(R_2/R_1)A_pA_c(0)\beta$ is the dc voltage gain of $-V(100)/V(1)$ found in
procedure 1.
5. By comparing the open-loop steady-state line and load regulations
obtained in procedure 4 with the required specifications of the reg-

ulator, we can determine from Eq. (8.5) the minimum dc loop-gain
required in the feedback loop:

$$A_L(0) \geq \frac{\Delta V_{oo}}{\Delta V_{oc}} - 1 \qquad (8.8)$$

where ΔV_{oo} is the change in V_o under the open-loop condition, as
found from procedure 4, and ΔV_{oc} is the allowable change in V_o
under the closed-loop condition (i.e., the regulation specification).

6. Assume that the transient-response specifications require the loop-
    gain characteristic to have

    Unity-gain bandwidth $\geq B_w$ (8.9)
    Phase margin $\geq \theta_m$ (8.10)

    We can then use Eqs. (8.8–8.10) to set the limits of the loop-gain
    characteristic, as illustrated in Fig. 8.4. Based on these constraints,
    and with the aid of further ac simulations, we can try various com-
    pensation methods, such as dominant-pole, single-pole single-zero,
    double-pole double-zero, and so forth, in an iteration process to
    achieve an acceptable design.

7. For each iteration in the design process, transient simulations as-
    suming step changes in input voltage and output current are carried
    out to ensure that the steady-state and transient responses of the
    regulator meet the required specifications.

When a design is evaluated, it is acceptable only if all worst-case combinations of input and loading conditions (such as maximum dc input voltage
together with maximum loading resistance, or minimum dc input voltage
together with minimum loading resistance) are tried out and verified.

While designing the feedback and compensation circuit, the designer
should also be aware of the following:

1. The settling time in the transient response of a regulator is closely
    related to the bandwidth and phase margin of the loop-gain char-
    acteristic $A_L(s)$. A larger phase margin and a wider bandwidth would
    result in a shorter settling time.
2. The worst-case phase margin of the feedback circuit should be larger
    than 45°.
3. The maximum unity-gain bandwidth of the loop-gain $A_L(s)$ should
    be smaller than one-quarter of the output ripple frequency of the

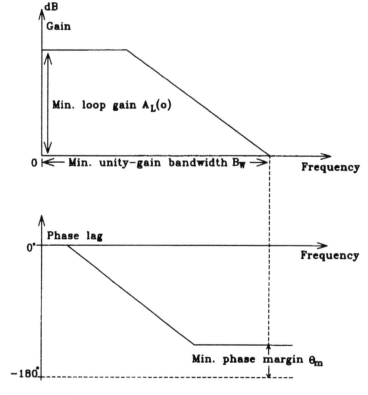

FIGURE 8.4   Allowable limits of loop-gain characteristic $A_L(s)$.

converter. This condition is necessary to ensure that the switching spikes can be sufficiently attenuated in the feedback circuit.

For the purpose of illustration, a regulator design example will be given in Section 8.4 after a discussion in the following subsection of the potential problems in a CAD exercise.

8.3.4 Commonly Encountered Problems

In this subsection, two problems that often appear in the computer-aided design of regulators will be discussed. The first is the multiple-solution problem in simulations involving the low-frequency behavior models of converters or regulators. The second is the apparently stable transient-simulation results for regulators that are actually unstable.

Let us consider the multiple-solution problem first. It is known that when an ac simulation, or a transient simulation without UIC (use initial conditions) option, is invoked for the low-frequency behavior model of a converter or regulator (actually for any circuit), a dc analysis will first be performed to find the biasing point of the circuit. If the behavior model contains nonlinear element(s) and feedback path(s), either explicitly or implicitly, it is possible for the dc analysis to converge to a false biasing point, which can exist mathematically but has no practical meaning. Typical indicators of such a false biasing point are:

1. An output voltage with a reversed polarity or an abnormally large or small amplitude
2. A negative value of D_2 (duty cycle of flywheel diode), which cannot exist practically
3. Unexpectedly large or small node voltages or branch currents

When any such phenomena occur, a possible method for solving the problem is to use a .NODESET statement to specify a more reasonable set of initial guesses of node voltages for the Newton–Raphson iteration process in SPICE to restart the dc biasing point analysis (instead of using computer-generated guesses). A transient simulation with UIC option and properly assigned initial conditions may also be used to help determine the desired biasing point. The dc biasing voltages obtained from the transient simulation can then be used as nodeset voltages for subsequent simulations.

The second problem to be discussed in this subsection is related to the transient simulation of circuits that are themselves unstable. SPICE transient simulations of such circuits often give apparently stable but untrue results. The following is a simple example that can be used to illustrate the nature and cause of the problem.

Consider the circuit shown in Fig. 8.5(a), which is a simple amplifier with positive feedback. It is assumed that the amplifier has a very high open-loop gain (10^5). A circuit designer can easily recognize that it is a Schmitt trigger. However, a SPICE transient simulation based on a sinusoidal input V_s gives the unexpected output waveform V_o as shown in Fig. 8.5(b). It is interesting to note that when the input terminals of the amplifier are swapped (so that the feedback becomes negative), as shown in Fig. 8.5(c), the simulated results remain virtually the same. From these results, it appears that the simulator has mistaken the positive feedback in the circuit as negative feedback.

It can be found that although the simulation results appear to be abnormal, there is actually nothing wrong with SPICE. The following analysis will help to trace the real cause of the problem.

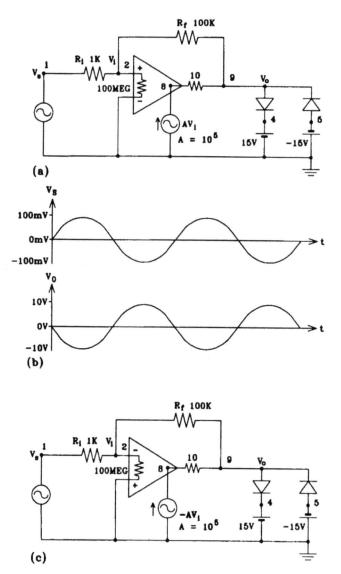

FIGURE 8.5 Amplifier with feedback: (a) Schmitt trigger; (b) simulated waveform of Schmitt trigger; (c) amplifier with negative feedback.

If we carry out a mathematical analysis on the circuit shown in Fig. 8.5(c), we can derive the mathematical equations given in the left-hand column of Table 8.1. If a similar analysis is repeated, but this time assuming a positive feedback as shown in Fig. 8.5(a), we shall get the expressions given in the right-hand column of Table 8.1. In these two analyses, the following assumptions are made:

1. The shunting effect of the 100-MEGΩ resistance (the input resistance of the amplifier) and the voltage drop of the 10-Ω resistor, in both Figs. 8.5(a) and 8.5(c), are negligible.
2. Both circuits operate in the linear region.

It is found that both analyses give similar expressions for the output voltage V_o! The conclusion from such a comparison is that SPICE is not at fault; it is the imperfect model that causes the problem.

A careful study should reveal that the mathematical solution of

$$V_o = V_s \frac{-R_f}{R_i} \tag{8.11}$$

can, in fact, exist in the Schmitt trigger circuit only as an unstable equilibrium state. In practice, however, even the slightest excitation in the feedback loop will result in a positive feedback to cause the amplifier to become saturated. It is actually the lack of elements-to-model excitations, such as thermal noise, dc drift, and interference, that results in an abnormal simulation of the circuit operation.

Table 8.1 Comparison of Analytical Results between Positive and Negative Feedback

For circuit with negative feedback	For circuit with positive feedback (Schmitt trigger)
$\dfrac{V_s - V_i}{R_i} = \dfrac{V_i - V_o}{R_f}$	$\dfrac{V_s - V_i}{R_i} = \dfrac{V_i - V_o}{R_f}$
$\dfrac{V_s - V_o/(-A)}{R_i} = \dfrac{V_o/(-A) - V_o}{R_f}$	$\dfrac{V_s - (V_o/A)}{R_i} = \dfrac{(V_o/A) - V_o}{R_f}$
$V_o = \dfrac{V_s}{R_i} \dfrac{1}{-1/(AR_f) - (1/R_f) - 1/(AR_i)}$	$V_o = \dfrac{V_s}{R_i} \dfrac{1}{1/(AR_f) - (1/R_f) + 1/(AR_i)}$
$V_o \approx V_s \dfrac{-R_f}{R_i}$ if $A \to \infty$	$V_o \approx V_s \dfrac{-R_f}{R_i}$ if $A \to \infty$

Similar kinds of erroneous results can also appear in the transient simulation of regulator circuits. The regulator model shown in Fig. 8.6 is a
typical example. The intended use of the model is to study the effect of a
change in the dc input voltage V_i on the output voltage V_o. An example
of the transient simulation is shown in Fig. 8.7, which appears to be quite
acceptable. However, when an ac simulation is performed to find the loop-
gain characteristic $-V(100)/V(1)$, the results are as shown in Fig. 8.8,
which indicates that the feedback loop has a negative phase margin (and
therefore should be unstable). The message is clear: The simulated waveform of V_o as shown in Fig. 8.7 is unreliable!

Possible methods that can be used to help the circuit designer uncover
the instability problem are:

1. Inspect the loop-gain characteristic $A_L(s)$, and determine the true
   phase or gain margin, as described above.
2. If a transient simulation is suspected to be unreliable, repeat one
   more simulation but, this time, use a .IC statement to specify a

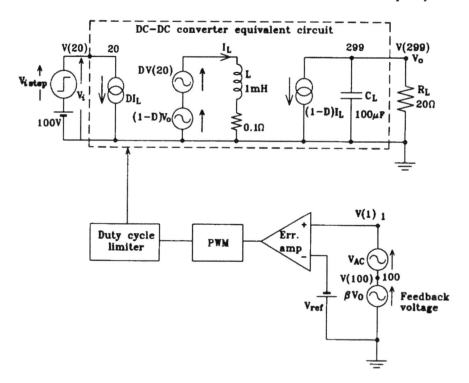

FIGURE 8.6 SPICE model of switching regulator.

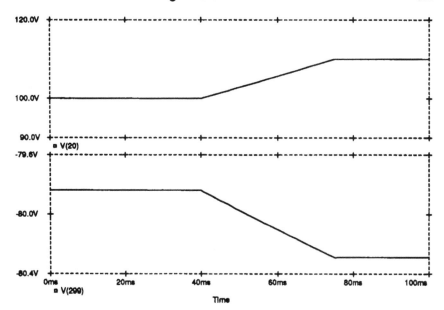

FIGURE 8.7 Simulated transient response of switching regulator: V(20) = V_i; V(299) = V_o.

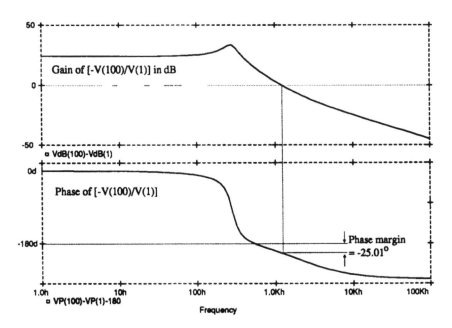

FIGURE 8.8 Simulated loop-gain characteristic of switching regulator.

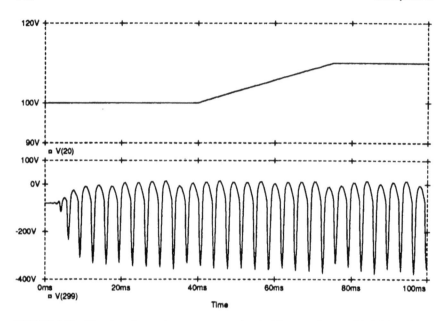

FIGURE 8.9   Simulated transient response of switching regulator with initial offset added: $V(20) = V_i$; $V(299) = V_o$.

slightly different output voltage as the initial condition. (This is equivalent to adding a perturbation to the circuit.) Consider the example shown in Fig. 8.6. The initial simulation in Fig. 8.7 gives an apparently stable dc biasing point at $V_o = -79.835$. However, if we repeat the simulation with an assigned initial V_o of -79.9 V (about 0.1% offset from the original output would be sufficient), we get the new waveform as shown in Fig. 8.9, which clearly shows the instability of the circuit. It should be noted that while this small offset is very effective in unveiling the oscillatory nature of an unstable circuit, the spurious effect it introduces will be hardly noticeable if the circuit is stable.

8.4   CONVERTER AND REGULATOR DESIGN EXAMPLE

In this section, a design example for a two-output half-bridge switching regulator, the specifications of which are given in Subsection 8.4.1, will be used to illustrate the practical techniques of using SPICE to help design switch-mode power supplies. The example actually consists of two parts. The first part is the design of the converter, which will be discussed in

Subsection 8.4.2. The second part is the design of the feedback compensation circuit for the regulator, which will be discussed in Subsection 8.4.3.

8.4.1 Regulator Specifications and Selection of Circuit Topology

The regulator to be designed is to meet the following specifications:

1. Allowable ac mains voltage (rms):
   195 V to 265 V (230 V ± 15%), 50/60 Hz
2. dc outputs and steady-state regulation:
   +5 V ± 0.2%, 5 to 25 A, ripple ≤ 100 mV peak to peak
   +15 V ± 15%, 1.4 to 7 A, ripple ≤ 150 mV peak to peak
3. Transient load regulation:
   For step changes in the 5-V output current between 5 and 12.5 A, or between 12.5 and 25 A, with 10-μs rise/fall time, the maximum allowable overshoots in the transient responses of the 5- and 15-V outputs are 200 mV and 1 V, respectively.
4. Transient line regulation:
   For changes in the dc input voltage between 193.5 and 374 V, at the rate of 70 V/ms, the allowable overshoots in the transient responses of the 5- and 15-V outputs are 50 mV.

Since the required output power is quite large (230 W), a half-bridge converter topology is considered appropriate. The switching frequency of each transistor is selected to be 100 kHz, which gives an output ripple frequency of 200 kHz. To obtain optimum feedback compensation, a two-pole, two-zero compensation network will be employed, as shown in Fig. 8.10.

8.4.2 Design of the Converter

In this design example, the selected converter topology is half-bridge, as shown in Fig. 8.10. The design procedures to be performed are outlined as follows (with details given later):

1. Based on the voltage and current requirements, the sizes of transformer cores, the number of turns of the windings, the required filtering inductances, capacitances, and so forth, will first be calculated. Together with the converter topology given in Fig. 8.11, this completes the preliminary design of the converter stage.

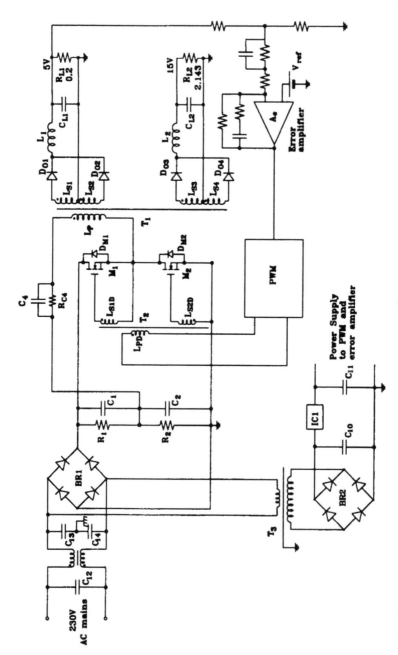

FIGURE 8.10 Basic half-bridge regulator circuit.

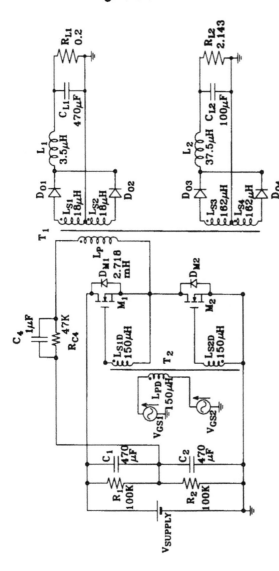

FIGURE 8.11 Preliminary design of half-bridge converter.

2. Based on the component values obtained in procedure 1, and with the assumption of near-ideal components and operating conditions, cycle-by-cycle simulations will be carried out for the converter circuit. The objectives of these near-ideal simulations are to test the reliability of the model, to verify the feasibility of the circuit under near-ideal operating conditions, and to extract design data such as voltage/current waveforms and stresses.
3. Based on the information obtained in procedure 2, a selection of diodes and transistors will be performed. The SPICE models of the selected devices will be entered into the simulation file.
4. A series of iterations will then be started to improve the paper design until it meets all requirements. The final simulation results will be used as guidelines for the practical implementation of the converter.

According to the procedures outlined above, the detailed design steps for the converter circuit and the related computer simulations will be elaborated in the following paragraphs.

First, consider the design of the power transformer T_1. The required $A_c A_w$ product of the magnetic core can be estimated from Eq. (10.53):

$$A_c A_w = \left[\frac{10^4 \, V_{max} I_{rms} T_{ON}}{120.3 \, K} \right]^{1.585} [K_H f + K_E f^2]^{0.6606} \text{ cm}^4 \qquad (8.12)$$

where

A_c = magnetic core area in cm²
A_w = window area in cm²
$V_{max} = \frac{1}{2}(265 \times 1.414) = 187.4 \text{ V}$ $\qquad\qquad (8.13)$
$T_{ON} = 2.2 \text{ μs}$ $\qquad\qquad\qquad\qquad\qquad\qquad\qquad (8.14)$

(The estimation of T_{ON} is based on the assumption that, for a dc input V_{max} of 187.4 V, the required duty cycle of each switching transistor is 0.22. The simulation result in Fig. A8.7.2 will confirm that this assumption is acceptable because the actual duty cycle required for each transistor is 0.16 only.)

$$I_{rms} = \text{dc input power} \frac{1}{0.5 \times \text{min dc input voltage}} \text{ form factor}$$

$$= \frac{\text{output power}}{\text{conversion efficiency}} \frac{2 \times \text{form factor}}{\text{min dc input voltage}}$$

$$\approx \frac{5 \times 25 + 15 \times 7}{0.75} \times \frac{2 \times 1.2}{195 \times 1.414 \times 0.85} = 3.14 \text{ A}$$

$$(8.15)$$

(The estimation of the form factor is based on the assumption that each switching transistor has a steady-state duty cycle of 0.35 when the dc input voltage is at its minimum. The simulation result in Fig. A8.7.2 will show that the actual duty cycle for each transistor is $0.5 \times D = 0.5 \times 0.714 = 0.357$, which implies that the assumption is acceptable.)

$$K = K_u K_p = 0.4 \times 0.41 = 0.164 \text{ [from Eq. (10.34)]} \tag{8.16}$$

$$K_H = 4 \times 10^{-5} \tag{8.17}$$

$$K_E = 4 \times 10^{-10} \tag{8.18}$$

$$f = 10^5 \tag{8.19}$$

Substitution of the preceding estimated values of V_{max}, T_{ON}, I_{rms}, K, K_H, and K_E into Eq. (8.12) gives

$$A_c A_w = 2.03 \text{ cm}^4 \tag{8.20}$$

In the actual circuit, we choose a magnetic core with

$$A_c = 1.42 \text{ cm}^2 \tag{8.21}$$

$$A_w = 1.496 \text{ cm}^2 \tag{8.22}$$

$$A_c A_w = 2.12 \text{ cm}^4 \tag{8.23}$$

The number of turns of the primary winding can now be determined from Eq. (10.30).

$$N_p = \frac{V_{max} T_{ON}}{A_c \Delta B_m} \tag{8.24}$$

where
$$A_c = 1.42 \times 10^{-4} \text{ (m}^2) \tag{8.25}$$
$$\Delta B_m = 0.4052 \, (A_c A_w)^{-0.1292} [K_H f + K_E f^2]^{-0.4167} \text{ teslas} \tag{8.26}$$
$$\text{[obtained from Eq. (10.50)]}$$
$$A_c A_w = 2.12 \text{ (cm}^4)$$

Substitution of Eqs. (8.25) and (8.26) into Eq. (8.24) gives

$$N_p = \frac{187.4 \times 2.2 \times 10^{-6}}{1.42 \times 10^{-4} \times 0.4052 \times 2.12^{-0.1292}[4 + 4]^{-0.4167}} = 18.8$$

The actual number of turns used for the primary winding is 22.

$$\therefore N_p = 22 \tag{8.27}$$

Since a step-down ratio of 11 is considered appropriate for the 5-V output, each of L_{S1} and L_{S2} will have two turns. Correspondingly, L_{S3} and L_{S4} will each have six turns.

Based on the number of turns of the windings and the A_L value (inductance of a single turn) of the magnetic core, the values of L_P, L_{S1}, L_{S2}, L_{S3}, and L_{S4}, and the coupling coefficients among them, are estimated in the following table.

Parameter of T_1	L_P	L_{S1}	L_{S2}	L_{S3}	L_{S4}	Coupling coefficient between each pair of inductances
Value	2.178 mH	18 μH	18 μH	162 μH	162 μH	0.9985

The parameters in the table are absolutely necessary for the simulation of the converter circuit.

The detailed design procedures for the gate-driving transformer T_2 will not be given here. However, the important transformer parameters in the following table will be used in later simulations.

Parameter of T_2	L_{PD}	L_{S1D}	L_{S2D}	Coupling coefficient between each pair of inductances
Value	150 μH	150 μH	150 μH	0.9985

The filtering inductance L_1 for the +5-V output is estimated from Eq. (1.21). By assuming that the converter remains in continuous-mode operation for an output current of 5 A ($R_{L1} \le 1\ \Omega$), we have

$$L_1 = \tfrac{1}{4} R_{L1}(1 - D)T \qquad (8.28)$$

Note that in calculating the required values of the output filtering components for push-pull converters (such as L_1, L_2, C_{L1}, and C_{L2} in Fig. 8.11), the parameter T should be interpreted as the period of the output ripple and the parameter D should be twice the duty cycle of each switching transistor. We therefore have

$$L_1 = \tfrac{1}{4} \times 1 \times (1 - 0.3) \times 5 \times 10^{-6}$$
$$= 1.75\ \mu H \text{ (assuming a minimum } D \text{ of 0.3)}$$

An actual value of 3.5 μH is chosen.

$$\therefore L_1 = 3.5 \ \mu H \tag{8.29}$$

The required A_cA_w product, number of turns, and width of the air gap of the inductor can be determined by direct application of Eqs. (10.19), (10.11), and (10.16).

The filtering capacitance C_{L1} is found from Eq. (1.24):

$$\Delta V_o = \frac{T^2 V_o}{8 C_{L1} L_1} (1 - D)$$

Assuming a minimum D of 0.3 and a ΔV_o of 10 mV, we have

$$C_{L1} = \frac{(5 \times 10^{-6})^2 \times 5}{8 \times 10 \times 10^{-3} \times 3.5 \times 10^{-6}} \times (1 - 0.3) = 312.5 \ \mu F \tag{8.30}$$

(To keep the total ΔV_o below 100 mV, the ripple voltage due to the effective series resistance of C_{L1} and noise pickup should be less than 90 mV).

The actual value selected is 470 μF.

$$\therefore C_{L1} = 470 \ \mu F \tag{8.31}$$

The filtering inductance L_2 for the +15-V output is estimated from Eq. (1.21) by assuming that the continuous-mode operation is maintained for an output current of 1.4 A ($R_{L2} \le 10.714 \ \Omega$):

$$L_2 = \tfrac{1}{2} R_{L2}(1 - D)T$$
$$= \tfrac{1}{2} \times 10.714 \times (1 - 0.3) \times 5 \times 10^{-6} = 18.75 \ \mu H \tag{8.32}$$

An actual value of 37.5 μH is chosen.

$$\therefore L_2 = 37.5 \ \mu H \tag{8.33}$$

Again, the A_cA_w product, number of turns, and width of air gap of the inductor can be found by direct application of Eqs. (10.19), (10.11), and (10.16).

The filtering capacitance C_{L2} is estimated from Eq. (1.24):

$$\Delta V_o = \frac{T^2 V_o}{8 C_{L2} L_2} (1 - D) \tag{8.34}$$

Assuming a minimum D of 0.3 and a ΔV_o of 15 mV, we have

$$C_{L2} = \frac{(5 \times 10^{-6})^2 \times 15}{8 \times 15 \times 10^{-3} \times 37.5 \times 10^{-6}} \times (1 - 0.3) = 58.3 \ \mu F$$

The actual value selected is 100 μF.

$$\therefore C_{L2} = 100 \ \mu F \tag{8.35}$$

(To keep the total ΔV_o below 150 mV, the ripple voltage due to the effective series resistance of C_{L2} and noise pickup should be less than 135 mV.)

The ac mains rectifier filtering capacitances C_1 and C_2 are found from Eq. (10.2):

$$C = \tfrac{1}{2}C_1 = \tfrac{1}{2}C_2 = \tfrac{1}{2}PT\frac{1}{0.51 \ (V_{acmin})^2} \tag{8.36}$$

where

$$P = \frac{\text{output power}}{\text{conversion efficiency}} = \frac{230 \ W}{0.75} = 306.7 \ W$$

$$T = \frac{1}{50} \ \text{sec}$$

$$V_{acmin} = 195 \ V$$

Substitution of the preceding parameters into Eq. (8.36) yields

$$C_1 = C_2 = 2C = 316 \ \mu F \tag{8.37}$$

The actual value selected is 470 μF. The bleeder resistances R_1 and R_2 are chosen to be 100 kΩ.

The coupling capacitance C_4 is assumed to be 1 μF, and its parallel resistance R_{C4} is assumed to be 47 kΩ.

With all key component values now estimated, we have actually completed the preliminary design stage of the converter circuit. The next step in the design process is to develop a model for the preliminary converter design (shown in Fig. 8.11) and start from there a series of iterations to obtain a practically achievable design with well-predicted performance. For the purpose of illustration, the simulation file for each typical iteration will be given in a separate attachment at the end of this chapter.

As a starting point, Fig. A8.1.1 of Attachment 8.1 shows schematically the near-ideal model of the preliminary converter design. The listing of the model is also given in the attachment. In this model the following assumptions are made:

1. The effective resistances of inductors, transformer windings, and filtering capacitors are very small (e.g., 1 $\mu\Omega$).
2. All transformers have very tight coupling (e.g., $K = 0.999999$).
3. All transistors and diodes have very fast switching speed and low forward voltage drop.

Using the near-ideal converter model, as represented by the input listing given in Attachment 8.1 (at the end of this chapter), simulations are carried out to test the reliability of the model. Figure A8.1.2 of Attachment 8.1 shows an example of such simulations. The simulated waveforms can be compared with the idealized ones, such as those given in Fig. 4.12, to cross-check the validity of the converter model (to ensure that no careless mistakes are made). Other simulations assuming different input voltages, duty cycles, and loading resistances may also be performed to estimate the voltage and current stresses of circuit components.

Based on the first-round simulation results obtained from the near-ideal circuit model and other information obtained from data books, measurements, or experience, we have decided to

1. Use a pair of IRF740 as the power switches.
2. Use a pair of 1N5831 as D_{o1} and D_{o2}.
3. Use a pair of MUR810 as D_{o3} and D_{o4}.
4. Include the following effective series resistances into the model.

Component	L_P	L_{S1}	L_{S2}	L_{S3}	L_{S4}	L_{PD}	L_{S1D}	L_{S2D}	L_1	L_2	C_{L1}	C_{L2}
Effective series resistance in mΩ	180	3	3	30	30	1×10^3	1.5×10^3	1.5×10^3	3	30	15	40

5. Replace the unrealistic magnetic coupling coefficients by practical ones. In this example, we assume that $K = 0.9985$.

Subsequently, the converter model and its listing are modified as shown in Attachment 8.2. It should be noted that, in this simulation file, the .MODEL parameters for the MOS power transistor IRF740 are obtained from the component library of PSpice. A convenient feature of this MOS model is that it includes the inherent source-to-drain diode in it. However, when the input file is to be run by SPICE (instead of PSpice), a different modeling arrangement will be required. (This applies to both Attachment 8.2 and Attachment 8.3.)

Based on the more realistic converter model given in Attachment 8.2, a second round of simulations can be carried out for the purpose of evaluating practical performance and rectifying design pitfalls. The simulation results given in Figs. A8.2.2–A8.2.5 are examples. From these examples, the following messages are quite clear:

1. The severe ringing in the voltages V(7), V(15), V(21), shown in Fig. A8.2.2 (e), (g), and (h), respectively, indicates that snubber circuits have to be added to suppress such ringing.
2. The ringing/overshoot in the gate voltage V(9), shown in Fig. A8.2.2(b), indicates that a damping/clamping circuit is required.
3. The $I_{DS} - V_{DS}$ switching trajectories for the two transistors, which are shown in Figs. A8.2.4 and A8.2.5, show that they are not desirable for the transistors.

As an iteration in the design process, snubber circuits and gate damping/clamping circuits are now added to the converter circuit, as shown in Attachment 8.3.

Based on the new model shown in Fig. A8.3.1, a third round of simulations is performed. Some simulation examples are given in Figs. A8.3.2–A8.3.5. These results show that the modified circuit is now much improved.

Similar iterations can be repeated until an acceptable design is obtained. Normally, many iterations are required before a final solution can be obtained.

However, in this example, we shall assume that the circuit shown in Fig. A8.3.1 is already our final converter design and that the simulated results in Figs. A8.3.2–A8.3.5 are the guidelines for the practical implementation of the converter circuit.

8.4.3 Design of the Regulator

Using the half-bridge converter designed (on paper) in Attachment 8.3 as a power conversion engine, the design of the complete regulator will be discussed in this subsection. The three main tasks involved in the design work are:

1. Selection of a PWM IC, and design of an appropriate gate-driving circuit.

2. Design of a power supply for the PWM IC and the driving circuitry.
3. Design of a feedback compensation circuit for the regulator.

For details on the selection of a PWM IC and design of the driving circuit, the reader is referred to Section 10.6.

The dc power to the PWM and error amplifier shown in Fig. 8.10 is supplied by a small linear regulator consisting of transformer T_3 (a small-sized mains transformer), bridge-rectifier $BR2$, filtering capacitors C_{10} and C_{11}, and IC regulator $IC1$. The design of the linear regulator is quite straightforward and will not be discussed here.

In the design of the feedback compensation circuit, we shall follow the steps given in Subsection 8.3.3, as described below.

First, we develop a low-frequency behavior model of the complete regulator, as shown in Fig. 8.12. The expanded circuitry of the dual-output half-bridge converter is shown in Fig. 8.13. Further details of the half-bridge converter subcircuit (XCIR1 and XCIR2) are shown in Fig. 8.14. (Note that the low-frequency behavior model of the half-bridge converter

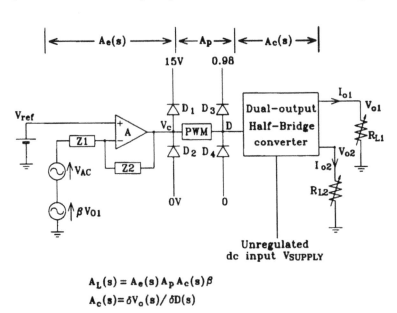

$$A_L(s) = A_e(s) A_p A_c(s) \beta$$
$$A_c(s) = \delta V_o(s) / \delta D(s)$$

$$D = 2 \times (\text{duty cycle of individual switch})$$
$$= \frac{1}{2.4} V_c - 0.375 \qquad 0 < D < 0.98$$

FIGURE 8.12   Complete regulator model for simulation and design purposes.

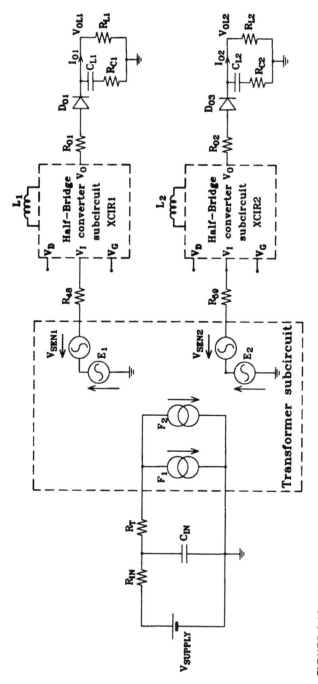

FIGURE 8.13 Expanded model of dual-output half-bridge converter. Note: (1) D_{o1} and D_{o3} are used to simulate the effects of the output rectifiers; (2) R_{o1} and R_{o2} are used to simulate the lossy components of the converter.

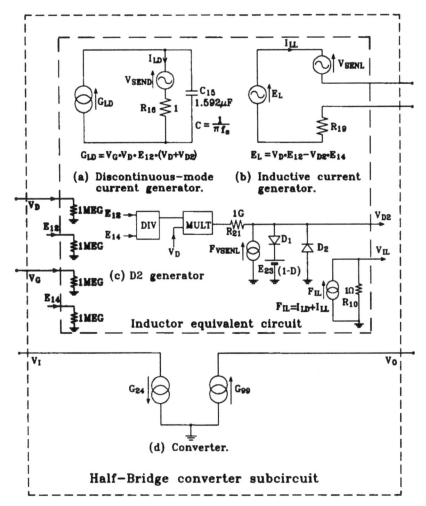

FIGURE 8.14 Details of half-bridge converter subcircuit.

is actually similar to that of the transformer-coupled push-pull converter given in Attachment 7.5 of Chapter 7 because both converters are push-pull versions of the buck converter.)

Particular attention should be paid to the resistances R_{o1} and R_{o2} shown in Fig. 8.13. They are used to represent the lossy components of the converter circuit, which include the loss in transistors, the loss in snubber circuits, the loss in the output power transformer, and the loss in the output

filtering inductors (but exclude the loss in the output rectifiers). In esti-
mating their equivalent resistance values, it is assumed that they represent
10% power loss under the full-load condition.

To begin the design iteration, we now start with the first low-frequency
behavior simulation, as shown in Attachment 8.4. There, based on the
regulator model shown in Figs. A8.4.1–A8.4.3, and assuming $R_1 = R_2 = 10\ k\Omega$, an ac simulation is carried out to determine the loop-gain characteristic $A_L(s)$ (for the feedback loop containing the 5-V output V_{oL1}, which
is fully loaded with an R_{L1} of 0.2 Ω).

$$A_L(s) = \frac{-V(100)}{V(1)} \tag{8.38}$$

The simulated result is shown in Fig. A8.4.4. From there, we find that the
circuit should be stable for the condition of $Z_1 = Z_2 = 10\ k\Omega$ because the
loop gain has a phase margin of about 55°.

Using the same circuit, dc analyses are carried out to determine the line
and load regulation characteristics (for $Z_1 = Z_2 = 10\ k\Omega$). Attachment
8.5 shows the simulation file and results.

From the ac and dc simulations in Attachments 8.4 and 8.5, the following
information is obtained (for the condition of $Z_1 = Z_2 = 10\ k\Omega$):

$$\frac{10K}{10K} A_p A_c(0)\beta = A_p A_c(0)\beta = \frac{-V(100)}{V(1)}$$

$$= 2.189 \text{ or } 6.805 \text{ dB} \tag{8.39}$$

(found from Fig. A8.4.4)

Worst-case closed-loop ΔV_{oc} (for the 5-V output) $= 6.29 - 5$

$$= 1.29 \text{ V}$$

$$\tag{8.40}$$

(found from Fig. A8.5.2 for the condition of $V_{supply} = 374$ V and
$I_{o1} = 5$ A)

Based on these parameters, we can use Eq. (8.7) to calculate the worst-
case open-loop ΔV_{oo} of the converter:

$$\Delta V_{oo} = \Delta V_{oc} \left[1 + \frac{R_2}{R_1} A_p A_c(0)\beta \right] = 1.29(2.189 + 1) = 4.1138 \text{ V}$$

$$\tag{8.41}$$

From Eq. (8.8), we have

$$\text{Required } A_L(0) = \frac{\Delta V_{oo}}{\text{Acceptable } \Delta V_o} - 1 \qquad (8.42)$$

$$= \frac{4.1138}{5 \times 0.1\%} - 1 = 821.76 \text{ or } 58.3 \text{ dB} \qquad (8.43)$$

(Note that although the ΔV_o specification is $\pm 0.2\%$, our design is aimed at $\pm 0.1\%$ for the nominal operating condition.)

In the design of the frequency compensation circuit, the following assumptions are made:

1. Two zeros at the frequency of

$$f_{z1} = f_{z2} = \frac{1}{2} \frac{1}{2\pi} [L_1 C_{L1}]^{-1/2} = 1962 \text{ Hz} \qquad (8.44)$$

   are required to compensate the phase lag of the output filtering circuit (L_1 and C_{L1}).

2. A pole at the frequency of

$$f_{p1} = \frac{1}{2\pi C_{L1} R_{C1}} = \frac{1}{2\pi \times 470 \times 10^{-6} \times 15 \times 10^{-3}} = 22575 \text{ Hz} \qquad (8.45)$$

   is required to compensate the zero caused by the effective series resistance (R_{C1}) of the output filtering capacitor C_{L1}.

3. A unity-gain bandwidth of 40 kHz (which is slightly less than one-quarter of the 200-kHz ripple frequency of the output) is required to maintain a good transient response.

   $$\text{Unity-gain bandwidth } B_w = 40 \text{ kHz} \qquad (8.46)$$

4. A dc gain (in decibels) of

   $$A_e(0) = A_L(0) - A_p A_c(0)\beta = 58.3 - 6.805 = 51.5 \text{ dB} \qquad (8.47)$$

   is required in the error amplifier to maintain the steady-state regulation specification.

In order to meet the requirements stated above, the following design steps are taken:

1. It is assumed that the compensation circuit as shown in Fig. 8.15 is used and that

$$R_{22} = 2 \text{ MEG}\Omega \qquad (8.48)$$

2. Since the dc gain required in the error amplifier is 51.5 dB (equivalent to a ratio of 375.8), we need to have

$$A_e(0) = \frac{R_{22}}{R_{11} + R_{12}} = 375.8 \qquad (8.49)$$

$$\therefore R_{11} + R_{12} = \frac{2 \times 10^6}{375.8} = 5322 \ \Omega \qquad (8.50)$$

From the expressions given in Fig. 8.15 and Eqs. (8.44) and (8.45), we have

$$\frac{f_{z1}}{f_{p1}} = \frac{1/(2\pi R_{12}C_{11})}{1/[2\pi C_{11}(R_{11}//R_{12})]} = \frac{R_{11}}{R_{11} + R_{12}} = \frac{1962}{22575} = 0.08691 \qquad (8.51)$$

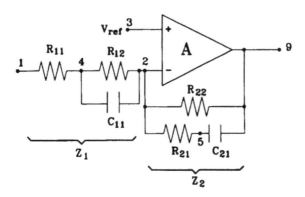

Poles: $\quad f_{p2} = \dfrac{1}{2\pi(R_{21}+R_{22})C_{21}} \qquad f_{p1} = \dfrac{1}{2\pi C_{11}(R_{11}//R_{12})}$

Zeros: $\quad f_{z2} = \dfrac{1}{2\pi R_{21} C_{21}} \qquad\qquad f_{z1} = \dfrac{1}{2\pi R_{12} C_{11}}$

FIGURE 8.15 Two-pole, two-zero compensation network.

The values of R_{11} and R_{12} can be found from Eqs. (8.50) and (8.51) as

$$R_{11} = 5322 \times 0.08691 = 463 \ \Omega$$
$$R_{12} = 5322 \times (1 - 0.08691) = 4859 \ \Omega$$

The actual values selected are

$$R_{11} = 470 \ \Omega \tag{8.52}$$
$$R_{12} = 4700 \ \Omega \tag{8.53}$$

3. The value of C_{11} is determined from Eq. (8.44):

$$f_{z1} = 1962 \ \text{Hz} = \frac{1}{2\pi R_{12}C_{11}} \tag{8.54}$$

$$C_{11} = \frac{1}{2\pi R_{12}f_{z1}} = \frac{1}{2\pi \times 4700 \times 1962} = 17.26 \ \text{nF}$$

The actual value selected is

$$C_{11} = 18 \ \text{nF} \tag{8.55}$$

4. By assuming

$$R_{21} = 15000 \ \Omega \tag{8.56}$$

$$C_{21} = \frac{1}{2\pi \times 1962 \times 15000} = 5.408 \ \text{nF} \tag{8.57}$$

$$\text{(so that } f_{z2} = 1962 \ \text{Hz)}$$

a trial ac simulation is now run to find the unity-gain bandwidth B_w. The B_w found from the simulated result is 64.7 kHz. This means that if the unity-gain bandwidth is to be reduced to 40 kHz, the capacitance C_{21} will have to be increased to (approximately)

$$C_{21} = \frac{64.7 \ \text{kHz}}{40 \ \text{kHz}} \times 5.408 \ \text{nF} = 8.75 \ \text{nF}$$

Correspondingly, the value of R_{21} will need to be changed to

$$R_{21} = \frac{1}{2\pi \times 1962 \times 8.75 \times 10^{-9}} = 9271 \ \Omega$$

The actual values selected for the iteration are:

$$C_{21} = 8.2 \text{ nF} \qquad (8.58)$$
$$R_{21} = 10 \text{ k}\Omega \qquad (8.59)$$

5. By using the new values of C_{21} and R_{21} as given by Eqs. (8.58) and (8.59), a further ac simulation is carried out to determine the exact value of the unity-gain bandwidth B_w. The new value obtained from simulation is 39.8 kHz, which is considered acceptable. (If this is not acceptable, further iterations based on steps 4 and 5 may be repeated until an acceptable, and practical, solution is obtained. It is worth mentioning that by using PSpice on an IBM-compatible PC with an 80486 processor operating at 33-MHz clock frequency, each $A_L(s)$ simulation takes only about 9 seconds to complete.)

The component values as given by Eqs. (8.48), (8.52), (8.53), (8.55), (8.58), and (8.59) are used in the final design:

$R_{22} = 2 \text{ MEG}\Omega$
$R_{11} = 470 \ \Omega$
$R_{12} = 4700 \ \Omega$
$C_{11} = 18 \text{ nF}$
$C_{21} = 8.2 \text{ nF}$
$R_{21} = 10 \text{ k}\Omega$

The complete regulator design is shown in Fig. A8.6.1.

The simulated loop-gain characteristic of the final design (including error amplifier, PWM, converter, and feedback ratio β) for an R_{L1} of 0.2 Ω is given as Fig. A8.6.2 in Attachment 8.6. The phase margin found there is 94°. However, when the load R_{L1} is increased to about 6 Ω (which is far beyond the normal operation range of 0.2 Ω $< R_{L1} < 1$ Ω), the phase margin falls to 45°. (This small phase margin is due to the nearly 90° phase lag for discontinuous-mode operation at low frequency and the existence of f_{p2} in the compensation circuit. The related simulation is not shown in the attachment.)

Based on the model of the final design shown in Fig. A8.7.1, simulations can be carried out to predict the performance of the regulator. Figures A8.7.2 and A8.7.3 are examples of such simulations. These simulations show that, under the normal operating conditions, the design would meet all steady-state and transient regulation specifications.

It should be interesting to note the following:

1. The parameter D (D being equal to twice the duty cycle of each switching transistor) is shown as $V(11)$ in Figs. A8.7.2 and A8.7.3. It is found that the required steady-state minimum and maximum values of D are 0.32 and 0.714, respectively. This information is very important because it confirms that the assumptions made for the value of D (or for the duty cycle of each switching transistor) in the design of the output transformer and filtering circuits, as described in Subsection 8.4.2, are acceptable.
2. Also from Fig. A8.7.2, it is found that when the 5-V output current changes from 12.5 to 25 A, the value of D will be clipped temporarily at the maximum limit of 0.98. Such a clipping is an important factor that limits the transient-response performance of the regulator. If the transient-response specification for such a step change is relaxed, we can actually design the regulator to operate at a larger steady-state duty cycle to improve the power conversion efficiency. (A larger duty cycle implies a smaller form factor of the current pulses and therefore a smaller conduction loss, as explained in Subsection 9.2.3.)
3. The simulations between 90 and 100 ms, in both Figs. A8.7.2 and A8.7.3, are actually out of the specified range of operating conditions. It is therefore reasonable to find that the 15-V output regulations within this period are worse than the specifications. These simulations are, however, still very useful because they show how much worse the performance is under such conditions.

It should be noted that the preceding example is not meant to be an optimized design. Rather, it is an illustration showing how CAD tools may be used to help produce a design that would meet specifications. Circuit designers should also realize that while it is of prime importance to meet the specifications, they should not fall into the trap of designing a circuit to outperform excessively any specification (knowingly or unknowingly) without knowing the hidden adverse side effects or implications of such a design.

8.5 PRACTICAL IMPLEMENTATION

When the paper design of a converter or regulator has been finalized by using CAD tools, the simulated performances can be used as objectives for the actual circuit to achieve. However, because of practical limitations, it may not be always possible to achieve all the simulated performances. Typical examples of such limitations are:

1. Transformers may not have coupling coefficients as close to 1 as we want.
2. Output filtering capacitors may not have effective series resistances as low as we need.
3. There are always noises and interferences that are not taken care of by simulations.
4. In addition to its direct adverse effect on the regulator's performance index, the presence of switching spikes and interferences may also imply an extra filtering requirement in the feedback circuit. This can result in a smaller phase margin in the control loop and poorer transient responses.
5. Any unexpected delay in the feedback loop that has not been included in the simulation will further reduce the practically achievable phase margin.

However, through design iterations and experience, circuit designers should be able to refine their circuit models and CAD techniques to ensure that their designs are realistic and easy to implement.

8.6 SUMMARY AND FURTHER REMARKS

In this chapter the philosophy of using the CAD tool to help design electronic circuits and the practical techniques of applying it to the design of converters and regulators have been discussed. Some potential problems that may be encountered in the design process have been identified. The solutions to these problems have also been discussed. A design example for a converter and a regulator has been given to illustrate in detail the design steps.

It is expected that when circuit designers learn to use a CAD tool, they may encounter many problems, such as being unfamiliar with the tool and unaware of its limitations, finding it difficult to develop suitable models for simulation purposes, being unable to interpret simulation results, not knowing what to do when problems occur, and not having sufficient libraries of component/behavior models. When these problems have been overcome, however, designers will find such a tool extremely useful. Users of CAD tools will soon realize that the capabilities of human designers and CAD tools are actually complementary: While human designers are very good at generating ideas, CAD tools are excellent at verifying good ideas, helping to correct wrong ideas, and documenting all ideas.

EXERCISES

1. In what ways can SPICE be used to help design a converter?
2. Comment on the differences among the following four sets of waveforms of the half-bridge converter:
   a. those shown in Fig. 4.12
   b. those shown in Fig. A8.1.2 of Attachment 8.1 at the end of this chapter
   c. those shown in Figs. A8.2.2 and A8.2.3 of Attachment 8.2
   d. those shown in Figs. A8.3.2 and A8.3.3 of Attachment 8.3
   What are the uses of each set of waveforms?
3. Based on the half-bridge converter designed in Section 8.4, design a full-bridge converter to double its output currents while maintaining the other specifications.
4. In what ways can SPICE be used to help design the feedback compensation circuit of a regulator?
5. Design a feedback compensation circuit for the full-bridge converter designed in question 3 that meets the regulation specifications given in Subsection 8.4.1. (Assume that all current values in the specifications are doubled.)

ATTACHMENT 8.1: SIMULATION OF NEAR-IDEAL*
HALF-BRIDGE CONVERTER

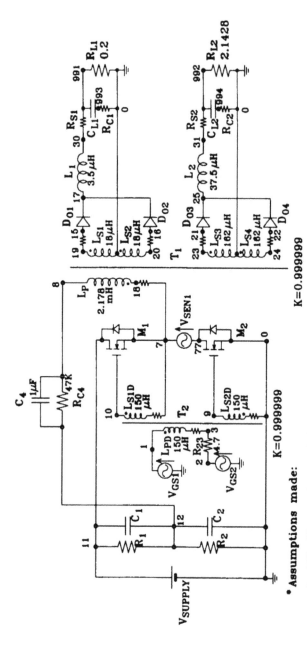

* Assumptions made:

1. Effective series resistances of inductors, transformer windings, and output filtering capacitors = 1 μΩ

2. K=0.999999

3. All transistors and diodes have near ideal characteristics

Fig. A8.1.1 Near-ideal* Half-Bridge converter model.

A8.1 SIMULATION OF NEAR-IDEAL HALF-BRIDGE CONVERTER

```
HALF BRIDGE CONVERTER (CYCLE-BY-CYCLE SIMULATION)
*NEAR-IDEAL SIMULATION
*
*DRIVING STAGE
VGS1 1 0 PULSE(0 15 0.1U 0.1U 0.1U 2U 10U)
VGS2 2 0 PULSE(0 15 5.1U 0.1U 0.1U 2U 10U)
R23 2 3 4.7
RLPD 4 3 1U
LPD 1 4 150U
LS1D 10 50 150U
RLS1D 50 7 1U
LS2D 60 9 150U
RLS2D 60 0 1U
K1D LPD LS1D 0.999999
K2D LPD LS2D 0.999999
K3D LS1D LS2D 0.999999
*
*
*POWER STAGE
M1 11 10 7 7 MOD1 (W=2)
M2 77 9 0 0 MOD1 (W=2)
VSEN1 7 77
VSUPPLY 11 0 270V
C1 11 12 470UF IC=135
R1 11 12 100K
C2 12 0 470UF IC=135
R2 12 0 100K
*
*
*POWER TRANSFORMER T1 AND ASSOCIATED CIRCUITS
RLP 7 18 1U
LP 18 8 2.178M IC=0.05
LS1 19 0 18U
RLS1 15 19 1U
LS2 0 20 18U
RLS2 20 16 1U
LS3 23 0 162U
RLS3 21 23 1U
LS4 0 24 162U
RLS4 24 22 1U
K1 LS1 LS2 0.999999
K2 LS1 LS3 0.999999
K3 LS1 LS4 0.999999
K4 LS2 LS3 0.999999
K5 LS2 LS4 0.999999
K6 LS3 LS4 0.999999
K7 LP LS1 0.999999
K8 LP LS2 0.999999
K9 LP LS3 0.999999
K10 LP LS4 0.999999
*
```

```
*COUPLING CIRCUIT
C4 8 12 1U
RC4 8 12 47K
*
*
*OUTPUT STAGE 1
DO1 15 17 DMOD
DO2 16 17 DMOD
L1 17 30 3.5U IC=25
RS1 30 991 1U
CL1 991 993 470U IC=5
RC1 993 0 1U
RL1 991 0 0.2
*
*
*OUTPUT STAGE 2
DO3 21 25 DMOD
DO4 22 25 DMOD
L2 25 31 37.5U IC=7
RS2 31 992 1U
CL2 992 994 100U IC=15
RC2 994 0 1U
RL2 992 0 2.1428
*
*
*
.MODEL MOD1 NMOS (VTO=3.5)
.MODEL DMOD D
.PROBE
*THE .PROBE STATEMENT IS FOR PSPICE ONLY.
*DELETE .PROBE STATEMENT IF RUN BY SPICE.
.OPTIONS ITL4=300
.TRAN 0.25U 50U 1U UIC
.PRINT TRAN I(VSEN1) V(7) V(991)
.PLOT TRAN I(VSEN1) V(7) V(991)
*MORE VARIABLES CAN BE ADDED TO THE .PRINT AND
*.PLOT LISTS AT THE DISCRETION OF THE DESIGNER.
.END
```

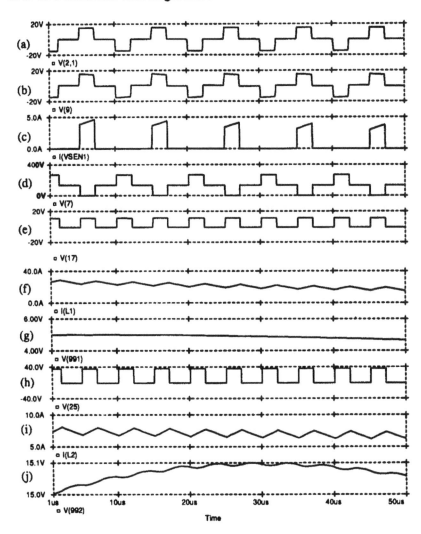

(a) V(2,1)=Driving voltage

(b) V(9)=Gate voltage of M2

(c) I(VSEN1)=Drain current of M2

(d) V(7)=Drain to source voltage of M2

(e) V(17)=Input voltage to L1

(f) I(L1)=Current in L1

(g) V(991)=Output voltage at node 991

(h) V(25)=Input voltage to L2

(i) I(L2)=Current in L2

(j) V(992)=Output voltage at node 992

Fig. A8.1.2   Simulated waveforms of near-ideal Half-Bridge converter.

ATTACHMENT 8.2: SIMULATION OF A MORE REALISTIC* HALF-BRIDGE CONVERTER

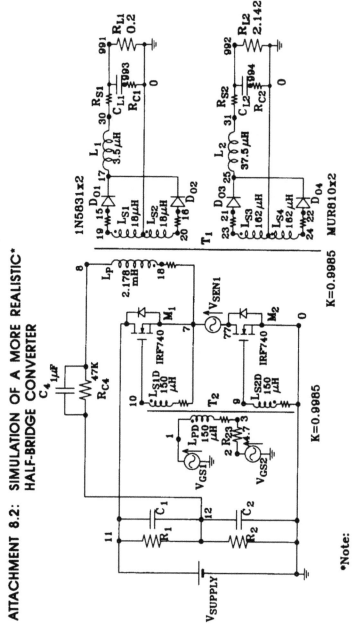

*Note:

For the actual values of effective series resistances of inductors, transformer windings, and output filtering capacitors, refer to the listing given in A8.2

Fig. A8.2.1 A more realistic* Half–Bridge converter model.

A8.2   SIMULATION OF A MORE REALISTIC HALF-BRIDGE CONVERTER

```
HALF BRIDGE CONVERTER (CYCLE-BY-CYCLE SIMULATION)
*MORE REALISTIC SIMULATION
*
*DRIVING STAGE
VGS1 1 0 PULSE(0 15 0.1U 0.1U 0.1U 2.2U 10U)
VGS2 2 0 PULSE(0 15 5.1U 0.1U 0.1U 2.2U 10U)
R23 2 3 4.7
RLPD 4 3 1
LPD 1 4 150U
LS1D 10 50 150U
RLS1D 50 7 1.5
LS2D 60 9 150U
RLS2D 60 0 1.5
K1D LPD LS1D 0.9985
K2D LPD LS2D 0.9985
K3D LS1D LS2D 0.9985
*
*
*POWER STAGE
M1 11 10 7 7 IRF740 (W=.78 L=2U)
M2 77 9 0 0 IRF740 (W=.78 L=2U)
VSEN1 7 77
VSUPPLY 11 0 270V
C1 11 12 470UF IC=135
R1 11 12 100K
C2 12 0 470UF IC=135
R2 12 0 100K
*
*
*POWER TRANSFORMER T1 AND ASSOCIATED CIRCUITS
RLP 7 18 180M
LP 18 8 2.178M
LS1 19 0 18U
RLS1 15 19 3M
LS2 0 20 18U
RLS2 20 16 3M
LS3 23 0 162U
RLS3 21 23 30M
LS4 0 24 162U
RLS4 24 22 30M
K1 LS1 LS2 0.9985
K2 LS1 LS3 0.9985
K3 LS1 LS4 0.9985
K4 LS2 LS3 0.9985
K5 LS2 LS4 0.9985
K6 LS3 LS4 0.9985
K7 LP LS1 0.9985
K8 LP LS2 0.9985
K9 LP LS3 0.9985
K10 LP LS4 0.9985
*
```

```
*COUPLING CIRCUIT
C4 8 12 1U IC=-5
RC4 8 12 47K
*
*
*OUTPUT STAGE 1
DO1 15 17 D1N5831
DO2 16 17 D1N5831
L1 17 30 3.5U IC=25
RS1 30 991 3M
CL1 991 993 470U IC=5
RC1 993 0 15M
RL1 991 0 0.2
*
*
*OUTPUT STAGE 2
DO3 21 25 MUR810
DO4 22 25 MUR810
L2 25 31 37.5U IC=7
RS2 31 992 30M
CL2 992 994 100U IC=15
RC2 994 0 40M
RL2 992 0 2.1428
*
*
*
.MODEL D1N5831   D(IS=89.43U RS=3.991M N=1 XTI=0 EG=1.11
+                CJO=3.488N M=.5044 VJ=.75 FC=.5)
.MODEL MUR810    D(IS=4.433P RS=5.91M N=1 XTI=3 EG=1.11
+                CJO=265.3P M=.4253 VJ=.75 FC=.5 TT=61.66N)
.MODEL IRF740    NMOS(LEVEL=3 GAMMA=0 DELTA=0 ETA=0 THETA=0
+                KAPPA=0 VMAX=0 XJ=0 TOX=100N UO=600
+                PHI=.6 RS=8.563M KP=20.59U VTO=3.657
+                RD=.3915 CBD=1.419N PB=.8 MJ=.5 FC=.5
+                CGSO=1.392N CGDO=146.6P IS=17.65P
+                RDS=1.778MEG  N=1 RG=.9088 TT=570N)
*NOTE: THE IRF740 MODEL USED IN THIS INPUT FILE IS
*      OBTAINED FROM THE COMPONENT LIBRARY OF PSpice.
*      THIS MODEL HAS INCLUDED THE SOURCE TO DRAIN
*      DIODE OF THE TRANSISTOR IN IT. IF RUN BY SPICE,
*      A DIFFERENT MODELING ARRANGEMENT IS REQUIRED.
.PROBE
*THE .PROBE STATEMENT IS FOR PSPICE ONLY.
*DELETE .PROBE STATEMENT IF RUN BY SPICE.
.OPTIONS ITL4=300 ITL5=0
.TRAN 0.25U 50U 1U UIC
.PRINT TRAN I(VSEN1) V(7) V(991)
.PLOT TRAN I(VSEN1) V(7) V(991)
*MORE VARIABLES CAN BE ADDED TO THE .PRINT AND
*.PLOT LISTS AT THE DISCRETION OF THE DESIGNER.
.END
```

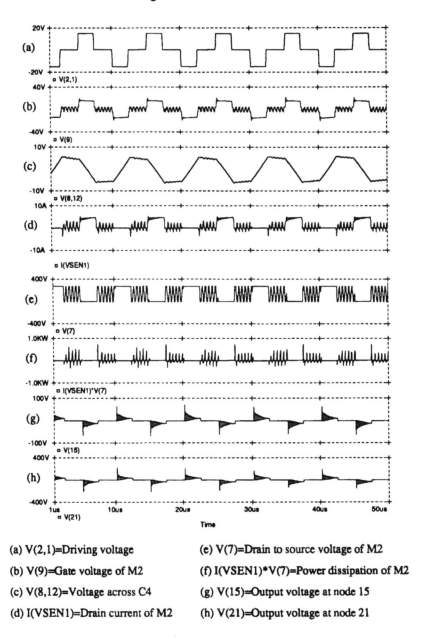

(a) V(2,1)=Driving voltage   (e) V(7)=Drain to source voltage of M2

(b) V(9)=Gate voltage of M2   (f) I(VSEN1)*V(7)=Power dissipation of M2

(c) V(8,12)=Voltage across C4   (g) V(15)=Output voltage at node 15

(d) I(VSEN1)=Drain current of M2   (h) V(21)=Output voltage at node 21

Fig. A8.2.2   Simulated waveforms of more realistic Half-Bridge converter.

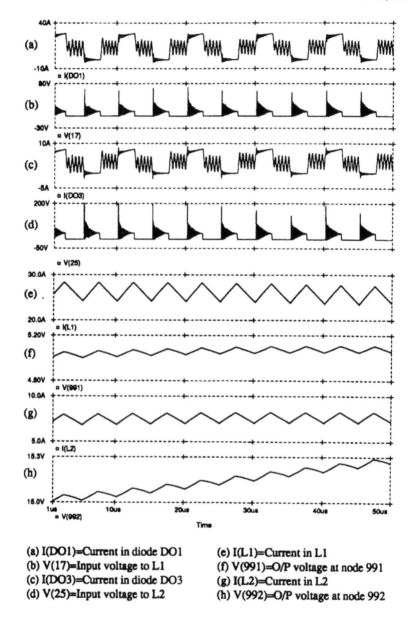

(a) I(DO1)=Current in diode DO1   (e) I(L1)=Current in L1
(b) V(17)=Input voltage to L1   (f) V(991)=O/P voltage at node 991
(c) I(DO3)=Current in diode DO3   (g) I(L2)=Current in L2
(d) V(25)=Input voltage to L2   (h) V(992)=O/P voltage at node 992

Fig. A8.2.3   Simulated waveforms of more realistic Half-Bridge converter.

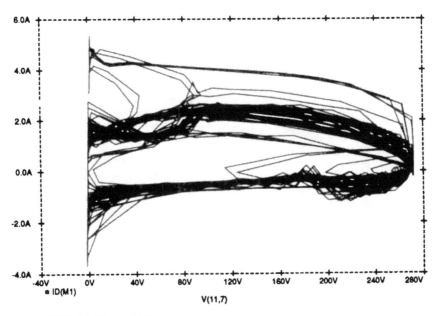

V(11,7)=V_{DS} of M_1
ID(M1)=I_{DS} of M_1
(I_{DS} includes current in source to drain diode.)

Fig. A8.2.4 I_{DS}-V_{DS} trajectory of power transistor M_1.

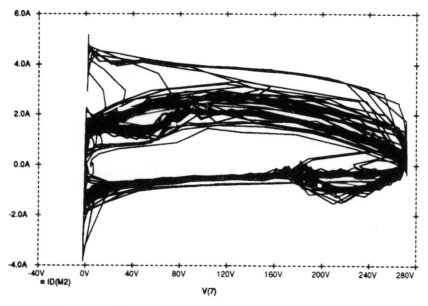

V(7)=V_{DS} of M_2
ID(M2)=I_{DS} of M_2
(I_{DS} includes current in source to drain diode.)

Fig. A8.2.5 I_{DS}-V_{DS} trajectory of power transistor M_2.

ATTACHMENT 8.3: SIMULATION OF FINAL DESIGN OF HALF-BRIDGE CONVERTER

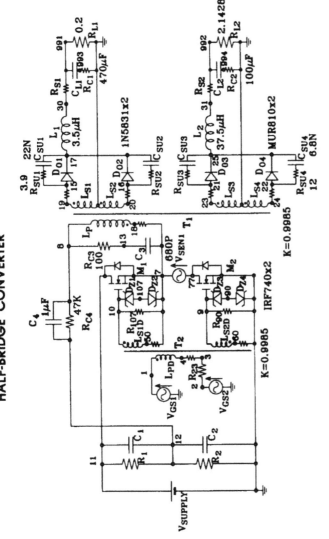

Fig. A8.3.1 Half–Bridge converter model with snubber and clamping circuits.

A8.3   <u>SIMULATION OF FINAL DESIGN OF HALF-BRIDGE
       CONVERTER</u>

```
HALF BRIDGE CONVERTER (CYCLE-BY-CYCLE SIMULATION)
*WITH SNUBBER AND CLAMPING CIRCUITS
*
*DRIVING STAGE
VGS1 1 0 PULSE(0 15 0.1U 0.1U 0.1U 2.2U 10U)
VGS2 2 0 PULSE(0 15 5.1U 0.1U 0.1U 2.2U 10U)
R23 2 3 4.7
RLPD 4 3 1
LPD 1 4 150U
LS1D 10 50 150U
RLS1D 50 7 1.5
LS2D 60 9 150U
RLS2D 60 0 1.5
K1D LPD LS1D 0.9985
K2D LPD LS2D 0.9985
K3D LS1D LS2D 0.9985
DZ1 107 10 D1N965
DZ2 107 7 D1N965
R107 10 7 270
DZ3 90 9 D1N965
DZ4 90 0 D1N965
R90 9 0 270
*
*
*POWER STAGE
M1 11 10 7 7 IRF740 (W=.78 L=2U)
M2 77 9 0 0 IRF740 (W=.78 L=2U)
VSEN1 7 77
VSUPPLY 11 0 270V
C1 11 12 470UF IC=135
R1 11 12 100K
C2 12 0 470UF IC=135
R2 12 0 100K
*
*
*POWER TRANSFORMER T1 AND ASSOCIATED CIRCUITS
RLP 7 18 180M
LP 18 8 2.178M
*
*SNUBBER CIRCUIT
C3 7 13 680P
RC3 13 8 100
*
*COUPLING CIRCUIT
C4 8 12 1U IC=-5
RC4 8 12 47K
*
LS1 19 0 18U
RLS1 15 19 3M
LS2 0 20 18U
```

```
RLS2 20 16 3M
LS3 23 0 162U
RLS3 21 23 30M
LS4 0 24 162U
RLS4 24 22 30M
K1 LS1 LS2 0.9985
K2 LS1 LS3 0.9985
K3 LS1 LS4 0.9985
K4 LS2 LS3 0.9985
K5 LS2 LS4 0.9985
K6 LS3 LS4 0.9985
K7 LP LS1 0.9985
K8 LP LS2 0.9985
K9 LP LS3 0.9985
K10 LP LS4 0.9985
*
*
*OUTPUT STAGE 1
DO1 15 17 D1N5831
RSU1 15 1517 3.9
CSU1 1517 17 22N
DO2 16 17 D1N5831
RSU2 16 1617 3.9
CSU2 1617 17 22N
L1 17 30 3.5U IC=25
RS1 30 991 3M
CL1 991 993 470U IC=5
RC1 993 0 15M
RL1 991 0 0.2
*
*
*OUTPUT STAGE 2
DO3 21 25 MUR810
RSU3 21 2125 12
CSU3 2125 25 6.8N
DO4 22 25 MUR810
RSU4 22 2225 12
CSU4 2225 25 6.8N
L2 25 31 37.5U IC=7
RS2 31 992 30M
CL2 992 994 100U IC=15
RC2 994 0 40M
RL2 992 0 2.1428
*
*
*
.MODEL D1N5831    D(IS=89.43U RS=3.991M N=1 XTI=0 EG=1.11
+                 CJO=3.488N M=.5044 VJ=.75 FC=.5)
.MODEL MUR810     D(IS=4.433P RS=5.91M N=1 XTI=3 EG=1.11
+                 CJO=265.3P M=.4253 VJ=.75 FC=.5 TT=61.66N)
.MODEL IRF740     NMOS(LEVEL=3 GAMMA=0 DELTA=0 ETA=0 THETA=0
+                 KAPPA=0 VMAX=0 XJ=0 TOX=100N UO=600
+                 PHI=.6 RS=8.563M KP=20.59U VTO=3.657
```

```
+                    RD=.3915 CBD=1.419N PB=.8 MJ=.5 FC=.5
+                    CGSO=1.392N CGDO=146.6P IS=17.65P
+                    RDS=1.778MEG  N=1 RG=.9088 TT=570N)
*NOTE: THE IRF740 MODEL USED IN THIS INPUT FILE IS
*      OBTAINED FROM THE COMPONENT LIBRARY OF PSpice.
*      THIS MODEL HAS INCLUDED THE SOURCE TO DRAIN
*      DIODE OF THE TRANSISTOR IN IT. IF RUN BY SPICE,
*      A DIFFERENT MODELING ARRANGEMENT IS REQUIRED.
.MODEL D1N965    D(IS=0.05U BV=15 IBV=0.05U)
.PROBE
*THE .PROBE STATEMENT IS FOR PSPICE ONLY.
*DELETE .PROBE STATEMENT IF RUN BY SPICE.
.OPTIONS ITL4=300 ITL5=0
.TRAN 0.25U 50U 1U UIC
.PRINT TRAN I(VSEN1) V(7) V(991)
.PLOT TRAN I(VSEN1) V(7) V(991)
*MORE VARIABLES CAN BE ADDED TO THE .PRINT AND
*.PLOT LISTS AT THE DISCRETION OF THE DESIGNER.
.END
```

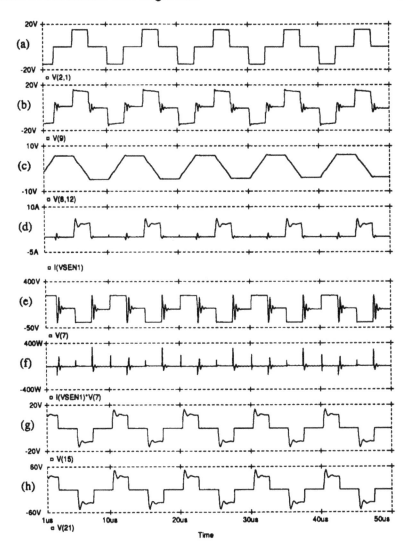

(a) V(2,1)=Driving voltage   (e) V(7)=Drain to source voltage of M2

(b) V(9)=Gate voltage of M2   (f) I(VSEN1)*V(7)=Power dissipation of M2

(c) V(8,12)=Voltage across C4   (g) V(15)=Output Voltage at node 15

(d) I(VSEN1)=Drain current of M2   (h) V(21)=Output Voltage at node 21

Fig. A8.3.2   Simulated waveforms of Half-Bridge converter with snubber and clamping circuits.

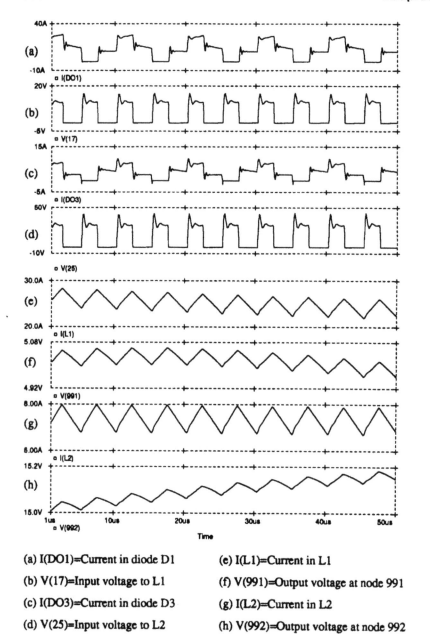

(a) I(DO1)=Current in diode D1   (e) I(L1)=Current in L1

(b) V(17)=Input voltage to L1   (f) V(991)=Output voltage at node 991

(c) I(DO3)=Current in diode D3   (g) I(L2)=Current in L2

(d) V(25)=Input voltage to L2   (h) V(992)=Output voltage at node 992

Fig. A8.3.3   Simulated waveforms of Half-Bridge converter with
                snubber and clamping circuits.

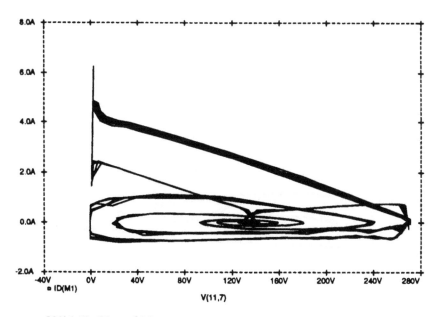

V(11,7)=V$_{DS}$ of M$_1$
ID(M1)=I$_{DS}$ of M$_1$
(I$_{DS}$ includes current in source to drain diode.)

Fig. A8.3.4   I$_{DS}$-V$_{DS}$ trajectory of power transistor M$_1$.

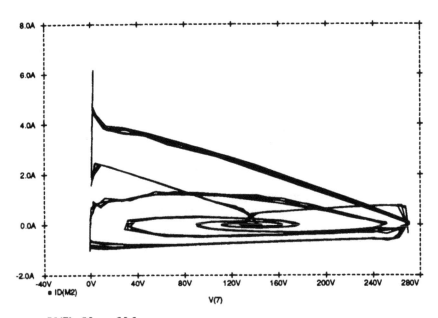

V(7)=V$_{DS}$ of M$_2$
ID(M2)=I$_{DS}$ of M$_2$
(I$_{DS}$ includes current in source to drain diode.)

Fig. A8.3.5   I$_{DS}$-V$_{DS}$ trajectory of power transistor M$_2$.

ATTACHMENT 8.4: SIMULATION OF LOOP-GAIN CHARACTERISTIC $A_L(s)$

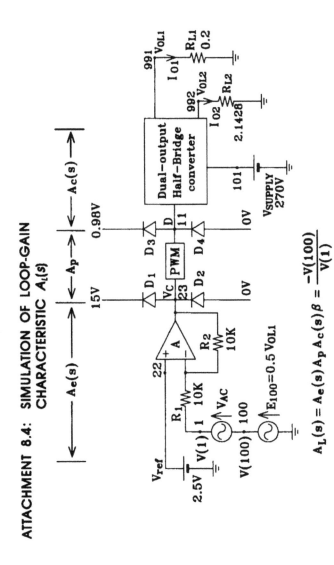

$$A_L(s) = A_e(s)\, A_p\, A_c(s)\, \beta = \frac{-V(100)}{V(1)}$$

Effective $D = 2 \times$ duty cycle of each switching transistor

Fig. A8.4.1 Regulator model for simulation of $A_L(s)$ characteristic.

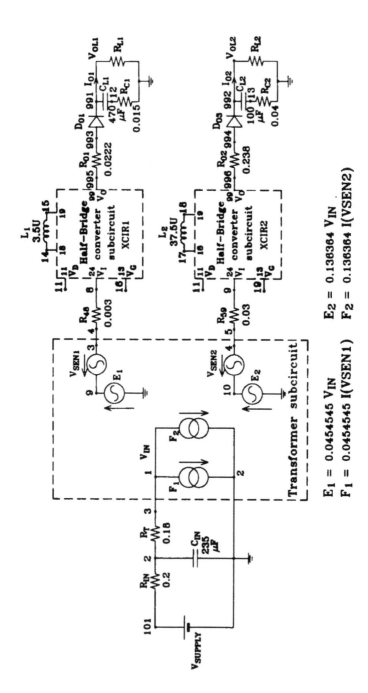

Fig. A8.4.2 Expanded SPICE model of dual−output Half−Bridge converter.

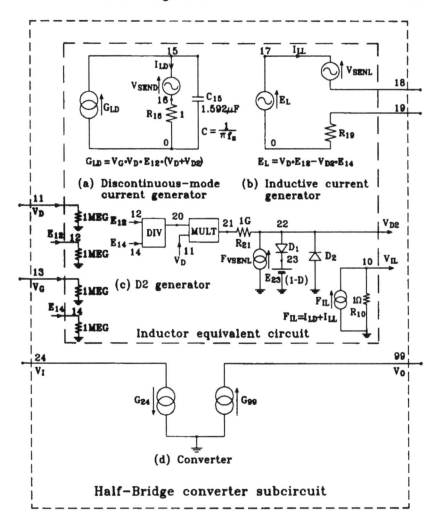

Fig. A8.4.3 Detailed SPICE model of Half–Bridge
converter subcircuit.

A8.4 SIMULATION OF THE LOOP-GAIN CHARACTERISTIC AL(S) (FOR THE FEEDBACK LOOP CONTAINING VO1)

```
SIMULATION OF LOOP-GAIN CHARACTERISTIC AL(S) FOR REGULATOR
*
*
*
.SUBCKT DIV 3 1 9
*DIVIDER SUBCIRCUIT
*V(9) = V(3)/V(1)
R30 3 0 1MEG
R10 1 0 1MEG
E20 2 0 POLY(2) 1 0 9 0  0 0 0 0 1
R20 2 0 1MEG
E90 9 0 POLY(1) 2 3  0 -1MEG
R90 9 0 1MEG
.ENDS DIV
*
*
*
.SUBCKT COMPEN 1 3 9
*V(1) = SIGNAL INPUT
*V(3) = REF VOLTAGE INPUT
*V(9) = OUTPUT
*V70 = VMAX LIMIT
*V60 = VMIN LIMIT
*AMPLIFIER
RIN1 2 0 1G
RIN2 3 0 1G
E8 8 0 POLY(1) 2 3  0 -10K
R89 8 9 50
D1 9 7 DIDEAL
V70 7 0 DC 15
D2 6 9 DIDEAL
V60 6 0 DC 0
*COMPENSATION NETWORK
R1 1 2 10K
R2 2 9 10K
.ENDS COMPEN
*
*
*
.SUBCKT PWM 1 9
* PULSE WIDTH MODULATOR
R10 1 0 1MEG
E8 8 0 POLY(1) 1 0  -0.375 0.4167
R89 8 9 1K
D3 9 4 DIDEAL
VDMAX 4 0 0.98
D4 0 9 DIDEAL
.ENDS PWM
*
*
```

```
*
.SUBCKT TRANSFORMER 1 2 3 4
*V(1,2)=INPUT, V(3)=OUTPUT 1, V(4)=OUTPUT 2
F1 1 2 POLY(1) VSEN1   0 0.0454545
*0.5*NS1/NP = 0.0454545
F2 1 2 POLY(1) VSEN2   0 0.136364
*0.5*NS2/NP = 0.136364
E1 9 0 POLY(1) 1 2   0 0.0454545
VSEN1 9 3
E2 10 0 POLY(1) 1 2   0 0.136364
VSEN2 10 4
.ENDS TRANSFORMER
*
*
*
.SUBCKT CONVERTER 11 13 18 19 24 99
*V(11) = DUTY CYCLE INPUT
*THE EFFECTIVE DUTY CYCLE FOR PUSH-PULL
*CIRCUITS (INCLUDING HALF-BRIDGE AND
*FULL-BRIDGE) IS EQUAL TO TWICE THE
*DUTY CYCLE OF EACH SWITCHING TRANSISTOR
*V(13) = VG INPUT = T/(2L)
*WHERE T IS ONE HALF OF THE SWITCHING PERIOD OF EACH
*SWITCHING TRANSISTOR FOR PUSH-PULL CIRCUITS (INCLUDING
*HALF-BRIDGE AND FULL-BRIDGE)
*V(24) = INPUT VOLTAGE
*V(99) = OUTPUT VOLTAGE
*18,19 = INDUCTOR TERMINALS
*
*START OF INDUCTOR EQUIVALENT CIRCUIT
*V(11) = DUTY CYCLE INPUT
R11 11 0 1MEG
*
*V(12) = VI-VO
R12 12 0 1MEG
*
*V(13) = VG
R13 13 0 1MEG
*
*V(14) = VO
R14 14 0 1MEG
*
*DISCONTINUOUS-MODE CURRENT GENERATOR
GLD 0 15 POLY(1) 100 0   0 1
*GLD = VG*VD*(VI-VO)*(VD+VD2)
*V(15) = VILD
VSEND 15 16
R16 16 0 1
C15 15 0 1.592U
*
E103 103 0 POLY(2) 11 0 12 0   0 0 0 0 1
*V(11) = VD
```

```
*V(12) = (VI-VO)
*V(103) = VD*(VI-VO)
R103 103 0 1MEG
*
E101 101 0 POLY(2) 22 0 11 0  0 1 1
*V(22) = VD2, V(11) = VD
*V(101) = VD+VD2
R101 101 0 1MEG
*
E100 100 0 POLY(3) 13 0 103 0 101 0
+0 0 0 0 0 0 0 0 0 0 0 0 1
*V(100) = VG*VD*(VI-VO)*(VD+VD2)
R100 100 0 1MEG
*
*INDUCTIVE CURRENT GENERATOR
EL 17 0 POLY(1) 110 0   0 1
*V(17) = V(110) = VD*(VI-VO)-VD2*VO
*
VSENL 17 18
R19 19 0 1M
*
E104 104 0 POLY(2) 22 0 14 0   0 0 0 0 1
*V(22) = VD2
*V(14) = VO
*V(104) = VD2*VO
R104 104 0 1MEG
*
E110 110 0 POLY(2) 103 0 104 0   0 1 -1
*V(110) = EL = VD*(VI-VO)-VD2*VO
R110 110 0 1MEG
*
*TOTAL INDUCTOR CURRENT GENERATOR
FIL 0 10 POLY(2) VSEND VSENL   0 1 1
R10 10 0 1
*V(10) = VILD+VILL = VIL(REPRESENTING IL)
*
*D2 GENERATOR
X1 12 14 20 DIV
*V(12) = VI-VO
*V(14) = VO
*V(20) = (VI-VO)/VO
E21 21 0 POLY(2) 20 0 11 0   0 0 0 0 1
*V(21) = VD*(VI-VO)/VO
R21 21 22 1G
FVSENL 0 22 POLY(1) VSENL   0 1
*FVSENL = ILL
D1 22 23 DIDEAL
E23 23 0 POLY(1) 11 0   1 -1
*V(11) = VD
*V(23) = 1-VD
D2 0 22 DIDEAL
*V(22) = VD2
*END OF INDUCTOR EQUIVALENT CIRCUIT
```

```
*
*HALF-BRIDGE CONVERTER
*INPUT CIRCUIT
X2 11 101 300 DIV
*V(11) = VD, V(101) = VD+VD2
*V(300) = VD/(VD+VD2)
*
G24 24 0 POLY(2) 10 0 300 0  0 0 0 0 1
*V(10) = VIL
*G24 = VIL*VD/(VD+VD2)
*
*OUTPUT CIRCUIT
G99 0 99 POLY(1) 10 0  0 1
*V(10) = VIL
*
*V(12) INPUT
E12 12 0 POLY(2) 24 0 99 0  0 1 -1
*
*V(14) INPUT
E14 14 0 POLY(1) 99 0  0 1
*
.ENDS CONVERTER
*
*
*
*MAIN CIRCUIT
VSUPPLY 101 0 DC 270V
*V(101) = SUPPLY VOLTAGE
RIN 101 2 0.2
CIN 2 0 235U
RT 2 3 180M
X1 3 0 4 5 TRANSFORMER
*V(3,0) = TRANSFORMER PRIMARY INPUT
*V(4), V(5) = TRANSFORMER SECONDARY OUTPUTS
R48 4 8 3M
R59 5 9 30M
*
*
*HALF-BRIDGE CONVERTER CIRCUIT 1
XCIR1 11 16 14 15 8 995 CONVERTER
*V(11) = DUTY CYCLE INPUT
*V(16) = VG INPUT
*14, 15 = INDUCTOR TERMINALS
*V(8) = CONVERTER INPUT VOLTAGE
*V(995) = CONVERTER OUTPUT VOLTAGE
L1 14 15 3.5U
VG1 16 0 0.714285714
RO1 995 993 22.2M
DO1 993 991 D1N5831
CL1 991 12 470U
RC1 12 0 15M
RL1 991 0 0.2
*V(991) = OUTPUT 1 = VOL1
```

```
*
*
* FEEDBACK LOOP
E100 100 0 POLY(1) 991 0  0 0.5
*V(100) = 0.5*VOL1
VAC 1 100 AC 1
XCOMP 1 22 23 COMPEN
VREF 22 0 DC 2.5V
*V(22) = REFERENCE VOLTAGE
XPWM 23 11 PWM
*
*
*HALF-BRIDGE CONVERTER CIRCUIT 2
XCIR2 11 19 17 18 9 996 CONVERTER
*V(11) = DUTY CYCLE INPUT
*V(19) = VG INPUT
*17, 18 = INDUCTOR TERMINALS
*V(9) = CONVERTER INPUT VOLTAGE
*V(996) = CONVERTER OUTPUT VOLTAGE
L2 17 18 37.5U
VG2 19 0 0.0666666667
RO2 996 994 0.238
DO3 994 992 MUR810
CL2 992 13 100U
RC2 13 0 40M
RL2 992 0 2.1428
*V(992) = OUTPUT 2 = VOL2
*
*
*
.MODEL DIDEAL   D(N=1M)
.MODEL D1N5831   D(IS=89.43U RS=3.991M N=1 XTI=0 EG=1.11
+                CJO=3.488N M=.5044 VJ=.75 FC=.5)
.MODEL MUR810   D(IS=4.433P RS=5.91M N=1 XTI=3 EG=1.11
+                CJO=265.3P M=.4253 VJ=.75 FC=.5 TT=61.66N)
.PROBE
*THE .PROBE STATEMENT IS FOR PSPICE ONLY.
*DELETE .PROBE STATEMENT IF RUN BY SPICE.
.OPTIONS ITL1=300
.AC DEC 10 1 100K
.PRINT AC VDB(100) VDB(1) VP(100) VP(1)
.PLOT AC VDB(100) VDB(1) VP(100) VP(1)
*THE OUTPUT SHOULD BE PLOTTED AS :
*AMPLITUDE OF AL(s)=VDB(100)-VDB(1)
*PHASE LAG OF AL(s)=VP(100)-VP(1)-180 DEGREES
.END
```

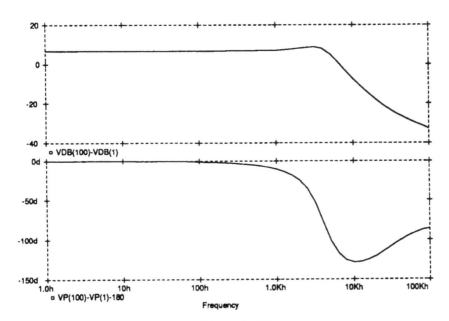

VDB(100)-VDB(1)=Loop gain of $A_L(s)$ in dB

VP(100)-VP(1)-180=Phase lag of $A_L(s)$ in degree

Fig. A8.4.4   Simulated result of the loop-gain characteristic $A_L(s)$
            (for the feedback loop containing V_{OL1}).

ATTACHMENT 8.5:   SIMULATION OF STEADY-STATE LINE AND
LOAD REGULATION CHARACTERISTICS

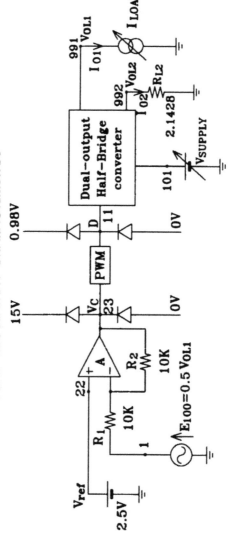

Fig. A8.5.1   Simulation for line and load regulation
characteristics for R1 = R2 = 10K.

A8.5 <u>SIMULATION OF STEADY-STATE LINE AND LOAD REGULATION</u>
<u>CHARACTERISTICS</u>

```
SIMULATION OF STEADY-STATE LINE AND LOAD REGULATION
*(FOR R1=R2=10K)
*THIS SIMULATION FILE IS FOR VSUPPLY=374V ONLY
*A SIMILAR SIMULATION SHOULD ALSO BE PERFORMED
*FOR VSUPPLY=193.5V
*
*
*
.SUBCKT DIV 3 1 9
*DIVIDER SUBCIRCUIT
*V(9) = V(3)/V(1)
R30 3 0 1MEG
R10 1 0 1MEG
E20 2 0 POLY(2) 1 0 9 0   0 0 0 0 1
R20 2 0 1MEG
E90 9 0 POLY(1) 2 3  0 -1MEG
R90 9 0 1MEG
.ENDS DIV
*
*
*
.SUBCKT COMPEN 1 3 9
*V(1) = SIGNAL INPUT
*V(3) = REF VOLTAGE INPUT
*V(9) = OUTPUT
*V70 = VMAX LIMIT
*V60 = VMIN LIMIT
*AMPLIFIER
RIN1 2 0 1G
RIN2 3 0 1G
E8 8 0 POLY(1) 2 3  0 -10K
R89 8 9 50
D1 9 7 DIDEAL
V70 7 0 DC 15
D2 6 9 DIDEAL
V60 6 0 DC 0
*COMPENSATION NETWORK
R1 1 2 10K
R2 2 9 10K
.ENDS COMPEN
*
*
*
.SUBCKT PWM 1 9
* PULSE WIDTH MODULATOR
R10 1 0 1MEG
E8 8 0 POLY(1) 1 0  -0.375 0.4167
R89 8 9 1K
D3 9 4 DIDEAL
VDMAX 4 0 0.98
```

```
D4 0 9 DIDEAL
.ENDS PWM
*
*
*
.SUBCKT TRANSFORMER 1 2 3 4
*V(1,2)=INPUT, V(3)=OUTPUT 1, V(4)=OUTPUT 2
F1 1 2 POLY(1) VSEN1  0 0.0454545
*0.5*NS1/NP = 0.0454545
F2 1 2 POLY(1) VSEN2  0 0.136364
*0.5*NS2/NP = 0.136364
E1 9 0 POLY(1) 1 2  0 0.0454545
VSEN1 9 3
E2 10 0 POLY(1) 1 2  0 0.136364
VSEN2 10 4
.ENDS TRANSFORMER
*
*
*
.SUBCKT CONVERTER 11 13 18 19 24 99
*V(11) = DUTY CYCLE INPUT
*THE EFFECTIVE DUTY CYCLE FOR PUSH-PULL
*CIRCUITS (INCLUDING HALF-BRIDGE AND
*FULL-BRIDGE) IS EQUAL TO TWICE THE
*DUTY CYCLE OF EACH SWITCHING TRANSISTOR
*V(13) = VG INPUT = T/(2L)
*WHERE T IS ONE HALF OF THE SWITCHING PERIOD OF EACH
*SWITCHING TRANSISTOR FOR PUSH-PULL CIRCUITS (INCLUDING
*HALF-BRIDGE AND FULL-BRIDGE)
*V(24) = INPUT VOLTAGE
*V(99) = OUTPUT VOLTAGE
*18,19 = INDUCTOR TERMINALS
*
*START OF INDUCTOR EQUIVALENT CIRCUIT
*V(11) = DUTY CYCLE INPUT
R11 11 0 1MEG
*
*V(12) = VI-VO
R12 12 0 1MEG
*
*V(13) = VG
R13 13 0 1MEG
*
*V(14) = VO
R14 14 0 1MEG
*
*DISCONTINUOUS-MODE CURRENT GENERATOR
GLD 0 15 POLY(1) 100 0  0 1
*GLD = VG*VD*(VI-VO)*(VD+VD2)
*V(15) = VILD
VSEND 15 16
R16 16 0 1
C15 15 0 1.592U
```

```
*
*
E103 103 0 POLY(2) 11 0 12 0   0 0 0 0 1
*V(11) = VD
*V(12) = (VI-VO)
*V(103) = VD*(VI-VO)
R103 103 0 1MEG
*
E101 101 0 POLY(2) 22 0 11 0   0 1 1
*V(22) = VD2, V(11) = VD
*V(101) = VD+VD2
R101 101 0 1MEG
*
E100 100 0 POLY(3) 13 0 103 0 101 0
+0 0 0 0 0 0 0 0 0 0 0 0 0 0 1
*V(100) = VG*VD*(VI-VO)*(VD+VD2)
R100 100 0 1MEG
*
*INDUCTIVE CURRENT GENERATOR
EL 17 0 POLY(1) 110 0   0 1
*V(17) = V(110) = VD*(VI-VO)-VD2*VO
*
VSENL 17 18
R19 19 0 1M
*
E104 104 0 POLY(2) 22 0 14 0   0 0 0 0 1
*V(22) = VD2
*V(14) = VO
*V(104) = VD2*VO
R104 104 0 1MEG
*
E110 110 0 POLY(2) 103 0 104 0   0 1 -1
*V(110) = EL = VD*(VI-VO)-VD2*VO
R110 110 0 1MEG
*
*TOTAL INDUCTOR CURRENT GENERATOR
FIL 0 10 POLY(2) VSEND VSENL   0 1 1
R10 10 0 1
*V(10) = VILD+VILL = VIL(REPRESENTING IL)
*
*D2 GENERATOR
X1 12 14 20 DIV
*V(12) = VI-VO
*V(14) = VO
*V(20) = (VI-VO)/VO
E21 21 0 POLY(2) 20 0 11 0   0 0 0 0 1
*V(21) = VD*(VI-VO)/VO
R21 21 22 1G
FVSENL 0 22 POLY(1) VSENL   0 1
*FVSENL = ILL
D1 22 23 DIDEAL
E23 23 0 POLY(1) 11 0   1 -1
*V(11) = VD
```

```
*V(23) = 1-VD
D2 0 22 DIDEAL
*V(22) = VD2
*END OF INDUCTOR EQUIVALENT CIRCUIT
*
*HALF-BRIDGE CONVERTER
*INPUT CIRCUIT
X2 11 101 300 DIV
*V(11) = VD, V(101) = VD+VD2
*V(300) = VD/(VD+VD2)
*
G24 24 0 POLY(2) 10 0 300 0  0 0 0 0 1
*V(10) = VIL
*G24 = VIL*VD/(VD+VD2)
*
*OUTPUT CIRCUIT
G99 0 99 POLY(1) 10 0  0 1
*V(10) = VIL
*
*V(12) INPUT
E12 12 0 POLY(2) 24 0 99 0  0 1 -1
*
*V(14) INPUT
E14 14 0 POLY(1) 99 0  0 1
*
.ENDS CONVERTER
*
*
*
*MAIN CIRCUIT
VSUPPLY 101 0 DC 374V
*V(101) = SUPPLY VOLTAGE
RIN 101 2 0.2
CIN 2 0 235U
RT 2 3 180M
X1 3 0 4 5 TRANSFORMER
*V(3,0) = TRANSFORMER PRIMARY INPUT
*V(4), V(5) = TRANSFORMER SECONDARY OUTPUTS
R48 4 8 3M
R59 5 9 30M
*
*
*HALF-BRIDGE CONVERTER CIRCUIT 1
XCIR1 11 16 14 15 8 995 CONVERTER
*V(11) = DUTY CYCLE INPUT
*V(16) = VG INPUT
*14, 15 = INDUCTOR TERMINALS
*V(8) = CONVERTER INPUT VOLTAGE
*V(995) = CONVERTER OUTPUT VOLTAGE
L1 14 15 3.5U
VG1 16 0 0.714285714
RO1 995 993 22.2M
DO1 993 991 D1N5831
```

```
CL1 991 12 470U
RC1 12 0 15M
RL1 991 0 1K
ILOAD 991 0
*V(991) = OUTPUT 1 = VOL1
*
*
*FEEDBACK LOOP
E100 100 0 POLY(1) 991 0  0 0.5
*V(100) = 0.5*VOL1
VAC 1 100
XCOMP 1 22 23 COMPEN
VREF 22 0 DC 2.5V
*V(22) = REFERENCE VOLTAGE
XPWM 23 11 PWM
*
*
*HALF-BRIDGE CONVERTER CIRCUIT 2
XCIR2 11 19 17 18 9 996 CONVERTER
*V(11) = DUTY CYCLE INPUT
*V(19) = VG INPUT
*17, 18 = INDUCTOR TERMINALS
*V(9) = CONVERTER INPUT VOLTAGE
*V(996) = CONVERTER OUTPUT VOLTAGE
L2 17 18 37.5U
VG2 19 0 0.0666666667
RO2 996 994 0.238
DO3 994 992 MUR810
CL2 992 13 100U
RC2 13 0 40M
RL2 992 0 2.1428
*V(992) = OUTPUT 2 = VOL2
*
*
*
.MODEL DIDEAL D (N=1M)
.MODEL D1N5831   D(IS=89.43U RS=3.991M N=1 XTI=0 EG=1.11
+                CJO=3.488N M=.5044 VJ=.75 FC=.5)
.MODEL MUR810    D(IS=4.433P RS=5.91M N=1 XTI=3 EG=1.11
+                CJO=265.3P M=.4253 VJ=.75 FC=.5 TT=61.66N)
.PROBE
*THE .PROBE STATEMENT IS FOR PSPICE ONLY.
*DELETE .PROBE STATEMENT IF RUN BY SPICE.
.DC ILOAD 0 25 0.25
.PRINT DC V(991)
.PLOT DC V(991)
.OPTIONS ITL1=300
.END
```

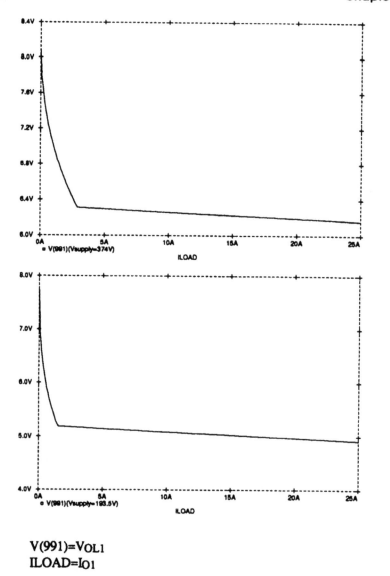

$$V(991) = V_{OL1}$$
$$ILOAD = I_{O1}$$

Fig. A8.5.2 Line and load regulation characteristics for $R1 = R2 = 10K$.

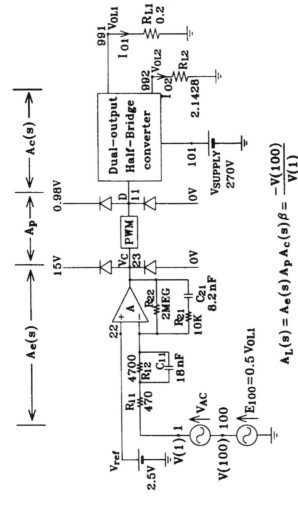

Fig. A8.6.1 Regulator with a 2-pole 2-zero compensation
network.

A8.6 <u>SIMULATION OF THE LOOP-GAIN CHARACTERISTIC WITH</u>
     <u>2-POLE 2-ZERO COMPENSATION</u>

```
SIMULATION OF THE LOOP-GAIN CHARACTERISTIC
*WITH 2-POLE 2-ZERO COMPENSATION
*
*
*
.SUBCKT DIV 3 1 9
*DIVIDER SUBCIRCUIT
*V(9) = V(3)/V(1)
R30 3 0 1MEG
R10 1 0 1MEG
E20 2 0 POLY(2) 1 0 9 0  0 0 0 0 1
R20 2 0 1MEG
E90 9 0 POLY(1) 2 3  0 -1MEG
R90 9 0 1MEG
.ENDS DIV
*
*
*
.SUBCKT COMPEN 1 3 9
*V(1) = SIGNAL INPUT
*V(3) = REF VOLTAGE INPUT
*V(9) = OUTPUT
*V70 = VMAX LIMIT
*V60 = VMIN LIMIT
*AMPLIFIER
RIN1 2 0 1G
RIN2 3 0 1G
E8 8 0 POLY(1) 2 3  0 -10K
R89 8 9 50
D1 9 7 DIDEAL
V70 7 0 DC 15
D2 6 9 DIDEAL
V60 6 0 DC 0
*COMPENSATION NETWORK
R11 1 4 470
C11 4 2 18N
R12 4 2 4700
R21 2 5 10K
C21 5 9 8.2N
R22 2 9 2MEG
.ENDS COMPEN
*
*
*
.SUBCKT PWM 1 9
* PULSE WIDTH MODULATOR
R10 1 0 1MEG
E8 8 0 POLY(1) 1 0  -0.375 0.4167
R89 8 9 1K
D3 9 4 DIDEAL
```

```
VDMAX 4 0 0.98
D4 0 9 DIDEAL
.ENDS PWM
*
*
*
.SUBCKT TRANSFORMER 1 2 3 4
*V(1,2)=INPUT, V(3)=OUTPUT 1, V(4)=OUTPUT 2
F1 1 2 POLY(1) VSEN1  0 0.0454545
*0.5*NS1/NP = 0.0454545
F2 1 2 POLY(1) VSEN2  0 0.136364
*0.5*NS2/NP = 0.136364
E1 9 0 POLY(1) 1 2  0 0.0454545
VSEN1 9 3
E2 10 0 POLY(1) 1 2  0 0.136364
VSEN2 10 4
.ENDS TRANSFORMER
*
*
*
.SUBCKT CONVERTER 11 13 18 19 24 99
*V(11) = DUTY CYCLE INPUT
*THE EFFECTIVE DUTY CYCLE FOR PUSH-PULL
*CIRCUITS (INCLUDING HALF-BRIDGE AND
*FULL-BRIDGE) IS EQUAL TO TWICE THE
*DUTY CYCLE OF EACH SWITCHING TRANSISTOR
*V(13) = VG INPUT = T/(2L)
*WHERE T IS ONE HALF OF THE SWITCHING PERIOD OF EACH
*SWITCHING TRANSISTOR FOR PUSH-PULL CIRCUITS (INCLUDING
*HALF-BRIDGE AND FULL-BRIDGE)
*V(24) = INPUT VOLTAGE
*V(99) = OUTPUT VOLTAGE
*18,19 = INDUCTOR TERMINALS
*
*START OF INDUCTOR EQUIVALENT CIRCUIT
*V(11) = DUTY CYCLE INPUT
R11 11 0 1MEG
*
*V(12) = VI-VO
R12 12 0 1MEG
*
*V(13) = VG
R13 13 0 1MEG
*
*V(14) = VO
R14 14 0 1MEG
*
*
*DISCONTINUOUS-MODE CURRENT GENERATOR
GLD 0 15 POLY(1) 100 0  0 1
*GLD = VG*VD*(VI-VO)*(VD+VD2)
*V(15) = VILD
VSEND 15 16
```

```
R16 16 0 1
C15 15 0 1.592U
*
E103 103 0 POLY(2) 11 0 12 0   0 0 0 0 1
*V(11) = VD
*V(12) = (VI-VO)
*V(103) = VD*(VI-VO)
R103 103 0 1MEG
*
E101 101 0 POLY(2) 22 0 11 0   0 1 1
*V(22) = VD2, V(11) = VD
*V(101) = VD+VD2
R101 101 0 1MEG
*
E100 100 0 POLY(3) 13 0 103 0 101 0
+0 0 0 0 0 0 0 0 0 0 0 0 1
*V(100) = VG*VD*(VI-VO)*(VD+VD2)
R100 100 0 1MEG
*
*INDUCTIVE CURRENT GENERATOR
EL 17 0 POLY(1) 110 0   0 1
*V(17) = V(110) = VD*(VI-VO)-VD2*VO
*
VSENL 17 18
R19 19 0 1M
*
E104 104 0 POLY(2) 22 0 14 0   0 0 0 0 1
*V(22) = VD2
*V(14) = VO
*V(104) = VD2*VO
R104 104 0 1MEG
*
E110 110 0 POLY(2) 103 0 104 0   0 1 -1
*V(110) = EL = VD*(VI-VO)-VD2*VO
R110 110 0 1MEG
*
*TOTAL INDUCTOR CURRENT GENERATOR
FIL 0 10 POLY(2) VSEND VSENL   0 1 1
R10 10 0 1
*V(10) = VILD+VILL = VIL(REPRESENTING IL)
*
*D2 GENERATOR
X1 12 14 20 DIV
*V(12) = VI-VO
*V(14) = VO
*V(20) = (VI-VO)/VO
E21 21 0 POLY(2) 20 0 11 0   0 0 0 0 1
*V(21) = VD*(VI-VO)/VO
R21 21 22 1G
FVSENL 0 22 POLY(1) VSENL   0 1
*FVSENL = ILL
D1 22 23 DIDEAL
E23 23 0 POLY(1) 11 0   1 -1
```

```
*V(11) = VD
*V(23) = 1-VD
D2 0 22 DIDEAL
*V(22) = VD2
*END OF INDUCTOR EQUIVALENT CIRCUIT
*
*HALF-BRIDGE CONVERTER
*INPUT CIRCUIT
X2 11 101 300 DIV
*V(11) = VD, V(101) = VD+VD2
*V(300) = VD/(VD+VD2)
*
G24 24 0 POLY(2) 10 0 300 0  0 0 0 0 1
*V(10) = VIL
*G24 = VIL*VD/(VD+VD2)
*
*OUTPUT CIRCUIT
G99 0 99 POLY(1) 10 0  0 1
*V(10) = VIL
*
*V(12) INPUT
E12 12 0 POLY(2) 24 0 99 0  0 1 -1
*
*V(14) INPUT
E14 14 0 POLY(1) 99 0  0 1
*
.ENDS CONVERTER
*
*
*
*MAIN CIRCUIT
VSUPPLY 101 0 DC 270V
*V(101) = SUPPLY VOLTAGE
RIN 101 2 0.2
CIN 2 0 235U
RT 2 3 180M
X1 3 0 4 5 TRANSFORMER
*V(3,0) = TRANSFORMER PRIMARY INPUT
*V(4), V(5) = TRANSFORMER SECONDARY OUTPUTS
R48 4 8 3M
R59 5 9 30M
*
*
*HALF-BRIDGE CONVERTER CIRCUIT 1
XCIR1 11 16 14 15 8 995 CONVERTER
*V(11) = DUTY CYCLE INPUT
*V(16) = VG INPUT
*14, 15 = INDUCTOR TERMINALS
*V(8) = CONVERTER INPUT VOLTAGE
*V(995) = CONVERTER OUTPUT VOLTAGE
L1 14 15 3.5U
VG1 16 0 0.714285714
RO1 995 993 22.2M
```

```
DO1 993 991 D1N5831
CL1 991 12 470U
RC1 12 0 15M
RL1 991 0 0.2
*V(991) = OUTPUT 1 = VOL1
*
*
*FEEDBACK LOOP
E100 100 0 POLY(1) 991 0  0 0.5
*V(100) = 0.5*VOL1
VAC 1 100 AC 1
XCOMP 1 22 23 COMPEN
VREF 22 0 DC 2.5V
*V(22) = REFERENCE VOLTAGE
XPWM 23 11 PWM
*
*
*HALF-BRIDGE CONVERTER CIRCUIT 2
XCIR2 11 19 17 18 9 996 CONVERTER
*V(11) = DUTY CYCLE INPUT
*V(19) = VG INPUT
*17, 18 = INDUCTOR TERMINALS
*V(9) = CONVERTER INPUT VOLTAGE
*V(996) = CONVERTER OUTPUT VOLTAGE
L2 17 18 37.5U
VG2 19 0 0.0666666667
RO2 996 994 0.238
DO3 994 992 MUR810
CL2 992 13 100U
RC2 13 0 40M
RL2 992 0 2.1428
*V(992) = OUTPUT 2 = VOL2
*
*
*
.MODEL DIDEAL  D(N=1M)
.MODEL D1N5831   D(IS=89.43U RS=3.991M N=1 XTI=0 EG=1.11
+              CJO=3.488N M=.5044 VJ=.75 FC=.5)
.MODEL MUR810   D(IS=4.433P RS=5.91M N=1 XTI=3 EG=1.11
+              CJO=265.3P M=.4253 VJ=.75 FC=.5 TT=61.66N)
.AC DEC 10 10M 100K
.PROBE
*THE .PROBE STATEMENT IS FOR PSPICE ONLY.
*DELETE .PROBE STATEMENT IF RUN BY SPICE.
.PRINT AC VDB(100) VDB(1) VP(100) VP(1)
.PLOT AC VDB(100) VDB(1) VP(100) VP(1)
*THE OUTPUT SHOULD BE PLOTTED AS :
*AMPLITUDE OF AL(s)=VDB(100)-VDB(1)
*PHASE LAG OF AL(s)=VP(100)-VP(1)-180 DEGREES
.OPTIONS ITL1=300
.END
```

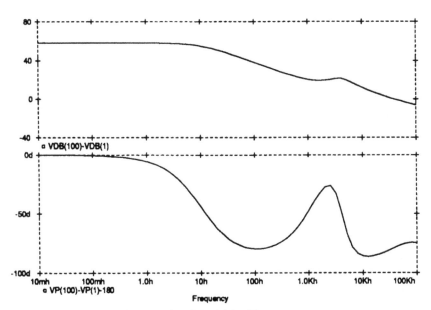

VDB(100)-VDB(1)=Loop gain of $A_L(s)$ in dB

VP(100)-VP(1)-180=Phase lag of $A_L(s)$ in degree

Fig. A8.6.2 Simulated result of the loop-gain characteristic $A_L(s)$
(with compensation network and $R_{L1}=0.2\Omega$).

ATTACHMENT 8.7: TRANSIENT SIMULATION OF FINAL DESIGN
OF REGULATOR

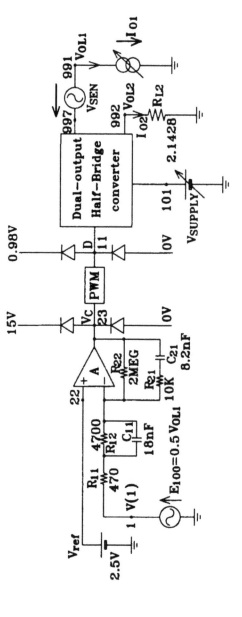

Fig. A8.7.1 Final regulator design and transient simulation.

A8.7 <u>TRANSIENT SIMULATION OF FINAL DESIGN OF REGULATOR</u>
<u>(WITH ACTIVE LOAD AT VOL1)</u>

```
TRANSIENT SIMULATION FOR THE FINAL REGULATOR DESIGN
*(WITH ACTIVE LOAD AT VOL1)
*(RESULTS ARE SHOWN IN FIG. A8.7.2)
*A SIMILAR SIMULATION SHOULD BE REPEATED WITH AN ACTIVE
*LOAD AT VOL2 IN ORDER TO GET THE RESULTS SHOWN IN
*FIG. A8.7.3
*
.SUBCKT DIV 3 1 9
*DIVIDER SUBCIRCUIT
*V(9) = V(3)/V(1)
R30 3 0 1MEG
R10 1 0 1MEG
E20 2 0 POLY(2) 1 0 9 0  0 0 0 0 1
R20 2 0 1MEG
E90 9 0 POLY(1) 2 3  0 -1MEG
R90 9 0 1MEG
.ENDS DIV
*
*
*
.SUBCKT COMPEN 1 3 9
*V(1) = SIGNAL INPUT
*V(3) = REF VOLTAGE INPUT
*V(9) = OUTPUT
*V70 = VMAX LIMIT
*V60 = VMIN LIMIT
*AMPLIFIER
RIN1 2 0 1G
RIN2 3 0 1G
E8 8 0 POLY(1) 2 3  0 -10K
R89 8 9 50
D1 9 7 DIDEAL
V70 7 0 DC 15
D2 6 9 DIDEAL
V60 6 0 DC 0
*COMPENSATION NETWORK
R11 1 4 470
C11 4 2 18N
R12 4 2 4700
R21 2 5 10K
C21 5 9 8.2N
R22 2 9 2MEG
.ENDS COMPEN
*
*
*
.SUBCKT PWM 1 9
* PULSE WIDTH MODULATOR
R10 1 0 1MEG
E8 8 0 POLY(1) 1 0  -0.375 0.4167
```

```
R89 8 9 1K
D3 9 4 DIDEAL
VDMAX 4 0 0.98
D4 0 9 DIDEAL
.ENDS PWM
*
*
*
.SUBCKT TRANSFORMER 1 2 3 4
*V(1,2)=INPUT, V(3)=OUTPUT 1, V(4)=OUTPUT 2
F1 1 2 POLY(1) VSEN1  0 0.0454545
*0.5*NS1/NP = 0.0454545
F2 1 2 POLY(1) VSEN2  0 0.136364
*0.5*NS2/NP = 0.136364
E1 9 0 POLY(1) 1 2  0 0.0454545
VSEN1 9 3
E2 10 0 POLY(1) 1 2  0 0.136364
VSEN2 10 4
.ENDS TRANSFORMER
*
*
*
.SUBCKT CONVERTER 11 13 18 19 24 99
*V(11) = DUTY CYCLE INPUT
*THE EFFECTIVE DUTY CYCLE FOR PUSH-PULL
*CIRCUITS (INCLUDING HALF-BRIDGE AND
*FULL-BRIDGE) IS EQUAL TO TWICE THE
*DUTY CYCLE OF EACH SWITCHING TRANSISTOR
*V(13) = VG INPUT = T/(2L)
*WHERE T IS ONE HALF OF THE SWITCHING PERIOD OF EACH
*SWITCHING TRANSISTOR FOR PUSH-PULL CIRCUITS (INCLUDING
*HALF-BRIDGE AND FULL-BRIDGE)
*V(24) = INPUT VOLTAGE
*V(99) = OUTPUT VOLTAGE
*18,19 = INDUCTOR TERMINALS
*
*START OF INDUCTOR EQUIVALENT CIRCUIT
*V(11) = DUTY CYCLE INPUT
R11 11 0 1MEG
*
*V(12) = VI-VO
R12 12 0 1MEG
*
*V(13) = VG
R13 13 0 1MEG
*
*V(14) = VO
R14 14 0 1MEG
*
*
*DISCONTINUOUS-MODE CURRENT GENERATOR
GLD 0 15 POLY(1) 100 0  0 1
```

```
*GLD = VG*VD*(VI-VO)*(VD+VD2)
*V(15) = VILD
VSEND 15 16
R16 16 0 1
C15 15 0 1.592U
*
E103 103 0 POLY(2) 11 0 12 0   0 0 0 0 1
*V(11) = VD
*V(12) = (VI-VO)
*V(103) = VD*(VI-VO)
R103 103 0 1MEG
*
E101 101 0 POLY(2) 22 0 11 0   0 1 1
*V(22) = VD2, V(11) = VD
*V(101) = VD+VD2
R101 101 0 1MEG
*
E100 100 0 POLY(3) 13 0 103 0 101 0
+0 0 0 0 0 0 0 0 0 0 0 0 0 0 1
*V(100) = VG*VD*(VI-VO)*(VD+VD2)
R100 100 0 1MEG
*
*INDUCTIVE CURRENT GENERATOR
EL 17 0 POLY(1) 110 0   0 1
*V(17) = V(110) = VD*(VI-VO)-VD2*VO
*
VSENL 17 18
R19 19 0 1M
*
E104 104 0 POLY(2) 22 0 14 0   0 0 0 0 1
*V(22) = VD2
*V(14) = VO
*V(104) = VD2*VO
R104 104 0 1MEG
*
E110 110 0 POLY(2) 103 0 104 0   0 1 -1
*V(110) = EL = VD*(VI-VO)-VD2*VO
R110 110 0 1MEG
*
*TOTAL INDUCTOR CURRENT GENERATOR
FIL 0 10 POLY(2) VSEND VSENL   0 1 1
R10 10 0 1
*V(10) = VILD+VILL = VIL(REPRESENTING IL)
*
*D2 GENERATOR
X1 12 14 20 DIV
*V(12) = VI-VO
*V(14) = VO
*V(20) = (VI-VO)/VO
E21 21 0 POLY(2) 20 0 11 0   0 0 0 0 1
*V(21) = VD*(VI-VO)/VO
R21 21 22 1G
FVSENL 0 22 POLY(1) VSENL   0 1
```

```
*FVSENL = ILL
D1 22 23 DIDEAL
E23 23 0 POLY(1) 11 0  1 -1
*V(11) = VD
*V(23) = 1-VD
D2 0 22 DIDEAL
*V(22) = VD2
*END OF INDUCTOR EQUIVALENT CIRCUIT
*
*HALF-BRIDGE CONVERTER
*INPUT CIRCUIT
X2 11 101 300 DIV
*V(11) = VD, V(101) = VD+VD2
*V(300) = VD/(VD+VD2)
*
G24 24 0 POLY(2) 10 0 300 0  0 0 0 0 1
*V(10) = VIL
*G24 = VIL*VD/(VD+VD2)
*
*OUTPUT CIRCUIT
G99 0 99 POLY(1) 10 0  0 1
*V(10) = VIL
*
*V(12) INPUT
E12 12 0 POLY(2) 24 0 99 0  0 1 -1
*
*V(14) INPUT
E14 14 0 POLY(1) 99 0  0 1
*
.ENDS CONVERTER
*
*
*
*MAIN CIRCUIT
VSUPPLY 101 0 PWL(0 270  2M 270  3.49M 374  40M 374
+ 42.58M 193.5  75M 193.5  77.58M 374  100M 374)
*V(101) = SUPPLY VOLTAGE
RIN 101 2 0.2
CIN 2 0 235U
RT 2 3 180M
X1 3 0 4 5 TRANSFORMER
*V(3,0) = TRANSFORMER PRIMARY INPUT
*V(4), V(5) = TRANSFORMER SECONDARY OUTPUTS
R48 4 8 3M
R59 5 9 30M
*
*
*HALF-BRIDGE CONVERTER CIRCUIT 1
XCIR1 11 16 14 15 8 995 CONVERTER
*V(11) = DUTY CYCLE INPUT
*V(16) = VG INPUT
*14, 15 = INDUCTOR TERMINALS
```

```
*V(8) = CONVERTER INPUT VOLTAGE
*V(995) = CONVERTER OUTPUT VOLTAGE
L1 14 15 3.5U
VG1 16 0 0.714285714
RO1 995 993 22.2M
DO1 993 997 D1N5831
CL1 997 12 470U
RC1 12 0 15M
R991 997 0 1K
VSEN 997 991
IO1 991 0 PWL(0 25  17M 25  17.01M 12.5  27M 12.5  27.01M 5
+ 58M 5  58.01M 12.5  68M 12.5  68.01M 25  90M 25  90.05M 1
+ 100M 1)
*V(991) = OUTPUT 1 = VOL1
*
*FEEDBACK LOOP
E100 100 0 POLY(1) 991 0  0 0.5
*V(100) = 0.5*VOL1
VAC 1 100 AC 1
XCOMP 1 22 23 COMPEN
VREF 22 0 DC 2.5V
*V(22) = REFERENCE VOLTAGE
XPWM 23 11 PWM
*
*
*HALF-BRIDGE CONVERTER CIRCUIT 2
XCIR2 11 19 17 18 9 996 CONVERTER
*V(11) = DUTY CYCLE INPUT
*V(19) = VG INPUT
*17, 18 = INDUCTOR TERMINALS
*V(9) = CONVERTER INPUT VOLTAGE
*V(996) = CONVERTER OUTPUT VOLTAGE
L2 17 18 37.5U
VG2 19 0 0.0666666667
RO2 996 994 0.238
DO3 994 992 MUR810
CL2 992 13 100U
RC2 13 0 40M
RL2 992 0 2.1428
*V(992) = OUTPUT 2 = VOL2
*
*
*
.MODEL DIDEAL   D(N=1M)
.MODEL D1N5831   D(IS=89.43U RS=3.991M N=1 XTI=0 EG=1.11
+                CJO=3.488N M=.5044 VJ=.75 FC=.5)
.MODEL MUR810   D(IS=4.433P RS=5.91M N=1 XTI=3 EG=1.11
+                CJO=265.3P M=.4253 VJ=.75 FC=.5 TT=61.66N)
.TRAN .5M 100M
.PROBE
*THE .PROBE STATEMENT IS FOR PSPICE ONLY.
*DELETE .PROBE STATEMENT IF RUN BY SPICE.
.PRINT TRAN V(101) I(VSEN) V(991) V(992)
.PLOT TRAN V(101) I(VSEN) V(991) V(992)
*MORE VARIABLES CAN BE ADDED TO THE .PRINT AND
*.PLOT LISTS AT THE DISCRETION OF THE DESIGNER.
.OPTIONS ITL1=300
.END
```

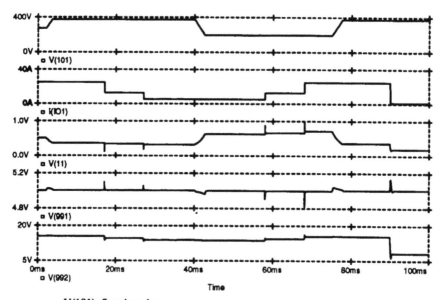

V(101)=Supply voltage
I(IO1)=Active load current
V(11)=Effective duty cycle D (2 x duty cycle of each switching transistor)
V(991)=V$_{OL1}$
V(992)=V$_{OL2}$

Fig. A8.7.2   Transient Simulation for the complete regulator
              design (with active load at V$_{OL1}$).

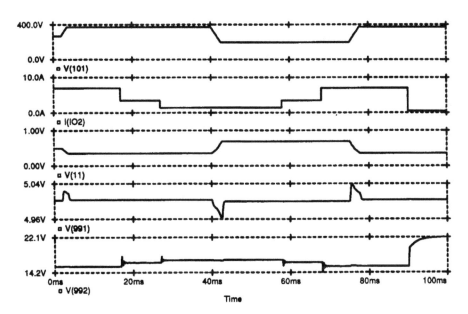

V(101)=Supply voltage
I(IO2)=Active load current
V(11)=Effective duty cycle D (2 x duty cycle of each switching transistor)
V(991)=V$_{OL1}$
V(992)=V$_{OL2}$

Fig. A8.7.3   Transient Simulation for the complete regulator
            design (with active load at V$_{OL2}$).

9

High-Frequency Quasi-Resonant and Resonant Converters

This chapter identifies the problems encountered in the design of high-frequency converters and examines some techniques for solving these problems. Typical examples of high-frequency quasi-resonant and resonant converters with good potential for further development will be studied.

The emphasis in this chapter is on understanding the fundamental principles. The objective is to equip readers with the necessary background to enable them to carry out computer-aided analysis and design work on high-frequency converters.

9.1 PROBLEMS ENCOUNTERED IN DESIGNING HIGH-FREQUENCY CONVERTERS

High-frequency converters can be designed to have low weight and small size because, at high frequencies, the transformers, inductors, and capacitors used have low weight and small size. Square-wave converters work well up to about 200 kHz [52]. When the frequency is pushed higher, however, a number of problems can occur. The main problems include the following:

1. Since the switching loss of transistors and diodes is proportional to the switching frequency, a higher frequency will result in a heavier switching loss and a lower conversion efficiency.
2. Since the hysteresis loss of inductors and transformers is proportional to the frequency, and the eddy-current loss is proportional to the square of frequency, such losses can become intolerable at high frequencies. The skin effect at high frequencies also results in additional copper loss.

3. The effect of leakage inductance in transformers can become very serious at a high switching frequency because of the fast rate of change of current (di/dt), which results in large voltage spikes. Snubber circuits with heavy loss may have to be used to overcome this problem. In addition, semiconductor devices with a higher voltage rating may be required. It is generally true that semiconductor devices with a higher voltage rating would have a larger on-state resistance and therefore a larger conduction loss.

The problems listed above indicate that the heavy losses at a high frequency are a major concern in the design of high-frequency converters. In the next section, various methods used to reduce such losses will be discussed.

9.2 MINIMIZATION OF LOSSES

It is generally true that the loss in high-frequency converters can be reduced by:

1. Using switching devices (transistors, diodes, etc.) with a faster switching speed
2. Using transformers and reactive components (magnetic components, in particular) that have lower high-frequency losses

There are limits, however, to the performance of these components. The cost consideration may also prevent the best components from being used for all applications.

The following subsections focus on the design aspects that can be used to reduce the losses in high-frequency converters.

9.2.1 Minimization of Turn-on Switching Loss

When a semiconductor device is being switched on, the finite switching speed of the device results in a dissipation of power. If the switching-on action can be carried out within a period in which the voltage across the switch is zero, however, the turn-on switching loss can be eliminated. In a class of converters known as quasi-resonant, zero-voltage turn-on converters, reactive components are used to create a zero-voltage condition for the switch to turn on. Such converters would have zero turn-on switching loss and can therefore operate at a higher switching frequency. Certain categories of resonant converters also possess the zero-voltage turn-on characteristic. For the purpose of illustration, an example of a quasi-resonant, zero-voltage turn-on converter will be studied in Subsection 9.3.2.

Examples of resonant converters with the zero-voltage turn-on character-istic will be described in Section 9.4.

9.2.2 Minimization of Turn-off Switching Loss

When a semiconductor device is being switched off, the finite switching speed of the device will also result in a dissipation of power. If, however, the switching off is arranged to take place during a time when the current is zero, we can eliminate the turn-off switching loss. In a class of converters known as quasi-resonant, zero-current turn-off converters, reactive com-ponents are employed to reduce the current to zero just before the switch is turned off. Such converters would therefore have a zero turn-off loss. Certain categories of resonant converters also possess this zero-current turn-off characteristic. An example of a quasi-resonant, zero-current turn-off converter will be studied in Subsection 9.3.1, and an example of a resonant converter with the zero-current turn-off characteristic will be ex-amined in Section 9.4.

9.2.3 Minimization of Conduction Loss

When a series of current pulses, denoted as i, is passed through a converter circuit component, the averaged power dissipation due to the conduc-tion loss in the effective series resistance R of the component is given by

$$P_d = \frac{1}{T} \int_t^{t+T} i^2 R \, dt \qquad (9.1)$$

$$P_d = I_{rms}^2 R \qquad (9.2)$$

where T is the switching period of the converter and I_{rms} is the rms value of the current. Whereas it is obvious that conduction loss can be reduced by minimizing the effective resistance R, it is not so obvious that, for a given averaged output current, the conduction loss can also be reduced by minimizing the form factor F of the current (the ratio of the rms value to the average value), as explained below.

Because of the output smoothing circuit, the dc output current of a dc-to-dc converter is proportional to only the averaged value of the current i. We therefore have the implied output power P_o given by

$$P_o = I_{av}^2 R_L = \left[\frac{I_{rms}}{F} \right]^2 R_L \qquad (9.3)$$

where I_{av} is the average value of the current i and R_L the effective load resistance, I_{rms} the rms value, and F the form factor, of the current i. Combining Eqs. (9.2) and (9.3), we have

$$P_d = F^2 \frac{P_o R}{R_L} \tag{9.4}$$

It is clear from Eq. (9.4) that the larger the form factor, the larger the conduction loss (for a given output power) will be.

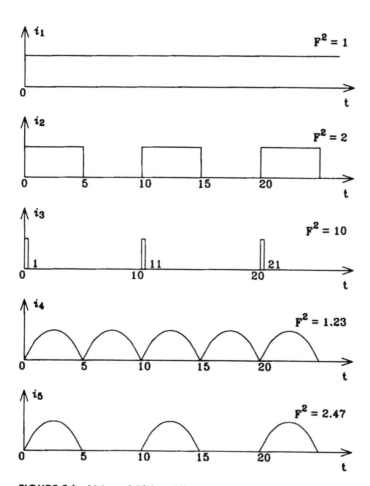

FIGURE 9.1   Value of F^2 for different waveforms.

For the purpose of illustration, the values of F^2 for various typical current waveforms encountered in power converters (square-wave and resonant types) are given in Fig. 9.1.

It should be noted that, in general, the smaller the percentage of time in which a current is allowed to flow, the larger the conduction loss will be.

9.2.4 Minimization of Losses in Transformers and Reactive Components

The losses in a transformer or inductor can be reduced by

1. Using a magnetic core that has lower hysteresis and eddy-current losses
2. Operating the magnetic core at a smaller flux-density level
3. Using winding wires with a larger effective cross-sectional area and a lower skin-effect resistance (litz wire, stranded wires, etc.)

The second and third requirements listed above would usually result in a larger size of the transformer or inductor.

In resonant-type converters, the voltage and current stresses of the resonant inductor or capacitor can be very large because of the resonance phenomenon. Such large stresses also imply heavy losses in the reactors. In the design of quasi-resonant and resonant converters, it is therefore desirable, for this and other obvious reasons, to keep such stresses to only the required minimum.

9.3 QUASI-RESONANT CONVERTERS

In this section, the principle of operation and the characteristics of quasi-resonant converters will be studied.

Quasi-resonant converters are a class of converters that make use of reactive components to create a zero-voltage condition for the switch to turn on or a zero-current condition for the switch to turn off. Quasi-resonant converters are, however, different from resonant converters in the following aspects:

1. Whereas the resonant circuit in a quasi-resonant converter is only a means to create a zero-voltage or zero-current condition for the power switch to turn on or off, the resonant circuit in a resonant converter is an integral part of the power conversion circuit.
2. The resonant circuit in a quasi-resonant converter can be associated with a power switch to form a resonant switch. Without such a

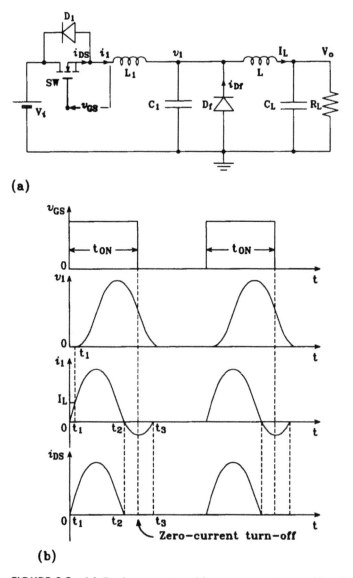

(a)

(b)

FIGURE 9.2 (a) Buck converter with zero-current turn-off and (b) voltage and current waveforms.

resonant circuit, a quasi-resonant converter will revert back to a normal square-wave converter. However, the same is not true for a resonant converter.

9.3.1 Quasi-Resonant Converter with Zero-Current Turn-off

As an example to illustrate the principle of zero-current turn-off, Fig. 9.2(a) shows the circuit of a buck converter in which the reactors L_1 and C_1 are added to create a zero-current condition for the switch SW to turn off. Based on the waveforms shown in Fig. 9.2(b), the operation of the circuit can be explained as follows:

1. It is assumed that the switching action of the circuit has reached a steady state and that it operates in the continuous mode.
2. At $t = 0$, the switch SW is turned on. Current i_1 starts to flow through L_1.
3. At $t = t_1$, i_1 increases to the value of I_L (the dc output current in the inductor L, which is assumed to be constant). Then i_1 replaces completely the diode current i_{Df}, and the diode D_f turns off. The equivalent circuit of the switching circuit at $t = t_1$ is shown in Fig. 9.3.
4. Assume that the conditions at $t = t_1$ are the initial conditions:

    $$i_1 = I_L \qquad v_1 = 0$$

    The current i_1 can then be found by applying the superposition theorem. The current component due to I_L is simply I_L. The current

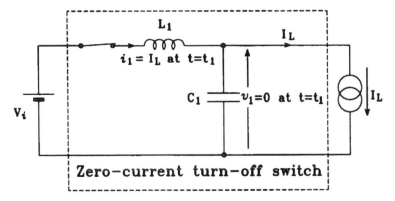

FIGURE 9.3 Equivalent circuit of zero-current turn-off switch shown in Fig. 9.2 for $t_1 < t < t_3$.

component due to V_i, which may be understood as a step input, is a sinusoidal current with amplitude $V_i \left[\dfrac{C_1}{L_1} \right]^{1/2}$. We therefore have i_1 given by

$$i_1 = I_L + V_i \left[\frac{C_1}{L_1} \right]^{1/2} \sin \left[\frac{1}{L_1 C_1} \right]^{1/2} (t - t_1)$$

(for $t_1 < t < t_3$) \hfill (9.5)

A possible waveform of i_1 for $t_1 < t < t_3$ is shown in Fig. 9.2(b).

5. It is obvious from Eq. (9.5) that if

$$V_i \left[\frac{C_1}{L_1} \right]^{1/2} \geq I_L \qquad (9.6)$$

we can create a temporary zero-current condition for the switch SW to turn off. (Between t_2 and t_3, i_{DS} is zero because i_1, having a negative value, flows through D_1 only.)

6. The other condition for zero-current switching is

$$t_2 < t_{ON} < t_3 \qquad (9.7)$$

as shown in Fig. 9.2(b).

As implied by Eq. (9.6), the condition for zero-current switching is actually not always obtainable because it depends on the loading current I_L and the input voltage V_i. It is generally true that, for most quasi-resonant converters, there is only a limited range of operating conditions in which zero-current (or zero-voltage) switching can be obtained. This is, in fact, one of the main disadvantages of quasi-resonant converters.

The circuit shown in Fig. 9.2(a) also suffers from the disadvantage that because of the constraint on t_{ON}, as implied by Eq. (9.7), the ordinary PWM (pulse-width modulation with constant switching frequency) cannot be used for the control of the output voltage. Instead, control is provided by varying the switching frequency f_s.

Because the output voltage V_o is equal to the averaged value of v_1, it is obvious from Fig. 9.2(b) that V_o should be proportional to the switching frequency f_s of the converter (under the continuous-mode operation).

9.3.2 Quasi-Resonant Converter with Zero-Voltage Turn-on

An example of a quasi-resonant flyback converter with zero-voltage turn-on is shown in Fig. 9.4(a). In this circuit, the capacitance C_1 and inductance

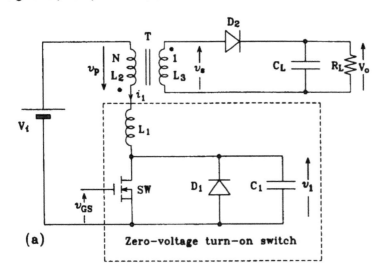

(a)

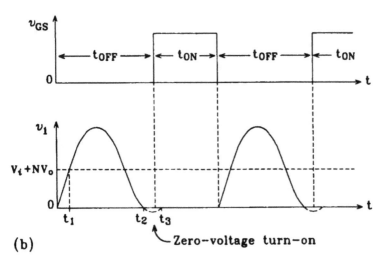

(b)

FIGURE 9.4 (a) Flyback converter with zero-voltage turn-on and (b) voltage waveforms.

L_1 are used as a resonant circuit to induce a zero-voltage condition for the transistor switch SW to turn on. The operation of the circuit can be explained as follows:

1. It is assumed that the switching action of the circuit has reached a steady state and that it operates in the continuous mode.

2. At $t = 0$, the switch SW is turned off. The primary winding of transformer T then acts as a constant current source to charge up C_1. When v_1 reaches the value of $(V_i + NV_o)$ at $t = t_1$, the primary voltage v_p is sufficient to produce a v_s to turn on the output rectifier D_2. (It is assumed that the voltage drop across L_1 at this moment is small because L_1 is normally much smaller than L_2.) The equivalent circuit of the zero-voltage turn-on switch at $t = t_1$ is shown in Fig. 9.5(a).

3. Assume that the conditions at $t = t_1$ are the initial conditions:

$$i_1 = I_L \tag{9.8}$$
$$v_1 = V_i + NV_o \tag{9.9}$$

where I_L is the current in L_1 just before diode D_2 is turned on. Based on these initial conditions, the following expression for v_1 can be found by applying the superposition theorem:

$$v_1 = (V_i + NV_o) + I_L \left[\frac{L_1}{C_1}\right]^{1/2} \sin \left[\frac{1}{L_1 C_1}\right]^{1/2} (t - t_1)$$
$$\text{(for } t_1 < t < t_2) \tag{9.10}$$

where the voltage component $I_L \left[\dfrac{L_1}{C_1}\right]^{1/2} \sin \left[\dfrac{1}{L_1 C_1}\right]^{1/2} (t - t_1)$, which is denoted as v_{11} in Fig. 9.5(b), is due to the initial inductor current I_L at $t = t_1$. It should be noted that the waveform of v_1 shown in dotted lines in Fig. 9.5(b) for $t > t_2$ does not actually exist because the diode D_1 will be turned on at $t = t_2$, reducing v_1 to practically zero.

4. It can be found from Eq. (9.10), or from the waveform of v_1 in Fig. 9.5(b), that if

$$I_L \left[\frac{L_1}{C_1}\right]^{1/2} \geq (V_i + NV_o) \tag{9.11}$$

we can create a zero-voltage condition for the switch SW to turn on. Note that the value of I_L in Eq. (9.11) needs to be estimated from the relationship

$$V_i I_L [\text{duty cycle of } SW] \approx \text{Input power to converter}$$
$$= \frac{\text{Output power}}{\text{Conversion efficiency}} \tag{9.12}$$

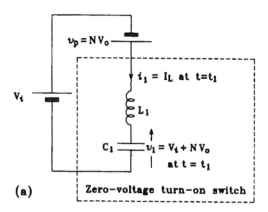

(a)

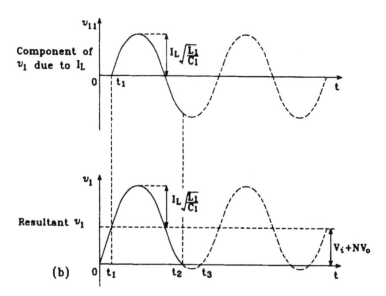

(b)

FIGURE 9.5 (a) Equivalent circuit of zero-voltage turn-on switch for $t_1 < t < t_2$ and (b) voltage waveforms.

$$I_L \approx \frac{\text{Output power}}{\text{Conversion efficiency}} \frac{1}{V_i[\text{duty cycle of } SW]} \tag{9.13}$$

5. In order to achieve zero-voltage switching, the turn-off period of SW is normally fixed. The output voltage is then controlled by varying the switching frequency. Because of the fixed turn-off time,

a lower switching frequency implies a larger duty cycle of *SW* and, therefore, a higher output voltage.

In the circuit shown in Fig. 9.4, the leakage inductance of the transformer *T* can be used as the inductance L_1 (or part of it), and the drain-to-source capacitance of *SW* can be used as the capacitor C_1 (or, again, part of it). In this way, the undesirable effects of leakage inductance and stray capacitance can be eliminated. Converters capable of incorporating such stray elements as components of the converter circuit have very good potential for high-frequency applications.

9.3.3 Characteristics of Quasi-Resonant Converters

There are many types of quasi-resonant converters in existence. Although only two types have been studied here, these are actually typical examples that can be used to illustrate the following special characteristics of such converters:

1. In a zero-current turn-off quasi-resonant converter, the electronic switch is effectively connected in series with a resonant circuit. When the resonant component of the current in the switch swings in a direction opposite to that of the normal current flow, the resultant current in the switch can be temporarily reduced to zero to allow for a zero-current turn-off. However, because of the additional resonance current, the peak current stress of the power switch is at least doubled.
2. Zero-current turn-off is particularly useful for switching devices with a fast turn-on speed but a slow turn-off speed, e.g., the bipolar transistor and IGBT (insulated-gate bipolar transistor), because zero-current turn-off minimizes only the turn-off switching loss.
3. In a zero-voltage turn-on quasi-resonant converter, the electronic switch is connected effectively in parallel with a resonant circuit. When the resonant component of the voltage across the switch swings in a direction opposite to that of the supply, the resultant voltage across the switch can be temporarily reduced to zero to allow for a zero-voltage turn-on. However, because of the additional resonance voltage, the peak voltage stress of the power switch is at least doubled.
4. Zero-voltage turn-on is particularly useful for power switches that have a fast turn-off speed but a slow turn-on speed.
5. While zero-current turn-off converters normally require a fixed turn-on time, zero-voltage turn-on converters normally require a fixed

turn-off time. It is actually possible to achieve both zero-current turn-off and zero-voltage turn-on in the same circuit, but then both the switching frequency and the duty cycle of the switching transistor have to be fixed, thus making it difficult to control the output voltage.

The resonance currents and voltages in quasi-resonant converters may create interference problems. Such problems can become more serious if frequency modulation is used to control or regulate the output voltage. Interferences with variable frequency make the design of filtering circuits more difficult. They may also result in objectionable effects in telecommunication systems and video displays.

9.4 RESONANT CONVERTERS

The resonant circuits used in the quasi-resonant converters discussed in Section 9.3 are used only to create a zero-voltage or zero-current condition for the power switch to turn on or turn off. These resonant circuits are not expected to take part actively in the power conversion process. In another type of converter, a resonant circuit is used to convert a chopped dc input to produce a sinusoidal current before the current is rectified in the output circuit. Such converters are known as resonant converters. Since the current in a resonant converter is of sinusoidal shape (or segments of sinusoidal waves), the di/dt rate is much reduced. This helps to minimize the switching loss of semiconductor devices because, during the switching period, the current and voltage levels are now much smaller. In a resonant converter, the transformer leakage inductance (if a transformer is used) is usually part of the resonant circuit. This helps to solve the problem of large voltage spikes encountered in square-wave converters as a result of leakage inductances.

Because of the above-mentioned advantages, resonant converters are very suitable for high-frequency applications.

9.4.1 Types of Resonant Converters

There are basically two classes of resonant dc-to-dc converters: the current-fed converter and the voltage-fed converter [58,71]. In a current-fed converter, an inductor is connected in series with the dc input to simulate a constant current source, which is switched between the ground and a parallel L-C circuit to induce a high-frequency resonance. The resonance current may then be transformer-coupled and rectified to provide the dc output. In a voltage-fed converter, the dc supply is first chopped into a high-frequency square-wave voltage, which is then fed into a series-resonant L-C circuit to produce a resonance. Similar to a current-fed converter, the resonance current can be transformer-coupled and rectified to provide the dc output. Note that although the current-fed converter requires an extra input inductor, it has the advantage that the current taken from the power source is smoothed.

Typical examples of current-fed and voltage-fed converters that have the potential for further development will be studied in the following two subsections.

9.4.2 The Class E Resonant Converter

The basic circuit of a class E resonant converter is shown in Fig. 9.6. It is an example of the current-fed resonant converter [46,56,68,70]. The waveform of the voltage v is shown in Fig. 9.7. It is found that, every time the electronic switch SW is turned off, the current I_{LP} will be forced into the resonant circuit to produce a resonance. During normal operation, the circulating current i_{LC} is built up to be so large that, within a fraction of the switching cycle, it is able to forward bias the diode D_1 (even in the presence of I_{LP}). It is within this fraction of time that SW is turned on again to allow L_P to store up inductive energy once more. Because of the near-zero forward voltage of diode D_1, the turn-on switching loss of SW is very small. Thus the class E converter also has the zero-voltage turn-on characteristic.

The output voltage of a class E converter is controlled by altering the switching frequency. A lower switching frequency gives a higher output voltage.

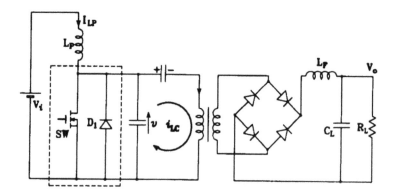

FIGURE 9.6 Class E resonant converter.

State of SW

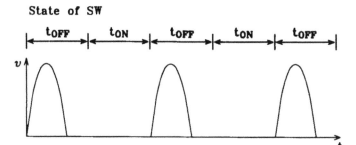

FIGURE 9.7 Voltage waveform of class E resonant converter.

The class E converter is simple and easy to drive and is capable of operating at very high frequencies [70]. However, it has the disadvantage that even the minimum loaded Q factor of the resonant circuit must be maintained high enough to enable the circulating current i_{LC} to build up adequately to produce the required zero voltage across the diode D_1 during the intervals when SW is to be turned on. It should be noted that high Q factors imply large reactive currents/voltages. Large "reactive" currents, when flowing through lossy reactors or electronic switches, result in heavy losses. Large "reactive" voltages also mean extra voltage stress and thus imply additional losses. In a class E converter, the voltage stress is normally much higher than the supply voltage.

9.4.3 Series-Resonant Converters

The basic circuit of a series-resonant converter, which is an example of the voltage-fed converter, is shown in Fig. 9.8. A typical output voltage versus switching frequency characteristic of the converter is given in Fig. 9.9. For this type of converter, two modes of operation are possible. They are defined here as:

1. The L-mode, in which the switching frequency f_s of the converter is always lower than the resonant frequency $1/(2\pi)\left[\dfrac{1}{LC}\right]^{1/2}$

2. The H-mode, in which the switching frequency is always higher than the resonant frequency.

The typical L-mode voltage and current waveforms are shown in Fig. 9.10. In this mode of operation, the current in the conducting switch (i_2 or i_1) falls sinusoidally to zero at the end of the half-cycle. SCRs can be used as

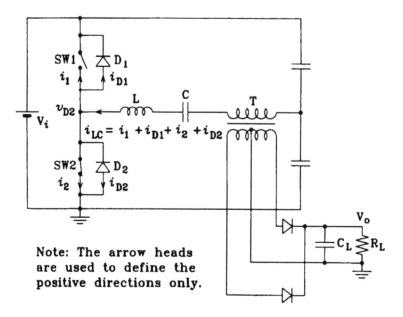

FIGURE 9.8 Series-resonant converter.

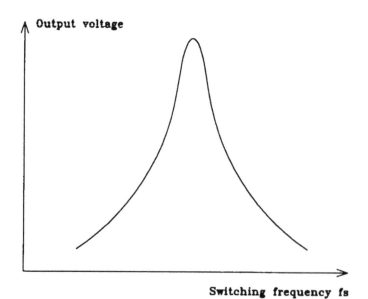

FIGURE 9.9 Output voltage versus switching frequency characteristic of resonant converter.

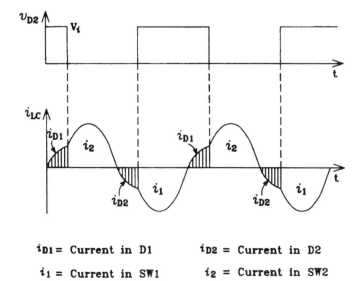

i_{D1} = Current in D1 i_{D2} = Current in D2

i_1 = Current in SW1 i_2 = Current in SW2

FIGURE 9.10 *L*-mode voltage and current waveforms.

the electronic switches *SW*1 and *SW*2 because the conducting SCR is allowed to turn off naturally at the end of the conduction period [48,49]. (It is effectively a zero-current turn-off converter.) Since SCRs have a larger power rating, this kind of converter is suitable for large power applications with relatively low operating frequency. Frequency modulation can be used to control the output voltage. The *L*-mode series-resonant converter has been used successfully in many applications [30,50,51,69].

In the *H*-mode operation, the switching frequency is always higher than the resonant frequency of the resonant circuit [3,6,42,58]. The voltage and current waveforms for the *H*-mode operation are given in Fig. 9.11. One important difference between the *L* and *H* modes of operation is the relative timing when *SW*1/*SW*2 is turned on. In the *L*-mode, this occurs while D_2/D_1 is still conducting. The switching loss in D_2/D_1 can become significant at high frequency because D_2/D_1 usually has a slow turn-off time. (The loss during the time when *SW*1/*SW*2 is turned on but D_2/D_1 has not recovered to the blocking state is significant.) In the *H*-mode, since *SW*1/*SW*2 is turned on while D_1/D_2 is conducting (effectively, zero-voltage turn-on), there is no problem at all: D_1/D_2 would have sufficient time to recover while *SW*1/*SW*2 is conducting. For this reason, the *H*-mode operation tends to have a higher conversion efficiency than the *L*-mode (particularly when the switching frequency is high). This is very obvious if

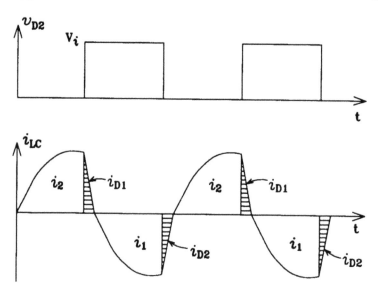

FIGURE 9.11 *H*-mode voltage and current waveforms.

power MOSFETs are used as switches *SW1* and *SW2* and their inherent diodes as D_1 and D_2 because, while MOSFETs have very fast switching times, in the order of tends of nanoseconds, their inherent diodes take hundreds of nanoseconds of recovery time to turn off.

Experimental works using MOSFETs as power switches have demonstrated that at high switching frequencies, the series-resonant converter with a series-connected load operating in the *H*-mode tends to give a higher conversion efficiency compared to many other types of resonant converters [3,6]. Such a converter suffers, however, from the disadvantage that there is a minimum load-current requirement if the output voltage is to be regulated by frequency modulation.

Series-resonant converters are now commonly used in many applications. To help designers analyze the circuit operation quantitatively, a graphical method of analysis will be introduced in the next subsection.

9.4.4 Graphical Analysis of Resonance Current in Series-Resonant Converters

In the design of resonant converters, it is necessary to determine the amplitude and shape of the transient currents in switching devices. The objective of this subsection is to introduce an easy-to-use graphical tool, based on the superposition theorem, for analysis of the resonance current in a series-resonant converter.

Refer to the converter circuit example shown in Fig. 9.12 and the H-mode waveforms of i_{LC} and Δv_{LC} shown in Fig. 9.13. It can be found that i_{LC} actually consists of a series of segments of sine waves, produced by step changes of Δv_{LC}. Representing each segment of i_{LC} by a current phasor, e.g., I_{OA} for $t_A < t < t_B$, I_{OB} for $t_B < t < t_C$, and I_{OC} for $t_C < t < t_D$, etc., it is found that these phasors can be combined in a phasor diagram in order to determine their relative phase relationships and amplitudes.

In the analysis that follows, it is assumed that the output transformer T_1, with a primary to secondary turns ratio of N, is ideal. Figure 9.14 is a phasor diagram for the current components of i_{LC}, which is found through the following steps:

1. It is assumed that, initially,

$$i_{LC} = 0, \qquad v_{D2} = V_i, \qquad V_{CO} = (-1/2)V_i$$

where V_{CO} is the dc voltage across the resonance capacitor C (the direction of which is defined in Fig. 9.12). It is also assumed that there is an initial dc voltage V_{o2} across the output filtering capacitor C_L. The gate-driving voltages v_{G1} and v_{G2} are applied at $t = t_A$, as shown in Fig. 9.13.

2. At $t = t_A$, SW2 is turned on. Sinusoidal current i_{LC} starts to flow in the positive direction through SW2. Effectively, this introduces a step voltage $\Delta v_{LC} = (V_i - V_{o1})$ across the LC circuit, resulting in

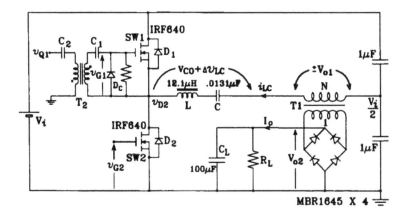

FIGURE 9.12 Series-resonant converter example.

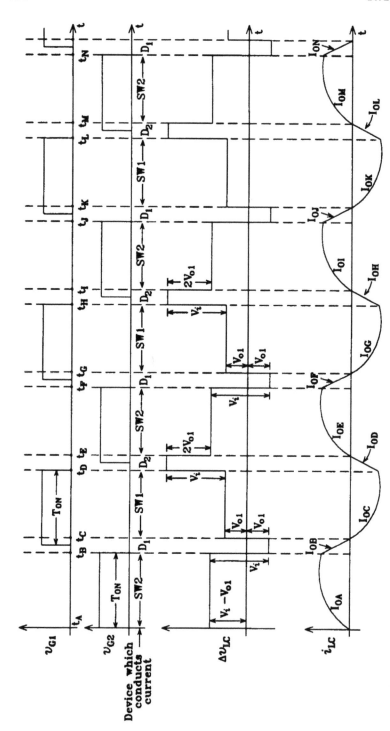

FIGURE 9.13 *H*-mode waveforms of series-resonant converter.

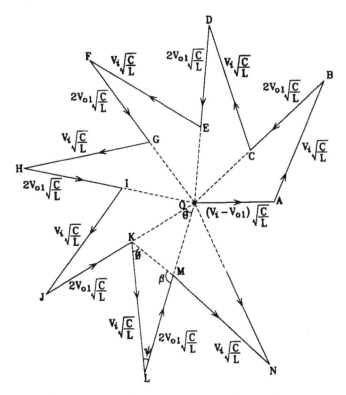

FIGURE 9.14 *H*-mode resonant current phasor diagram.

a sine wave current in the LC circuit, which is represented by the phasor OA in Fig. 9.14. The corresponding current phasor (in peak amplitude) will be referred to as I_{OA}.

$$\text{Amplitude of } I_{OA} = \left[\frac{C}{L}\right]^{1/2}(V_i - V_{o1}) \tag{9.14}$$

where $V_{o1} = NV_{o2}$. The phase angle of I_{OA} is taken as $0°$, as shown in Fig. 9.14.

3. At $t = t_B$, SW2 is turned off. The positive i_{LC} continues to flow, but now through D_1 (instead of SW2). Since v_{D2} changes from OV to V_i, a negative step voltage $-V_i$ appears in Δv_{LC} to produce an additional sinusoidal current component, which is represented by the phasor AB in Fig. 9.14.

$$\text{Amplitude of } AB = \left[\frac{C}{L}\right]^{1/2}V_i \tag{9.15}$$

The phase lag of AB due to the delay $(t_B - t_A)$ is

$$\left[\frac{T_{ON}}{T_r} \right] 360° \quad \text{or} \quad (T_{ON}f_r)360°$$

where T_{ON} is the on duration of the switching transistor, T_r is the natural oscillation period of the resonance current, and

$$f_r = \frac{1}{T_r} = \frac{1}{2\pi} \left[\frac{1}{LC} \right]^{1/2} \tag{9.16}$$

Together with the phase reversal (due to the negative polarity of the step), the resultant phase angle of AB is given by

$$\text{Phase angle of } AB = 180° - (T_{ON}f_r)360° \tag{9.17}$$

The phasor sum of OA and AB in Fig. 9.14 is OB, which is referred to as I_{OB}. Note that although I_{OB} appears to be larger than I_{OA}, its instantaneous value can actually be lower than that of I_{OA} because I_{OB} exists only in the near zero-crossing region, as can be seen in the waveform of i_{LC} in Fig. 9.13.

4. At $t = t_C$, i_{LC} reverses to the negative direction to flow through $SW1$. Because of the reversed voltage drop across T_1, a step change of $2V_{o1}$ appears in Δv_{LC}, resulting in an additional current, shown as BC in Fig. 9.14.

$$\text{Amplitude of } BC = \left[\frac{C}{L} \right]^{1/2} 2V_{o1} \tag{9.18}$$

It should be noted that the $2V_{o1}$ edge occurs at the time the instantaneous value of I_{OB} changes from positive to negative. This implies that the phase angle of BC is exactly opposite to that of I_{OB}, as shown in Fig. 9.14.

The current for $t_C < t < t_D$ is represented by OC in Fig. 9.14 and is referred to as I_{OC} in Fig. 9.13.

5. At $t = t_D$, $SW1$ is turned off. The negative i_{LC} continues to flow, but now through D_2. A step change of V_i thus appears in Δv_{LC}, introducing one more current component shown as CD in Fig. 9.14.

$$\text{Amplitude of } CD = \left[\frac{C}{L} \right]^{1/2} V_i \tag{9.19}$$

If the phase angle of CD with respect to AB (the phasor due to the previous step change of V_i) is denoted as α, we have

$$\alpha = \frac{(1/2)T_r - (1/2)T}{(1/2)T_r} 180° = \frac{T_r - T}{T_r} 180° \tag{9.20}$$

where T is the switching period of the converter. Or, alternatively,

$$\alpha = \frac{f_s - f_r}{f_s} 180° \tag{9.21}$$

where f_s is the switching frequency of the converter.

The current for $t_D < t < t_E$ is represented by OD in Fig. 9.14 and by I_{OD} in Fig. 9.13.

6. At $t = t_E$, i_{LC} reverses back to the positive direction to flow through $SW2$. A step change of $-2V_{ol}$ thus appears in Δv_{LC}, resulting in the next current component shown as DE in Fig. 9.14.

$$\text{Amplitude of } DE = \left[\frac{C}{L}\right]^{1/2} 2V_{ol} \tag{9.22}$$

The phase angle of DE is opposite to that of I_{OD}.

7. At $t = t_F$, t_J, and t_N: similar to $t = t_B$.
8. At $t = t_G$ and t_K: similar to $t = t_C$.

The plotting of current phasors can be continued until it reaches a steady state.

Assuming that $[(f_s - f_r)/f_s]180° = \alpha \le 90°$, which is true if the switching frequency is not larger than twice the resonance frequency, the amplitudes of the current phasors I_{OA}, I_{OC}, I_{OE}, I_{OG}, I_{OI}, I_{OK}, and I_{OM} will truly represent the actual peak amplitudes of currents in the switching circuits (as can be seen in the waveform of i_{LC} in Fig. 9.13). The currents I_{OB}, I_{OD}, I_{OF}, I_{OH}, I_{OJ}, I_{OL}, and I_{ON} have, however, never any chance for their instantaneous values to reach the peak values.

The value of I_{LC} (peak value of i_{LC} for $\alpha \le 90°$) in the steady state can be determined through the following steps:

1. Assume that I_{LC} reaches a steady state at $t = t_K$, so that, in Fig. 9.14, we have

$$I_{LC} = OM = OK \text{ and } \Delta OMN = \Delta OKL$$

$$\theta = \text{angle between } OM \text{ and } OK$$
$$= \text{angle between } MN \text{ and } KL = \alpha \tag{9.23}$$

where $\alpha = [(f_s - f_r)/f_s]180°$. From the isosceles triangle OKM, we have also

$$\beta = 180° - \frac{1}{2}(180° - \theta) = 90° + \frac{\theta}{2} \tag{9.24}$$

2. In triangle KLM,

$$\frac{KL}{\sin \beta} = \frac{LM}{\sin \phi} \tag{9.25}$$

or

$$\frac{\left[\dfrac{C}{L}\right]^{1/2} V_i}{\sin \beta} = \frac{\left[\dfrac{C}{L}\right]^{1/2} 2V_{ol}}{\sin \phi} \tag{9.26}$$

Combining Eqs. (9.26) and (9.24) yields

$$\phi = \sin^{-1}\left[\frac{2V_{ol}}{V_i}\sin \beta\right] \tag{9.27}$$

$$= \sin^{-1}\left[\frac{2V_{ol}}{V_i}\cos \frac{\theta}{2}\right] \tag{9.28}$$

Also, from triangle KLM, we have

$$\psi = 180° - \beta - \phi \tag{9.29}$$

Substitution of Eqs. (9.24) and (9.28) into Eq. (9.29) gives

$$\psi = 90° - \frac{\theta}{2} - \sin^{-1}\left[\frac{2V_{ol}}{V_i}\cos \frac{\theta}{2}\right] \tag{9.30}$$

3. In triangle KLO

$$\frac{KL}{\sin \theta} = \frac{OK}{\sin \psi} \tag{9.31}$$

Knowing from Eq. (9.23) that $\theta = \alpha$, we then have

$$\frac{\left[\frac{C}{L}\right]^{1/2} V_i}{\sin \alpha} = \frac{I_{OK}}{\sin \psi} \tag{9.32}$$

From Eqs. (9.23), (9.30), and (9.32), the steady-state I_{LC} at $t = t_K$ is found as

$$I_{LC} = I_{OK} = \frac{1}{2 \sin \frac{\alpha}{2}} \left[\frac{C}{L}\right]^{1/2} \left\{V_i^2 - \left[2V_{o1} \cos \frac{\alpha}{2}\right]^2\right\}^{1/2}$$

$$- \left[\frac{C}{L}\right]^{1/2} V_{o1} \tag{9.33}$$

where $V_{o1} = NV_{o2}$. \hfill (9.34)

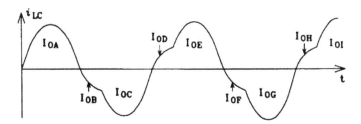

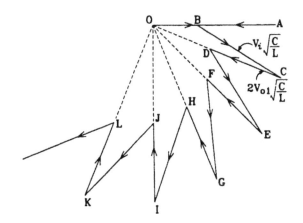

FIGURE 9.15 *L*-mode resonant current and phasor diagram.

Assuming that $\alpha \leq 90°$, the I_{LC} in Eq. (9.33) is the steady-state value of the peak amplitude of i_{LC}. The corresponding value of the dc output current I_o is then approximately equal to

$$I_o = N \frac{2}{\pi} I_{LC} \qquad (9.35)$$

It should be interesting to note that, by applying the techniques described in Chapter 7 for the simulation of algebraic expressions, the low-frequency behavior of the complete converter can be modeled as a single current source whose amplitude is given by Eqs. (9.35) and (9.33) [7].

The graphical analysis given above is for a series-resonant converter in the H-mode operation. A similar analysis can be carried out for the L-mode operation. The resultant current waveform and its graphical representation (for the L-mode operation) are shown in Fig. 9.15. The lengths OA, OC, OE, OG, OI, and OK in Fig. 9.15 truly represent the amplitudes of the currents.

9.5 THE ROLE OF CAD TOOLS IN THE DESIGN OF HIGH-FREQUENCY CONVERTERS

The operation of high-frequency quasi-resonant and resonant converters is not difficult to understand. Even the waveforms and converter characteristics (such as the relationship between switching frequency and output voltage) can often be roughly estimated by the circuit designer. However, when it comes to an accurate and detailed mathematical analysis, which possibly takes into account the stray components, difficulties may arise as a result of the switching nature of the circuit. The CAD techniques introduced in Chapters 6–8 can, in general, also be applied to the design of quasi-resonant and resonant converters and regulators. By combining the experience of a circuit designer and the analytical power of computer simulation, the design process can be made much easier.

9.6 SUMMARY AND FURTHER REMARKS

In this chapter, the common problems encountered in the design of high-frequency converters have been identified. Possible solutions to these problems have been suggested. The operation of typical examples of quasi-resonant and resonant converters that have good potential for further development have also been studied.

The techniques for designing quasi-resonant and resonant converters and regulators are actually much less mature than those of their square-wave counterparts. The intention of this chapter is to familiarize readers with the basic principles to enable them to carry out further analysis, design, and development work using CAD tools.

EXERCISES

1. Describe the methods that can be used to reduce
   a. the turn-on switching loss
   b. the turn-off switching loss
   c. the conduction loss
   of the semiconductor devices in power converters.
2. What are the differences between a quasi-resonant converter and a resonant converter?
3. Why is the current stress of the switching transistor in a zero-current turn-off quasi-resonant converter higher than that of an ordinary square-wave converter?
4. Why is the voltage stress of the switching transistor in a zero-voltage turn-on quasi-resonant converter higher than that of an ordinary square-wave converter?
5. Comment on the effects of variable load on the operation of
   a. an ordinary square-wave converter
   b. a quasi-resonant converter
   c. a series-resonant converter in L-mode operation
   d. a series-resonant converter in H-mode operation
6. Assume that the circuit shown in Fig. 9.12 has the following operational conditions:

   $V_i = 140$ V
   $N = 1$
   Switching frequency $f_s = 600$ kHz
   Gate-driving pulse width $T_{ON} = 0.73$ μs
   Waveforms of v_{G1} and v_{G2}: as shown in Fig. 9.13
   Initial $v_{D2} = 0$
   $V_{co} = 1/2\ V_i = 70$ V
   Initial $i_{LC} = 0$
   $V_{o2} = 40$ V

   a. Draw to scale a phasor diagram for the current i_{LC}.
   b. Based on the phasor diagram found, sketch the waveform of i_{LC} for the first three switching cycles.
7. Repeat Question 6, assuming a switching frequency of 300 kHz and a gate-driving pulse width T_{ON} of 1.4 μs.

10
Practical Techniques of Design of Switch-Mode Power Supplies

This chapter examines the various practical techniques that have not been covered in previous chapters for designing switch-mode power supplies. It describes the design of the input stage, the selection of the converter topology, the design of inductors, the design of transformers, the design of output circuits, the design of controller circuits, and the minimization of electromagnetic interference. The role of CAD tools in the design of such circuits will also be outlined. The discussion will be based on a mains-operated off-line switching regulator.

10.1 DESIGN OF THE INPUT STAGE

The function of the input stage of an off-line regulator is to convert an ac mains into a dc supply in order to provide input power to the dc-to-dc converter in the switch-mode power supply. Normally, no mains-frequency transformer is used here because such a transformer is usually bulky. The galvanic isolation and voltage stepping, up or down, can be provided by a high-frequency transformer in the dc-to-dc converter.

The basic requirements of the input stage are:

1. The maximum ripple of its dc output voltage must be within the allowable range acceptable by the dc-to-dc converter.
2. The input inrush current, which is the input current when the ac mains switch is turned on, must be limited to a safe value so that it will not damage the circuit.
3. The conducted interference that gets into the ac mains should be minimized to satisfy various regulations, such as those of the FCC

(Federal Communications Commission of the United States) or the VDE (Verband Deutscher Elektrotechniker of Germany).

The above-mentioned three design requirements will be discussed in the following three subsections.

10.1.1 Filtering the Ripple

For each switching regulator, there is always a limited range of allowable dc input voltage, beyond which the regulator will not function properly. This sets a limit on the ripple in the dc input voltage of the converter. To maintain a low ripple voltage, we need a large input filtering capacitance C (shown in Fig. 10.1). However, too large a value of capacitance will result in an unnecessarily high cost, a larger size, a narrower charging current pulse, and a larger peak charging current. A common practice is to allow for a maximum ripple equal to approximately 30% of the peak value of the minimum allowable ac mains voltage. Based on this empirical rule, we can estimate the minimum capacitance required, as described below.

Assume that

$$f = \text{ac mains frequency}$$

$$T = \frac{1}{f}$$

$$V_{acmin} = \text{minimum allowable ac mains voltage (rms value)}$$

$$P = \text{output power of the rectifying circuit}$$

$$C = \text{required filtering capacitance}$$

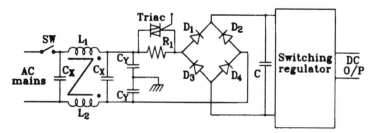

FIGURE 10.1   Input circuit of off-line regulator.

From the energy equation over a half-cycle, we have

$$\frac{T}{2} P = \frac{1}{2} C \text{ (max capacitor voltage)}^2 - \frac{1}{2} C \text{ (min capacitor voltage)}^2$$

$$\frac{T}{2} P = \frac{1}{2} C [1.414 \, V_{acmin}]^2 - \frac{1}{2} C [(1 - 30\%) \, 1.414 \, V_{acmin}]^2 \qquad (10.1)$$

$$\therefore C = \frac{1}{2} PT \frac{1}{0.51 \, (V_{acmin})^2} \qquad (10.2)$$

Equation (10.2) therefore gives the minimum filtering capacitance required.

In addition to the 30% ripple voltage mentioned above, if the ac mains is allowed to change from a minimum of V_{acmin} to a maximum of

$$V_{acmax} = 130\% \, V_{acmin} \text{ (a typical allowance)} \qquad (10.3)$$

then the range of dc input voltage that the switching regulator must be designed to accept will be from a minimum of 70% × (1.414 V_{acmin}) to a maximum of 130% × (1.414 V_{acmin}). A typical switching regulator should therefore be designed to provide the required regulation within this variation range of dc input voltage.

10.1.2 Input Inrush Current Limiter

At the instant the ac power switch SW in Fig. 10.1 is turned on, the instantaneous charging current flowing into the input filtering capacitor C can become very large because of the possible large potential difference between the input voltage and the capacitor voltage. This may damage the rectifier circuit. In order to prevent such damage, a Triac in parallel with a current-limiting resistor R_1 may be inserted into the input circuit, as shown in Fig. 10.1, to limit the inrush current. During the turn-on switching, no gate drive is applied to the Triac, so that the Triac remains in the off state. The current-limiting resistance R_1 is then used to limit the maximum inrush current to a safe value. When the capacitor C has been charged up, a continuous gate drive is applied to the Triac to turn it on to shunt the current in R_1. Since the voltage drop of the Triac is low, the power dissipation in the Triac can be maintained very low during normal operation.

As an alternative method of reducing the inrush current, a thermistor (a resistor with a negative temperature coefficient) may be used instead of the parallel resistor–Triac combination. A thermistor has a large resistance when it is cold and a small resistance when it is hot. During the turn-on switching, the relatively large resistance of the thermistor can be used to limit the inrush current. The relatively low resistance when it is hot reduces its power dissipation to an acceptable level during normal operation. However, an undesirable feature of the thermistor method is that when the power switch is turned off, it should not be turned on again until the thermistor has cooled down.

10.1.3 Suppression of Conducted Interference

Switch-mode power supplies produce both radiated and conducted interferences. Since the input stage is one of the main paths through which the conducted interference can contaminate the ac mains, the methods of minimizing this interference will be discussed in this subsection.

The coupled inductors L_1 and L_2 and the capacitors C_X and C_Y shown in Fig. 10.1 may be used as a line filter to stop the conducted interference from polluting the ac mains. The polarity of coupling between L_1 and L_2 is such that the magnetic effects of the ac mains currents will cancel each other. This eliminates the possibility of magnetic saturation, even with a relatively small core size. In theory, such a polarity of coupling can provide only common-mode interference rejection. However, since L_1 and L_2 are wound in such a way that they both have significant leakage inductance, the leakage inductance can be used to reject the differential-mode interference.

Typical values of L_1, L_2, C_X, and C_Y are:

$$L_1 = L_2 = 2 \text{ to } 50 \text{ mH} \tag{10.4}$$
$$C_X = 0.1 \text{ to } 1 \text{ }\mu\text{F} \tag{10.5}$$
$$C_Y = 2200 \text{ pF to } 0.033 \text{ }\mu\text{F} \tag{10.6}$$

The actual values should depend on the range of frequency of the interference and the power rating of the converter.

10.1.4 Use of SPICE as a CAD Tool

The circuit shown in Fig. 10.1 can be directly simulated by the SPICE program. Both the switch SW and the Triac may be modeled as "near-ideal" switches, based on the method introduced in Section 6.2. For most types of simulations, the loading effect of the "switching regulator" in Fig. 10.1 can be simplified to a resistance.

The following simulations are particularly useful for the design of the input stage:

1. Transient simulation to predict the current and voltage waveforms, especially during the turn-on transient condition
2. Steady-state ac simulation to predict the frequency-response characteristics of the line filter

Repeated trial-and-error simulations can help designers to reach a design quickly and safely.

10.2 SELECTION OF THE CONVERTER TOPOLOGY

In selecting a converter topology for a switch-mode power supply, the following questions must be answered:

1. Is transformer isolation (including transformer step-up or step-down) between the input and the output required?
2. Are multiple outputs required?
3. What is the output power required?
4. What is the desired switching frequency?
5. Are there any special requirements?

Each of the points raised above will be discussed briefly below.

Regarding the need for isolating the input from the output, safety is the major consideration. If safety is not a problem and there is no real need for transformer voltage step-up or step-down, then transformerless converters should be used to eliminate the problems caused by the leakage and primary inductances of transformers. The elimination of transformers will also improve the conversion efficiency.

Regarding the need for multiple outputs, if there is such a requirement, a transformer-coupled converter with multiple output windings should be the best solution. Flyback converters, in this connection, tend to give a better cross-load regulation because of the absence of output filtering inductances, which isolate outputs from each other.

Regarding the output power requirement, this is often the most important factor in determining whether single-ended or push-pull circuits should be used. Normally, for circuits with less than about 200-W output power, simple single-end circuits, such as the buck, boost, buck-boost, Ćuk, forward, or flyback converters, are preferred. For circuits of between 200 and

2000 W, two-transistor forward/flyback and half-bridge push-pull circuits are often used. For circuits with outputs larger than about 2000 W, the full-bridge circuit would be a better choice.

Regarding the switching frequency, it is well known that the higher the switching frequency, the smaller the physical size of the inductors, transformers, and capacitors should be. There is always a limit, however, beyond which the advantages gained do not justify the associated disadvantages. For example, if the switching frequency is so high, and the switching loss is so large, that the increase in the heat-sink size is larger than the decrease in size of the inductor and the capacitor, what is the point of operating at such a high frequency? The optimum operating frequency depends, of course, on the speed of the switching devices, the characteristics of the components, and the circuit topology. Some converter circuits, including the simple forward converter, the simple flyback converter, the transformer-coupled Ćuk converter, and the transformer-coupled push-pull converter, in which the transformer leakage inductance can easily cause excessively high voltage spikes, are not very suitable for high-frequency operation. Circuits in which the voltage spikes are more easily clamped, such as the two-transistor forward, two-transistor flyback, half-bridge, and full-bridge circuits, are more suitable for high-frequency application. Quasi-resonant and resonant converters that make use of the leakage inductance as part of the resonance inductance are, of course, even better for high-frequency operation (e.g., higher than about 500 kHz).

Regarding special requirements, a specification may dictate the use of a certain type of converter. For example, if smooth input and output currents are essential, a Ćuk converter with coupled inductors would be an obvious choice.

10.3 DESIGN OF INDUCTORS

In the design of an inductor, the basic requirement is that it provide the required inductance L at the rated current I without magnetic saturation or overheating. In other words, it must have a large enough LI rating. To achieve this, we need to choose a magnetic core with a sufficiently large core area and to design a winding with a sufficient number of turns and sufficient current-handling capacity, based on the following considerations.

From basic electromagnetic theory, we have

$$L \frac{dI}{dt} = N \frac{d\phi}{dt} \tag{10.7}$$

$$L \, dI = N \, d\phi \tag{10.8}$$

$$LI = N\phi$$
$$LI = NA_cB \tag{10.9}$$

where A_c is the effective cross-sectional area of the magnetic core and B is the magnetic flux density of the inductor.

$$\therefore N = \frac{LI}{A_cB} \tag{10.10}$$

For a given peak current I_{pk}, and a maximum allowable magnetic flux density B_{max}, we have the required number of turns equal to

$$N = \frac{LI_{pk}}{A_cB_{max}} \tag{10.11}$$

where L is in henries, I_{pk} in amperes, A_c in square meters, and B_{max} in teslas. However, for a core with a given winding area A_w, the maximum number of ampere-turns is limited by the current density in the winding:

$$NI_{rms} = JK_uA_w$$
$$N = \frac{JK_uA_w}{I_{rms}} \tag{10.12}$$

where J is the maximum allowable current density of the copper wire in amperes (rms)/square meter, K_u the winding utilization factor (i.e., the percentage of winding area that can be occupied by copper), and K_uA_w the useful copper area of the inductor winding in square meters. (The term JK_uA_w in the numerator is actually the maximum ampere-turns that can be provided by the winding area.) The winding factor K_u for a single-winding inductor is typically 0.4.

Combining Eqs. (10.11) and (10.12), we have

$$\frac{LI_{pk}}{A_cB_{max}} = \frac{JK_uA_w}{I_{rms}} \tag{10.13}$$

$$A_cA_w = \frac{LI_{pk}I_{rms}}{K_uJB_{max}} \text{ m}^4$$

where L is in henries, I_{pk} and I_{rms} are in amperes, J is in amperes/square meter, and B_{max} is in teslas.

If the unit of J is changed to amperes/square centimeter, we have

$$A_c A_w = \frac{L I_{pk} I_{rms}}{K_u J B_{max}} 10^4 \text{ cm}^4 \tag{10.14}$$

Equation (10.14) sets a limit to the minimum size of the magnetic core. It may be understood that the requirement on the magnetic core area A_c is due to the limited maximum flux density of the magnetic core, and the requirement on the winding area A_w is due to the limited current density of the wire. Equation (10.14), together with either Eq. (10.11) or Eq. (10.12), defines the basic requirements of the inductor. (Note that a trade-off between A_c and A_w is possible for inductors and transformers.)

The following three subsections discuss the design of the inductor based on Eqs. (10.14) and (10.11), as well as three other, different constraints.

10.3.1 Design for a Given Current Density

If the current density J in Eq. (10.14) is given, the design will be straightforward. We need only select a magnetic core with a sufficiently large $A_c A_w$ product and then, based on the known values of L, I_{pk}, B_{max}, and A_c, calculate the required number of turns from Eq. (10.11). However, for inductors using magnetic cores with a high μ_r, an air gap in the magnetic core is often required to limit the magnetic flux density to the value of B_{max} at the rated value of peak current I_{pk}. If no air gap is provided, the magnetic core may enter into the saturation region because of the high value of μ_r.

Assuming that the reluctance of the air gap is much larger than that of the magnetic core, we shall have almost all the mmf (magnetomotive force) developed across the air gap. Correspondingly, the inductance is given by

$$L = \frac{\mu_o N^2 A_c}{l_g} \tag{10.15}$$

where l_g is the length of the air gap. From Eq. (10.15), the required air gap can be found as

$$l_g = \frac{\mu_o N^2 A_c}{L} \text{ m} \tag{10.16}$$

where A_c is in square meters and L is in henries.

Based on Eqs. (10.14), (10.11), and (10.16), and a given current density J, we can therefore design the inductor.

10.3.2 Design for a Given Temperature Rise

The design of the inductor described in Subsection 10.3.1 assumes that the current density J in Eq. (10.14) is given. However, in the actual implementation, the selection of J cannot be completely arbitrary. At least one of the limiting factors is the rise in temperature of the inductor. In this subsection, we shall consider a design in which the temperature rise due to copper loss is to be limited to a maximum of 30°C. The magnetic core losses are, for the time being, assumed to be negligible. This assumption is justified if the B_{max} chosen for Eq. (10.14) is sufficiently below the saturation level and the operating frequency of the inductor is below about 50 kHz.

With natural convection cooling, a core with an $A_c A_w$ product of 1 cm⁴, and an allowable temperature rise of 30°C, can have a maximum current density of 420 A/cm². However, the maximum current density decreases as the core size is increased. Empirically, we have [106]

$$J_{30} = 420 (A_c A_w)^{-0.24} \text{ A/cm}^2 \tag{10.17}$$

where J_{30} is the current density for a 30°C rise in temperature. Substitution of Eq. (10.17) into Eq. (10.14) gives

$$A_c A_w = \frac{L I_{pk} I_{rms} 10^4}{K_u B_{max} 420 (A_c A_w)^{-0.24}} \text{ cm}^4 \tag{10.18}$$

$$A_c A_w = \left[\frac{10^4 L I_{pk} I_{rms}}{420 K_u B_{max}} \right]^{1.316} \text{ cm}^4 \tag{10.19}$$

Based on Eq. (10.19), we can therefore select a magnetic core with a sufficiently large $A_c A_w$ product for the inductor. Once the value of A_c of the selected core is known, Eqs. (10.11) and (10.16) can be used to determine the number of turns and the length of the air gap required. The design of the inductor is thus complete. It should, however, be noted that since the derivation of Eq. (10.19) is based on the empirical assumption given in Eq. (10.17), a different expression for $A_c A_w$ may result if a different empirical assumption is made.

10.3.3 Design of the High-Frequency Inductor

In the design described in Subsection 10.3.2, an assumption is made that it is the copper loss that predominantly determines the minimum size of the inductor. This is true only when the operating frequency of the inductor is low, e.g., below 50 kHz. However, for inductors carrying high-frequency currents (e.g., 100 kHz or higher), another factor that will affect the minimum allowable size of the inductor is the core losses due to the magnetic hysteresis and eddy current. When such core losses are large, a standard design practice is to allow for a 15°C rise in temperature from core losses and another 15°C from copper loss.

Assuming such a criterion, the rise in temperature ΔT due to core losses will be equal to

$$\Delta T = 15 = P_c V_c R_{TH} \tag{10.20}$$

where P_c is the power loss per cubic centimeter of the magnetic core, V_c the volume of the core in cubic centimeters, and R_{TH} the thermal resistance of the inductor. Empirically, the values of P_c, V_c, and R_{TH} can be estimated from the following approximate expressions [106]:

$$P_c = \Delta B^{2.4} (K_H f + K_E f^2) \text{ W/cm}^3 \tag{10.21}$$
$$V_c = 5.7 (A_c A_w)^{0.68} \text{ cm}^3 \tag{10.22}$$
$$R_{TH} = 23 (A_c A_w)^{-0.37} \text{ °C/watt} \tag{10.23}$$

where ΔB is the peak-to-peak swing of the magnetic flux density, K_H (typically equal to 4×10^{-5}) the hysteresis loss coefficient, K_E (typically equal to 4×10^{-10}) the eddy-current loss coefficient, and f the operating frequency.

Substitution of Eqs. (10.21), (10.22), and (10.23) into Eq. (10.20) gives

$$15 = \Delta B^{2.4} (K_H f + K_E f^2) \; 5.7 \; (A_c A_w)^{0.68} \; 23 \; (A_c A_w)^{-0.37}$$
$$\therefore \max \Delta B = \Delta B_m = 0.4052 \; (A_c A_w)^{-0.1292} \; [K_H f + K_E f^2]^{-0.4167} \tag{10.24}$$

Since it is now the peak-to-peak swing of magnetic flux density ΔB_m (but not the maximum magnetic flux density B_{max}) that limits the size of the magnetic core, the B_{max} in Eq. (10.14) has to be replaced by ΔB_m. Correspondingly, the peak current I_{pk} has to be replaced by the maximum peak-to-peak swing ΔI_m. Subsequent substitutions of B_{max} by ΔB_m and I_{pk} by ΔI_m in Eq. (10.14) gives the following alternative expression of $A_c A_w$ for high-frequency inductors:

$$A_c A_w = \frac{L \, \Delta I_m \, I_{rms}}{K_u J \, \Delta B_m} 10^4 \text{ cm}^4$$

$$= \frac{L \, \Delta I_m \, I_{rms} \, 10^4}{K_u J \, 0.4052 (A_c A_w)^{-0.1292} \, [K_H f + K_E f^2]^{-0.4167}} \text{ cm}^4 \tag{10.25}$$

Since the current density J in Eq. (10.25) is assumed to be of such a value that it will cause a further increase in temperature equal to 15°C, we have, also empirically [106],

$$J_{15} = 297 \, (A_c A_w)^{-0.24} \text{ A/cm}^2 \qquad (10.26)$$

Substitution of Eq. (10.26) into Eq. (10.25) gives

$$A_c A_w = \frac{L \, \Delta I_m \, I_{rms} \, 10^4}{K_u \, 297 (A_c A_w)^{-0.24} \, 0.4052 (A_c A_w)^{-0.1292} \, [K_H f + K_E f^2]^{-0.4167}} \text{ cm}^4$$

$$A_c A_w = \left[\frac{10^4 L \, \Delta I_m \, I_{rms}}{120.3 \, K_u} \right]^{1.585} [K_H f + K_E f^2]^{0.6606} \text{ cm}^4 \qquad (10.27)$$

In designing inductors carrying high-frequency currents, Eq. (10.27) can be used to determine the $A_c A_w$ product required. Once a core has been selected, Eqs. (10.11) and (10.16) can be used to find the number of turns and the air gap required.

10.4 DESIGN OF TRANSFORMERS

In the design of a transformer, the two basic requirements are:

1. There is no magnetic flux saturation even when the transformer is subjected to the rated maximum voltage.
2. There is no overheating or excessive power loss under the full-load condition.

To meet the first requirement, we need to have a magnetic core with a sufficiently large core area so that

$$\frac{N \, \Delta \phi_m}{T_{ON}} = \frac{N \, \Delta B_m \, A_c}{T_{ON}} > V_{max} \qquad (10.28)$$

where N is the number of turns of the transformer winding, $\Delta \phi_m$ the maximum allowable peak-to-peak swing of the magnetic flux in the magnetic core, ΔB_m the maximum allowable peak-to-peak swing of the flux density of the magnetic core, A_c the effective cross-sectional area of the magnetic core, V_{max} the expected maximum voltage across the transformer winding, and T_{ON} the maximum time for which V_{max} is applied to the winding before a change of polarity takes place.

By assuming that $N \Delta B_m A_c / T_{ON} = V_{max}$, we can determine the magnetic core area required:

$$A_c = \frac{V_{max} T_{ON}}{N \Delta B_m} \qquad (10.29)$$

Or, for a given core area A_c, the required number of turns is

$$N = \frac{V_{max} T_{ON}}{A_c \Delta B_m} \qquad (10.30)$$

where V_{max} is in volts, T_{ON} in seconds, A_c in square meters, and ΔB_m in teslas.

Note that, in the calculation of the number of turns N, it has been assumed that the μ_r of the magnetic core is so high and, therefore, the inductance is so large that the resultant magnetizing current (if any) will be negligibly small.

In order to meet the second requirement stated above, we need to have a sufficiently large winding area so that, for every winding, we have

$$KA_w J = I_{rms} N \qquad (10.31)$$

where A_w is the total physical winding area of the transformer, KA_w the usable copper area of the winding concerned, and I_{rms} the maximum rms current in that winding. (Note that K is different from K_u, as explained later on.) The product $KA_w J$ is actually the maximum number of ampere-turns that can be provided by the winding area of the winding concerned.

From Eq. (10.31), we can determine the physical winding area required:

$$A_w = \frac{I_{rms} N}{KJ} \qquad (10.32)$$

Multiplication of Eq. (10.32) by Eq. (10.29) gives the required $A_c A_w$ product:

$$A_c A_w = \frac{V_{max} I_{rms} T_{ON}}{KJ \Delta B_m} \ \mathrm{m^4}$$

where J is in amperes/square meter and ΔB_m is in teslas.

If J is in amperes/square centimeter, we have

$$A_cA_w = \frac{V_{max}I_{rms}T_{ON}}{KJ\,\Delta B_m} 10^4 \text{ cm}^2 \tag{10.33}$$

Based on Eqs. (10.33) and (10.30), and for given values of V_{max}, I_{rms}, T_{ON}, J, and ΔB_m, the transformer can be designed. However, care must be taken in determining the factor K and in interpreting the current I_{rms} and the current density J.

When applied to the primary winding, the factor K in Eq. (10.32) or (10.33) is actually a product of two factors:

$$K = K_uK_p \tag{10.34}$$

where K_u is the window utilization factor and K_p the primary area factor of the transformer.

The window utilization factor K_u is the percentage of window area that can be usefully occupied by copper. For a transformer with good high-voltage insulation between the primary and secondary windings, the utilization factor is typically 0.4.

The primary area factor K_p in Eq. (10.34) is the percentage of the winding area that can be assigned to the primary winding. For circuits with both the primary and secondary windings single-ended, or both of them center-tapped, the K_p factor is 0.5 because the primary and secondary windings share the current equally (assuming that the additional magnetizing current in the primary winding is very small). However, for circuits with a single-ended primary winding and a center-tapped secondary winding, K_p will be reduced to 0.41. The half-bridge converter circuit shown in Fig. 10.2 can be used to illustrate how this factor of 0.41 is obtained. (In this example, both switches $SW1$ and $SW2$ are assumed to have a duty cycle of 0.5.) Referring to the current waveforms given there, it can be found that

rms value of $i_i = I_{irms} = I$ $\hspace{3cm}$ (10.35)

rms value of $i_{o1} = \dfrac{I}{1.414}$ $\hspace{3cm}$ (10.36)

rms value of $i_{o2} = \dfrac{I}{1.414}$ $\hspace{3cm}$ (10.37)

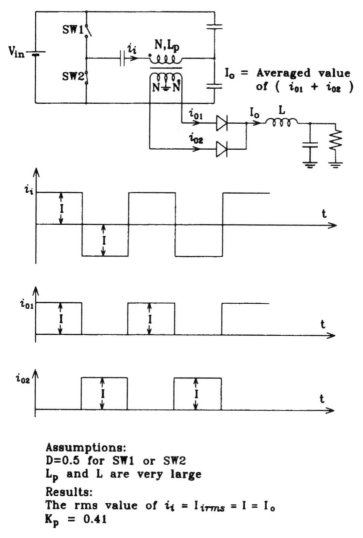

Assumptions:
D=0.5 for SW1 or SW2
L_p and L are very large

Results:
The rms value of $i_i = I_{irms} = I = I_o$
$K_p = 0.41$

FIGURE 10.2 Primary area factor for half-bridge converter.

Hence, we should have the primary-winding share of the ampere-turns, in a percentage, equal to

$$\frac{I}{I + (1/1.414)I + (1/1.414)I} = 41\% \tag{10.38}$$

Only 41% of the winding area should therefore be allocated to the primary winding. The primary area factor for both half-bridge and full-bridge converters is equal to 0.41.

The implication of current I_{rms} in Eq. (10.33) should also be noted with care. Since the allowable current density J in the winding is in rms value to reflect its heating effect, all calculations on winding areas should be based on rms currents. For half-bridge or full-bridge converters with a square-wave current, such as that given in Fig. 10.2, we have the rms value of i_i, denoted as I_{irms}, given by (assuming that $D = 0.5$ for $SW1$ or $SW2$):

$$I_{irms} = I = I_o \tag{10.39}$$

where I_o is the averaged value of the output current $(i_{o1} + i_{o2})$.

However, for single-ended two-transistor forward (or flyback) converters, such as that shown in Fig. 10.3, we have

$$I_{irms} = \frac{1}{1.414} I = \frac{1}{1.414} (2\,I_o) = 1.414\,I_o \tag{10.40}$$

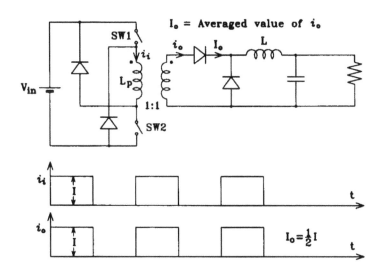

I_o = Averaged value of i_o

Assumptions:
D=0.5 for SW1 or SW2
L_p and L are very large
Results:
I_{irms} (rms value of i_i)=$\frac{I}{\sqrt{2}}$=$\sqrt{2}I_o$
K_p = 0.5

FIGURE 10.3 Root-mean-square current in single-ended two-transistor converter.

A comparison between Eqs. (10.39) and (10.40) indicates that, for a given averaged output current I_o, the half-bridge and full-bridge circuits can have a smaller-size copper wire for the primary winding (because they have a smaller I_{irms}), implying that these circuits make more efficient use of the primary winding than the single-ended two-transistor circuit shown in Fig. 10.3.

For the transformer-coupled push-pull converter shown in Fig. 10.4, we have the rms value of i_{i1} or i_{i2} given by

$$I_{irms} = \frac{1}{1.414} I = \frac{1}{1.414} I_o = 0.707\ I_o \tag{10.41}$$

Equation (10.41) indicates that the primary winding of a transformer-coupled push-pull converter needs to carry an rms current of only $0.707\ I_o$. However, since two windings are required for the primary, it is equally as

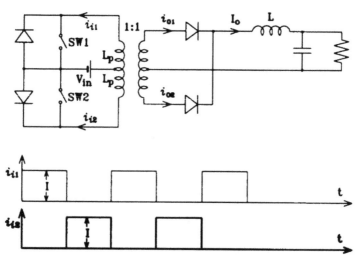

Assumptions:
D=0.5 for SW1 or SW2
L_p and L are very large

Results:
I_{irms} = rms value of i_{i1} (or i_{i2}) $= \dfrac{I}{\sqrt{2}} = \dfrac{I_o}{\sqrt{2}}$
$K_p = 0.5$

FIGURE 10.4 Root-mean-square current in transformer-coupled push-pull converter.

inefficient as the single-ended, two-transistor forward (or flyback) circuit shown in Fig. 10.3.

As a general statement, it may be said that the smaller the form factor of the current (I_{rms}/I_{av}), the better its ability to make use of the transformer (or inductor) winding.

Now, let us consider the current density J in Eq. (10.33). The maximum usable current density in a transformer winding is limited by the allowable temperature rise and, therefore, the power dissipation, of the transformer. In low-frequency transformers, say below 50 kHz, the copper loss in the winding is a main factor that determines the power dissipation and the $A_c A_w$ product. In high-frequency transformers, say above 100 kHz, both the copper loss and core loss (hysteresis loss plus eddy-current loss) are limiting factors. Because of this difference, the design methods for low-frequency and high-frequency transformers are also different, as described in the following two subsections.

10.4.1 Design of the Low-Frequency Transformer

At frequencies below about 50 kHz, a main factor that determines the minimum $A_c A_w$ product is the copper loss. Assuming that a 30°C rise in temperature due to copper loss is allowed, we have the maximum allowable current density J_{30} given empirically [106] by

$$J_{30} = 420(A_c A_w)^{-0.24} \text{ A/cm}^2 \tag{10.42}$$

Substitution of Eq. (10.42) into Eq. (10.33) gives

$$A_c A_w = \frac{V_{max} I_{rms} T_{ON}}{K \, 420(A_c A_w)^{-0.24} \, \Delta B_m} \, 10^4 \text{ cm}^4 \tag{10.43}$$

$$A_c A_w = \left[\frac{10^4 \, V_{max} I_{rms} T_{ON}}{420 \, K \, \Delta B_m} \right]^{1.316} \text{ cm}^4 \tag{10.44}$$

Based on Eq. (10.44), a magnetic core with a sufficiently large $A_c A_w$ product can be selected for the transformer. Once a core has been selected, the value of A_c is known. The number of turns of the primary winding can then be found from Eq. (10.30). The number of turns of other windings are found from their respective voltage ratios.

The window area of the transformer should be shared according to the maximum ampere-turns of the windings and, for each winding, the largest possible size of wire should be used to minimize the copper loss.

10.4.2 Design of the High-Frequency Transformer

At high frequencies, the hysteresis and eddy-current losses become significant. Empirically, we have the core loss per cm^3 given by [106]

$$P_c = \Delta B^{2.4} (K_H f + K_E f^2) \text{ W/cm}^3 \qquad (10.45)$$

where P_c is the total core loss in watts/cubic centimeter, ΔB the peak-to-peak swing of the magnetic flux density, K_H (typically equal to 4×10^{-5}) the hysteresis loss coefficient, and K_E (typically equal to 4×10^{-10}) the eddy-current loss coefficient. For high-frequency transformers, the design should be such that the core loss is equal to the copper loss. Assuming that a total rise in temperature of 30°C is allowed, we should have a rise of 15°C from the core losses and a further rise of 15°C from the copper loss. For the core losses, we have

$$\Delta T = 15 = P_c V_c R_{TH} \qquad (10.46)$$

where V_c is the volume of the magnetic core and R_{TH} the thermal resistance of the transformer. Empirically, V_c and R_{TH} are given by [106]

$$V_c = 5.7(A_c A_w)^{0.68} \text{ cm}^3 \qquad (10.47)$$
$$R_{TH} = 23(A_c A_w)^{-0.37} °C/W \qquad (10.48)$$

Substitution of Eqs. (10.45), (10.47), and (10.48) into Eq. (10.46) gives

$$15 = \Delta B^{2.4} [K_H f + K_E f^2] \, 5.7(A_c A_w)^{0.68} \, 23(A_c A_w)^{-0.37} \qquad (10.49)$$
$$\therefore \Delta B = \Delta B_m = 0.4052 \, (A_c A_w)^{-0.1292} [K_H f + K_E f^2]^{-0.4167} \qquad (10.50)$$

Equation (10.50) indicates that, in order to limit the core loss to the equivalent of a 15°C rise in temperature, the maximum peak-to-peak swing of the flux density ΔB_m needs to be confined to the value given in the equation.

Now, let us consider the 15°C rise in temperature due to the copper loss. We have, empirically, the current density given by [106]

$$J_{15} = 297 \, (A_c A_w)^{-0.24} \text{ A/cm}^2 \qquad (10.51)$$

Substitution of Eqs. (10.50) and (10.51) into Eq. (10.33) gives

$$A_c A_w = \frac{V_{max} I_{rms} T_{ON} \, 10^4}{K \, 297(A_c A_w)^{-0.24} \, 0.4052(A_c A_w)^{-0.1292} [K_H f + K_E f^2]^{-0.4167}} \text{ cm}^4 \qquad (10.52)$$

$$\therefore A_c A_w = \left[\frac{10^4 \ V_{max} I_{rms} T_{ON}}{120.3 \ K} \right]^{1.585} [K_H f + K_E f^2]^{0.6606} \ cm^4 \qquad (10.53)$$

Based on Eqs. (10.53), (10.50), and (10.30), the transformer can therefore be designed using the method described in the preceding subsection.

10.4.3 The Role of CAD Tools in the Design of Transformers and Inductors

The design of transformers/inductors is one of the areas in which CAD tools can be used efficiently to help designers. Some software packages may accept user-specified physical details of a transformer/inductor, such as the type of magnetic core (within the range supported by the model library), the number of turns of each winding, and the coupling coefficients, and produce a SPICE model for the transformer/inductor. This kind of soft model is very useful for trial-and-error simulations to optimize a transformer/inductor design on paper. Simulations based on these models are also valuable for the computer-aided analysis and design work described in Chapters 6, 8, and 9.

Software tools purely for the design of inductors and transformers for switch-mode power supplies are also available. Based on a given set of design rules, such programs provide an efficient guide for the user in arriving at a design.

It should be noted that, because of the different empirical assumptions, the final designs may also differ from one another. In practice, the design and construction work may take a few cycles of iteration and fine-tuning before an optimized solution can be reached.

10.5 DESIGN OF OUTPUT CIRCUITS

The output circuit, which we shall discuss in this section, includes output rectifiers and output filtering inductors and capacitors.

10.5.1 Rectifier Circuits

In the selection of a rectifier, the two most important requirements are:

1. It must have a sufficiently fast recovery time and a low forward voltage drop, both of which are necessary to maintain a high conversion efficiency.
2. It must have a sufficiently large reverse blocking voltage.

In order to meet the first requirement, Schottky diodes should be used if possible because they have practically zero recovery time and a low forward voltage drop when conducting. However, the disadvantages of the Schottky diode are its large junction capacitance and low reverse blocking voltage (up to about 100 V). When a large reverse voltage is present, fast recovery diodes, very fast recovery diodes, or ultrafast recovery diodes may have to be used. Although they are called fast, very fast, or ultrafast recovery diodes, these diodes actually have a longer recovery time than Schottky diodes. They also have a larger voltage drop when they conduct current in the forward direction.

For converters with a very low output voltage, e.g., below 5 V, the voltage drop across the rectifier becomes a significant factor in limiting the conversion efficiency of the converter. Under such conditions, the use of synchronous rectifiers should be considered. The so-called synchronous rectifier is actually a transistor that is driven to the fully on state when it is supposed to conduct. When the rectifier is supposed to be in the blocking state, the transistor is cut off by the gate/base control voltage. Because transistors can be designed to have a low conduction voltage (lower than the forward voltage of diodes), they can be used to improve the rectification efficiency. As examples, Fig. 10.5 gives the simplified circuits of two synchronous rectifiers, one using MOS power transistors and the other using bipolar transistors. It should be noted that, because of the existence of the inherent parallel diode, the source of the MOS power transistor has to be used as the anode of the synchronous rectifier. In the actual operation of the circuit, the inherent parallel diode should not conduct current because the voltage drop across the transistor is too small to turn it on.

The disadvantages of the synchronous rectifier include the following:

1. A more complex circuit
2. Require extra driving power for the gate/base of the transistor (synchronous rectifier)

During operation, if a large ringing voltage is present across a rectifier, an RC snubber circuit may have to be connected across the rectifier to prevent excessive voltage stress on the rectifier. Figure 10.6 shows a typical circuit arrangement and the expected voltage waveforms before and after a snubber circuit is connected.

10.5.2 Determination of the Filtering Inductance

For the buck family of converters (including buck, forward, two-transistor forward, transformer-coupled push-pull, half-bridge, and full-bridge), a

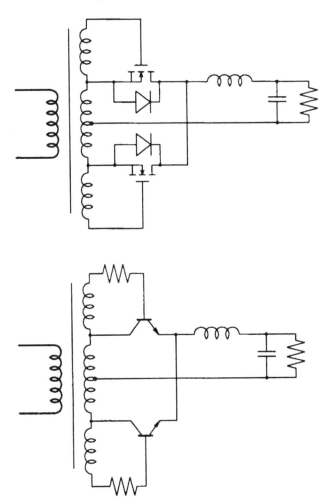

FIGURE 10.5 Synchronous rectifiers.

filtering inductor in the output circuit is necessary for proper circuit operation. Selection of the filtering inductance is commonly based on the
criterion that the converter should remain in continuous-mode operation
when the output current drops to about 20% of the rated maximum. That
being the case, the output inductor current waveform will be as shown in
Fig. 10.7. From there, we have

$$I_{omin} = \frac{1}{2}\frac{V_o}{L}(1 - D)T$$

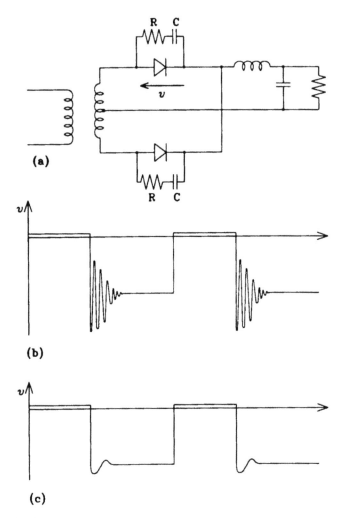

FIGURE 10.6 Use of RC snubber circuit for output rectifier. (a) RC snubber circuit, (b) voltage waveform without snubber, (c) voltage waveform when RC snubber is added.

$$L = \frac{1}{2} \frac{V_o}{I_{omin}} (1 - D)T \tag{10.54}$$

or

$$L = 2.5 \frac{V_o}{I_{omax}} (1 - D)T \tag{10.55}$$

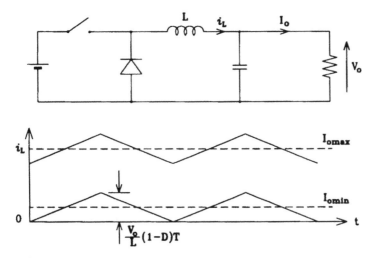

FIGURE 10.7 Output inductor current of buck family converters. Waveform of i_L for maximum and minimum load.

The required inductance L can therefore be found from Eq. (10.54) or Eq. (10.55).

Note that, for push-pull converters (including half-bridge and full-bridge), the effective T in Eqs. (10.54) and (10.55) is half the switching period of each transistor, and the effective D is twice the duty cycle of each transistor.

10.5.3   Selection of the Filtering Capacitor

In the selection of an output filtering capacitor for a converter, the main requirements are:

1. It must have a sufficiently large capacitance to maintain a sufficiently low ripple in the output voltage.
2. It must have a sufficiently small ESR (effective series resistance) to minimize the ripple voltage due to such resistance.

To meet the first requirement stated above, the expression found in Chapter 1 for the output ripple voltages may be used as guidelines in determining the necessary capacitance values. Care should be taken, however, in the interpretation of the parameters T and D in the ripple-voltage expressions for push-pull converters (including half-bridge and full-bridge). For these circuits, T is the period of the output ripple, and the effective D is twice the duty cycle of each switching transistor.

It should also be noted that because of the existence of the effective series resistance and inductance, the actual capacitance value to be used should be considerably larger than the calculated value. In the worst case, the ESR should be of such a value that

$$\text{ESR} \times \Delta I_o < \Delta V_o$$

where ΔI_o is the expected ripple current in the filtering capacitor and ΔV_o is the acceptable ΔV_o due to ESR alone.

In order to reduce the effective series resistance and inductance, a number of capacitors may be connected in parallel to provide the required filtering capacitance instead of using a single capacitor.

10.5.4 The Use of SPICE in the Design of Output Circuits

SPICE is very useful for the analysis and design of the output circuits of switch-mode power supplies. For example, the cycle-by-cycle simulation techniques described in Chapter 6 can be applied to predict the waveforms and performance of the circuits shown in Figs. 10.5 and 10.6. Fine-tuning the RC snubber circuits, shown in Fig. 10.6, based on iterated simulations, can be carried out to optimize their performance. If we have the SPICE models of commercially available rectifiers or synchronous rectifiers, they can also be tried out in simulations to assess their suitability for any particular application.

10.6 DESIGN OF CONTROLLER CIRCUITS

Almost all the control functions required for a converter can now be provided by IC (integrated circuit) controllers. However, circuit designers still have to select an appropriate controller and tailor an interface circuitry for any specific application. In the design and construction of a controller circuit, the basic requirements are:

1. The IC controller selected should be able to provide all (or at least most) of the control functions required, e.g., pulse-width modulation, slow start, and overload protection.
2. The controller circuit must be able to drive the switching transistor(s) properly and with a fast enough speed.
3. The controller circuit must be able to operate stably under all operating conditions. (Some controller circuits tend to produce ab-

normal output pulses when the error-amplifier input voltage or the output-loading capacitance is large.)

4. If galvanic isolation is required between the ac mains and the output circuit, it must be properly provided.

5. In the PCB (printed circuit board) layout, the controller circuitry must be carefully separated from the converter circuitry in order to minimize interference.

Except for special circuits, it is usually easy to find an IC controller to provide all or most of the control functions required by a designer. For converters with a large output power, the limited current driving capability of IC controllers may be insufficient to drive the large-power switching transistors. A buffer stage with a larger output current may therefore have to be added as an interface between the IC controller and the transistor switch.

Regarding the requirement for galvanic isolation, the need may arise from the following situations:

1. If, for safety reasons, there already exists an isolation transformer between the switching transistor and the regulator output circuitry, then, in order to maintain this safety aspect, a means of isolation in the feedback loop will also be necessary.

2. If the transistor switch of the converter is a floating component, whereas the pulse output of the controller circuit is using the ground as the reference, we also need an isolation between the floating transistor and the controller circuit.

Isolation using a pulse transformer between the controller output and the transistor switch is quite common. An example is shown in Fig. 10.8. However, compared with a direct drive, such pulse transformers normally have relatively slow rise and fall times. Clamping or clipping circuits may sometimes also be necessary in the secondary winding to shape the wave-forms.

Optical isolators are often used in the feedback-loop signal path to provide the isolation required. An example of such a circuit is shown in Fig. 10.9.

Isolation using a modulator and a demodulator in the feedback loop, with an RF (radio frequency) transformer acting as the isolator, may also be employed. An example is shown in Fig. 10.10, in which the RF transformer T provides the isolation in the feedback loop. The amplitude modulator in Fig. 10.10 uses a carrier frequency of 2 MHz. (Note that the power to the PWM is supplied by the winding L_1 of the output power transformer.)

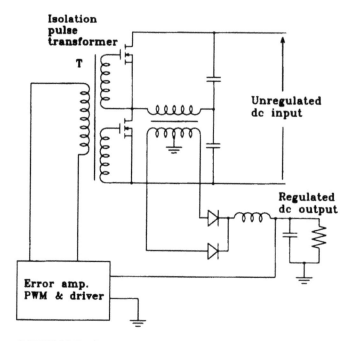

FIGURE 10.8 Regulator with transformer isolation.

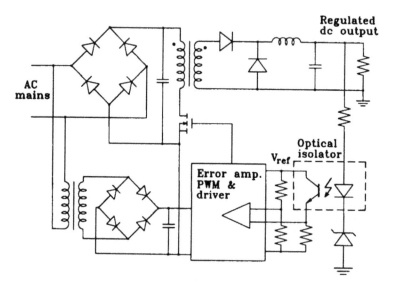

FIGURE 10.9 Regulator with optical isolator in feedback loop.

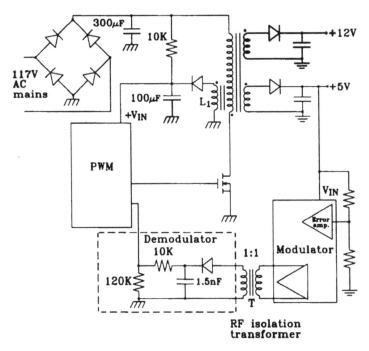

FIGURE 10.10 Regulator with modulator/demodulator in feedback loop.

It is interesting to note that, for converters in which no output filtering inductors are used, e.g., flyback converters, the output sample voltage can actually be transformer-coupled into the feedback loop without any metallic connection. An example of such a circuit is given in Fig. 10.11, in which the transformer winding L_1 is the output-voltage sampling element. At the same time, L_1 also provides power to the PWM IC.

Regarding the PCB layout of the controller circuitry, care should be taken to separate the controller circuit from the electronic switches and power transformer of the converter in order to minimize interference. Proper grounding and shielding should also be arranged to reduce unwanted coupling. PCBs with a ground plane are preferred for both the converter and controller circuits.

Special attention should be paid to the current feedback loop of current-mode-controlled switching regulators. Because of the relatively small current-control feedback voltage and the absence of a large time-constant circuit to remove the high-frequency interference, current-mode-controlled circuits are particularly susceptible to noise pickup and interference.

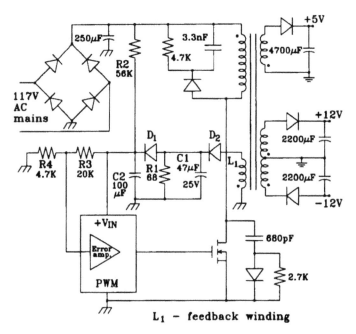

FIGURE 10.11   Flyback regulator with transformer-coupled feedback loop.

Control functions for protection purposes, such as soft start and current limiting (not including current detection), are often incorporated into the IC controllers for switching regulators. ICs designed specifically for supervisory and protection purposes, such as overvoltage sensing, undervoltage sensing, and crowbar protection, are also commonly available. Detailed application circuits can often be found from related application notes.

Regarding the design of the feedback and compensation circuit, SPICE is again a very powerful CAD tool for the designer, as explained in Chapter 8.

10.7   MINIMIZATION OF INTERFERENCE

Two kinds of interference are generated by switch-mode power supplies: conducted interference and radiated interference. Precautions should be taken to minimize both kinds of interference during the design stage of the switch-mode power supply.

Since conducted interference is transmitted mainly through the ac-mains cable, an effective method for stopping the conducted interference from contaminating the ac mains is to insert an appropriate low-pass line filter between the ac mains and the switch-mode power supply, as discussed in Subsection 10.1.3.

So-called radiated interference can actually be classified in the following three categories:

1. Interference through capacitive coupling
2. Interference through inductive coupling
3. Interference through electromagnetic (EM) wave

The first type of interference, due to capacitive coupling, is caused by a voltage source that produces a surrounding electric field. This type of interference can be minimized by inserting a grounded metal sheet between the source of the interference and the object to be shielded. The source of interference may also be totally enclosed within a grounded metal box to prevent the escape of the electric field.

The second type of interference, which is due to inductive coupling, is caused by the magnetic field generated by a current. For low-frequency magnetic fields, high-permeability magnetic materials can be used efficiently for shielding purposes. However, at high frequencies, e.g., 100 kHz or higher, good conducting materials such as copper can provide an equally effective, or even better, shielding. At such high frequencies, the eddy current induced by the magnetic interference produces an opposing magnetic field that cancels out the incident interference.

The third type of interference, which appears in the form of electromagnetic wave (EM wave), is caused by the antenna radiation effect of a current-carrying conductor. The interference due to the EM wave is significantly different from that due to the electric field alone or magnetic field alone. Whereas the electric field is only a storage of electric energy and the magnetic field a storage of magnetic energy, the EM wave is actually a radiation of energy. It is therefore natural that the interference due to an EM wave can extend to a much greater distance than that due to an electric field alone or a magnetic field alone.

The following precautions can be taken to reduce the interference due to EM wave and inductive coupling:

1. Minimize the length of conductors that carry large and high-frequency current.
2. Minimize the area of the current loops.
3. Use twisted pairs of wires if appropriate.
4. Use printed circuit boards with ground planes if necessary.
5. Shield the complete switching power supply unit, using a grounded metal box if necessary.

It should be noted that a lot of trial-and-error work is often required in order to minimize the interference.

10.8 SUMMARY AND FURTHER REMARKS

In this chapter, some practical techniques for designing switch-mode power supplies have been described. The topics covered include the design of the input stage, selection of the converter topology, design of inductors, design of transformers, design of output circuits, design of controller circuits, and the minimization of interference. The role of CAD tools in the design work has also been briefly discussed. Most of the materials presented in this chapter are complementary to those presented in earlier chapters. It is hoped that this chapter will enable the reader to gain a more complete understanding of the practical design aspects of switch-mode power supplies.

EXERCISES

1. In the design of an inductor or a transformer, a trade-off can be made between the magnetic core area and the winding area. Explain why such a trade-off is possible.
2. What are the basic requirements in the design of the following?
   a. low-frequency inductor
   b. low-frequency transformer
   c. high-frequency inductor
   d. high-frequency transformer
3. Compare the characteristics and performance of the Schottky diode with those of the fast-recovery diode.
4. What is a synchronous rectifier? What are the advantages and disadvantages of synchronous rectifiers compared with oridnary rectifiers?
5. Why are the effective series resistance and inductance of filtering capacitors undesirable? How can they be reduced?
6. What are the basic requirements in the design of a controller circuit for a power converter?
7. Describe the practical methods that can be used to reduce noise and interference in switch-mode power supplies.
8. Redesign the power transformer in the converter design example given in Section 8.4, assuming that the converter switching frequency is changed to
   a. 25 KHz
   b. 300 KHz

APPENDIX A

An Introduction to SPICE

A.1 INTRODUCTION

The four appendixes in this book are designed to help the reader learn SPICE (Simulation Program with Integrated Circuit Emphasis).

This first appendix is included for the information of circuit designers who have little or no experience with SPICE. The objective is to enable these designers to appreciate the origin, capability, and applications of SPICE, as well as its limitations.

SPICE was developed in the early 1970s, by the Electronics Research Laboratory of the University of California, Berkeley. The software is now in the public domain, which means that it may be used freely by U.S. citizens. Although the "emphasis" is on integrated circuits, SPICE has been widely used as an industry-standard tool for the design of a wide range of circuits, including discrete and integrated circuits, either analog or digital.

Various commercial versions of SPICE with improved graphics and interactive features, such as PSpice, HSPICE, and IGSPICE, are available. MicroSim Corporation, the software vendor of PSpice, is prepared to provide, free of charge, class instructors with evaluation versions of PSpice, which can be run on an IBM PC or Macintosh. This evaluation version is capable of simulating circuits with up to 10 active components and 64 nodes. Many of the example programs given in this book can be run by the evaluation version of PSpice. (For more details about PSpice, please see Appendix D.)

A.2 TYPES OF ANALYSIS AVAILABLE

The basic types of analysis that can be performed by SPICE include the following:

1. *dc analysis*: The dc analysis can be used to find the dc operating point (biasing point) of a circuit and to determine the dc output voltage (or current) versus dc input voltage (or current) transfer characteristic.
2. *ac analysis*: For a given operating point, the ac analysis can be used to determine the small-signal voltage gain of a circuit and to plot its gain versus frequency transfer function.
3. *Transient analysis*: For a given time-domain input at any node of a circuit, the transient analysis can be used to determine the time-domain response at any other node of the circuit.

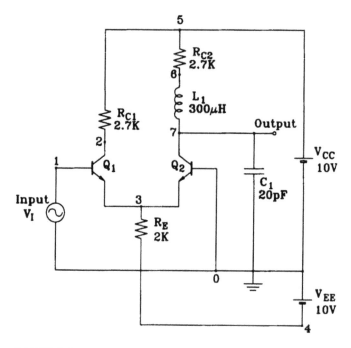

FIGURE A.1 Differential amplifier.

4. *dc sensitivity analysis*: The sensitivity analysis calculates the changes of the dc voltage at any circuit node due to changes of circuit parameters such as supply voltages and resistances.

5. *Fourier analysis*: The Fourier analysis determines the dc, fundamental, and harmonic components of the output of a transient analysis.

6. *Distortion analysis*: The distortion analysis calculates the intermodulation distortion of a circuit. However, this analysis is not available in PSpice.

7. *Noise analysis*: The noise analysis calculates the noise contribution from each component of a circuit and helps the designer to evaluate the noise performance of the circuit.

Of the seven basic types of analysis mentioned above, the first four are more commonly used. In the following sections, examples of these four types of analysis, all based on the circuit shown in Fig. A.1, will be studied.

A.3 DC ANALYSIS AND SENSITIVITY ANALYSIS

In this example, both the dc analysis and sensitivity analysis will be demonstrated. The circuit to be simulated is shown in Fig. A.1. The actual SPICE input file is given below:

```
DIFFERENTIAL AMPLIFIER
VI       1       0       DC       0
Q1       2       1       3       QMOD
RC1      5       2       2.7K
RE       3       4       2K
Q2       7       0       3       QMOD
L1       6       7       300U
RC2      5       6       2.7K
C1       7       0       20P
VCC      5       0       DC       10
VEE      0       4       DC       10
.MODEL   QMOD    NPN
.DC  VI  -0.05   0.15    0.005
.PRINT   DC   V(7,0)
.PLOT    DC   V(7,0)
.SENS    V(7,0)
.END
```

In the SPICE input file given above, the meanings of the statements are:

1. The first statement in the listing:

   ```
   DIFFERENTIAL AMPLIFIER
   ```

   is, by the SPICE language rule, the title of the input file.
2. The second statement,

   ```
   VI      1      0      DC      0
   ```

   specifies that a voltage source called V_I is connected between nodes 1 and 0. V_I is a dc voltage of 0 V. (In this example, since we run only the dc and sensitivity analyses, we do not have any signal input.)
3. The third statement,

   ```
   Q1      2      1      3         QMOD
   ```

   specifies that the bipolar transistor Q_1 is connected to nodes 2, 1, and 3, in the order of collector, base, and emitter. QMOD is the model type of transistor used.
4. The fourth statement,

   ```
   RC1    5      2      2.7K
   ```

   specifies that the resistance R_{C1} is connected between nodes 5 and 2. The resistance value is $2.7K$.
5. The seventh statement,

   ```
   L1     6      7      300U
   ```

   specifies that the inductance L_1 is connected between nodes 6 and 7. The inductance value is 300 μH.
6. The ninth statement,

   ```
   C1    7      0      20P
   ```

   specifies that the capacitance C_1 is connected between nodes 7 and 0. The capacitance value is 20 pF.
7. The tenth statement,

   ```
   VCC    5      0      DC      10
   ```

specifies that the voltage source V_{cc} is connected between nodes 5 and 0 and that V_{cc} is a dc voltage of 10 V. Node 5 is the positive node, and node 0 is the negative node.

8. The model statement,

       .MODEL   QMOD   NPN

defines that QMOD is the default NPN bipolar transistor in the SPICE program.

9. The control statement,

       .DC   VI   -0.05   0.15   0.005

specifies that dc transfer characteristic analyses for a range of V_I from -0.05 to 0.15 V, in steps of 0.005 V, are to be performed.

10. The control statement,

       .PRINT   DC   V(7,0)

specifies that the dc voltage between nodes 7 and 0, with node 0 as reference, is to be printed as output.

11. The control statement,

       .PLOT   DC   V(7,0)

specifies that the dc voltage between nodes 7 and 0, with node 0 as reference, is to be plotted as output.

12. The control statement,

       .SENS   V(7,0)

specifies that a sensitivity analysis is to be carried out to find the sensitivity of V(7,0) due to changes of circuit parameters.

13. The last statement,

       .END

is, by the SPICE language rule, the concluding statement of any SPICE input file.

When the input file mentioned above has been prepared using a text editor, it can be run in SPICE. (Be sure not to use document editors because they may introduce invisible characters not recognized by SPICE.) In a matter of seconds, results such as those given below and on the next few pages can be obtained. (Note that if the program is to be run under the PSpice environment, a .PROBE statement should be added before the .END statement.)

```
******* 08/21/  ************************************** 15:12:49 *******


DIFFERENTIAL AMPLIFIER


****    CIRCUIT DESCRIPTION


*******************************************************************************



VI 1 0 DC 0
Q1 2 1 3 QMOD
RC1 5 2 2.7K
RE 3 4 2K
Q2 7 0 3 QMOD
L1 6 7 300U
RC2 5 6 2.7K
C1 7 0 20P
VCC 5 0 DC 10
VEE 0 4 DC 10
.MODEL QMOD NPN
.DC VI -0.05 0.15 0.005
.PRINT DC V(7,0)
.PLOT DC V(7,0)
.SENS V(7,0)
.END


******* 08/21/  ************************************** 15:12:49 *******


DIFFERENTIAL AMPLIFIER


****    BJT MODEL PARAMETERS


*******************************************************************************



        QMOD
        NPN
   IS   100.000000E-18
   BF   100
   NF     1
   BR     1
   NR     1
```

DIFFERENTIAL AMPLIFIER

****   DC TRANSFER CURVES   TEMPERATURE -   27.000 DEG C

****

VI	V(7,0)
-5.000E-02	-6.720E-01
-4.500E-02	-4.502E-01
-4.000E-02	-1.286E-01
-3.500E-02	2.360E-01
-3.000E-02	6.445E-01
-2.500E-02	1.096E+00
-2.000E-02	1.589E+00
-1.500E-02	2.119E+00
-1.000E-02	2.677E+00
-5.000E-03	3.257E+00
0.000E+00	3.848E+00
5.000E-03	4.440E+00
1.000E-02	5.020E+00
1.500E-02	5.580E+00
2.000E-02	6.110E+00
2.500E-02	6.604E+00
3.000E-02	7.057E+00
3.500E-02	7.467E+00
4.000E-02	7.833E+00
4.500E-02	8.157E+00
5.000E-02	8.423E+00
5.500E-02	8.648E+00
6.000E-02	8.847E+00
6.500E-02	9.021E+00
7.000E-02	9.172E+00
7.500E-02	9.303E+00
8.000E-02	9.415E+00
8.500E-02	9.510E+00
9.000E-02	9.591E+00
9.500E-02	9.659E+00
1.000E-01	9.716E+00
1.050E-01	9.764E+00
1.100E-01	9.804E+00
1.150E-01	9.838E+00
1.200E-01	9.865E+00
1.250E-01	9.889E+00
1.300E-01	9.908E+00
1.350E-01	9.924E+00
1.400E-01	9.937E+00
1.450E-01	9.948E+00
1.500E-01	9.957E+00

DIFFERENTIAL AMPLIFIER

****   DC TRANSFER CURVES   TEMPERATURE =   27.000 DEG C

****


```
   VI          V(7,0)
 (*)----------  -5.0000E+00   0.0000E+00   5.0000E+00   1.0000E+01   1.5000E+01
               - - - - - - - - - - - - - - - - - - - - - - - - -
-5.000E-02 -6.720E-01 .              *  .          .            .            .
-4.500E-02 -4.502E-01 .            *.            .            .            .
-4.000E-02 -1.286E-01 .           *             .            .            .
-3.500E-02  2.360E-01 .            .*            .            .            .
-3.000E-02  6.445E-01 .            . *           .            .            .
-2.500E-02  1.096E+00 .            .  *          .            .            .
-2.000E-02  1.589E+00 .            .    *        .            .            .
-1.500E-02  2.119E+00 .            .      *      .            .            .
-1.000E-02  2.677E+00 .            .        *    .            .            .
-5.000E-03  3.257E+00 .   .        .          *  .            .            .
 0.000E+00  3.848E+00 .            .            *.            .            .
 5.000E-03  4.440E+00 .            .            .*.           .            .
 1.000E-02  5.020E+00 .            .            .  *          .            .
 1.500E-02  5.580E+00 .            .            .    *        .            .
 2.000E-02  6.110E+00 .            .            .      *      .            .
 2.500E-02  6.604E+00 .            .            .        *    .            .
 3.000E-02  7.057E+00 .            .            .          *  .            .
 3.500E-02  7.467E+00 .            .            .            *            .
 4.000E-02  7.833E+00 .            .            .            . *          .
 4.500E-02  8.157E+00 .            .            .            .   *        .
 5.000E-02  8.423E+00 .            .            .            .     *      .
 5.500E-02  8.648E+00 .            .            .            .      *      .
 6.000E-02  8.847E+00 .            .            .            .        * .  .
 6.500E-02  9.021E+00 .            .            .            .         * . .
 7.000E-02  9.172E+00 .            .            .            .          * .
 7.500E-02  9.303E+00 .            .            .            .          * .
 8.000E-02  9.415E+00 .            .            .            .          *  .
 8.500E-02  9.510E+00 .            .            .            .           *. .
 9.000E-02  9.591E+00 .            .            .            .           *. .
 9.500E-02  9.659E+00 .            .            .            .           *. .
 1.000E-01  9.716E+00 .            .            .            .           *. .
 1.050E-01  9.764E+00 .            .            .            .           *. .
 1.100E-01  9.804E+00 .            .            .            .           *. .
 1.150E-01  9.838E+00 .            .            .            .            * .
 1.200E-01  9.865E+00 .            .            .            .            * .
 1.250E-01  9.889E+00 .            .            .            .            * .
 1.300E-01  9.908E+00 .            .            .            .            * .
 1.350E-01  9.924E+00 .            .            .            .            * .
 1.400E-01  9.937E+00 .            .            .            .            * .
 1.450E-01  9.948E+00 .            .            .            .            * .
 1.500E-01  9.957E+00 .            .            .            .            * .
               - - - - - - - - - - - - - - - - - - - - - - - - -
```

```
****** 08/21/  ************************************* 15:12:49 ******

DIFFERENTIAL AMPLIFIER

****     SMALL SIGNAL BIAS SOLUTION      TEMPERATURE =   27.000 DEG C

*********************************************************************************


  NODE    VOLTAGE     NODE    VOLTAGE     NODE    VOLTAGE     NODE    VOLTAGE


(    1)    0.0000  (    2)    3.8485  (    3)    -.7955  (    4)  -10.0000

(    5)   10.0000  (    6)    3.8485  (    7)    3.8485



      VOLTAGE SOURCE CURRENTS
      NAME         CURRENT

      VI          -2.278E-05
      VCC         -4.557E-03
      VEE         -4.602E-03

      TOTAL POWER DISSIPATION   9.16E-02  WATTS
```

****** 08/21/ ************************************* 15:12:49 ******

DIFFERENTIAL AMPLIFIER

****   DC SENSITIVITY ANALYSIS   TEMPERATURE -   27.000 DEG C

****

DC SENSITIVITIES OF OUTPUT V(7,0)

	ELEMENT NAME	ELEMENT VALUE	ELEMENT SENSITIVITY (VOLTS/UNIT)	NORMALIZED SENSITIVITY (VOLTS/PERCENT)
	RC1	2.700E+03	-3.037E-12	-8.199E-11
	RE	2.000E+03	3.067E-03	6.134E-02
	RC2	2.700E+03	-2.278E-03	-6.152E-02
	VI	0.000E+00	1.186E+02	0.000E+00
	VCC	1.000E+01	1.000E+00	1.000E-01
	VEE	1.000E+01	-6.664E-01	-6.664E-02
Q1				
	RB	0.000E+00	0.000E+00	0.000E+00
	RC	0.000E+00	0.000E+00	0.000E+00
	RE	0.000E+00	0.000E+00	0.000E+00
	BF	1.000E+02	-3.037E-04	-3.037E-04
	ISE	0.000E+00	0.000E+00	0.000E+00
	BR	1.000E+00	3.599E-22	3.599E-24
	ISC	0.000E+00	0.000E+00	0.000E+00
	IS	1.000E-16	3.067E+16	3.067E-02
	NE	1.500E+00	0.000E+00	0.000E+00
	NC	2.000E+00	0.000E+00	0.000E+00
	IKF	0.000E+00	0.000E+00	0.000E+00
	IKR	0.000E+00	0.000E+00	0.000E+00
	VAF	0.000E+00	0.000E+00	0.000E+00
	VAR	0.000E+00	0.000E+00	0.000E+00
Q2				
	RB	0.000E+00	0.000E+00	0.000E+00
	RC	0.000E+00	0.000E+00	0.000E+00
	RE	0.000E+00	0.000E+00	0.000E+00
	BF	1.000E+02	-3.037E-04	-3.037E-04
	ISE	0.000E+00	0.000E+00	0.000E+00
	BR	1.000E+00	2.700E-13	2.700E-15
	ISC	0.000E+00	0.000E+00	0.000E+00
	IS	1.000E-16	-3.084E+16	-3.084E-02
	NE	1.500E+00	0.000E+00	0.000E+00
	NC	2.000E+00	0.000E+00	0.000E+00
	IKF	0.000E+00	0.000E+00	0.000E+00
	IKR	0.000E+00	0.000E+00	0.000E+00
	VAF	0.000E+00	0.000E+00	0.000E+00
	VAR	0.000E+00	0.000E+00	0.000E+00

JOB CONCLUDED

TOTAL JOB TIME   8.07

A.4 AC ANALYSIS

An example of the SPICE input file for an ac analysis is shown below:

```
DIFFERENTIAL AMPLIFIER, AC ANALYSIS
VI      1      0      AC       1
Q1      2      1      3        QMOD
RC1     5      2      2.7K
RE      3      4      2K
Q2      7      0      3        QMOD
L1      6      7      300U
RC2     5      6      2.7K
C1      7      0      20P
VCC     5      0      DC       10
VEE     0      4      DC       10
.MODEL  QMOD   NPN
.AC     DEC    10     100K    1G
.PRINT  AC     VDB(7,0)    VP(7,0)
.PLOT   AC     VDB(7,0)    VP(7,0)
.END
```

In the above SPICE input listing, attention should be paid to the usage of the following statements:

·1. The second statement,

```
VI      1      0      AC       1
```

specifies that an ac input voltage source called V_I is connected between nodes 1 and 0. The amplitude of V_I is 1 V. Since the ac analysis is based on the linearized small-signal equivalent circuit, the output voltage should be interpreted as the gain of the circuit rather than as the actual value given in the output file.

2. The control statement,

```
.AC    DEC    10    100K    1G
```

specifies that an ac analysis is to be carried out. The specifications for the analysis are:

i.   The sweep of frequency is by decades in logarithmic scale.
ii. Ten points of analysis for each decade of frequency are required.

    iii.   The start frequency is 100 kHz.
    iv.   The end frequency is 1 GHz.
3. The control statement,

      `.PRINT    AC    VDB(7,0)    VP(7,0)`

specifies that the ac output at node 7 with respect to node 0 is to be printed. Both the amplitude in the decibel scale, VDB(7,0), and the phase, VP(7,0), are required.

4. The control statement,

      `.PLOT    AC    VDB(7,0)    VP(7,0)`

specifies that both VDB(7,0) and VP(7,0) are also to be plotted as computer output.

After the SPICE simulation is run (in a matter of seconds), results such as those given below and on the next few pages can be obtained. (Note that if the program is to be run under the PSpice environment, a .PROBE statement should be added before the .END statement.)

```
******  08/21/  ************************************** 15:12:49 *******


DIFFERENTIAL AMPLIFIER, AC ANALYSIS


****     CIRCUIT DESCRIPTION


**************************************************************************



VI 1 0 AC 1
Q1 2 1 3 QMOD
RC1 5 2 2.7K
RE 3 4 2K
Q2 7 0 3 QMOD
L1 6 7 300U
RC2 5 6 2.7K
C1 7 0 20P
VCC 5 0 DC 10
VEE 0 4 DC 10
.MODEL QMOD NPN
.AC DEC 10 100K 1G
.PRINT AC VDB(7,0) VP(7,0)
.PLOT AC VDB(7,0) VP(7,0)
.END
```

```
******  08/21/  ************************************** 15:12:49 *******
```

DIFFERENTIAL AMPLIFIER, AC ANALYSIS

```
****     BJT MODEL PARAMETERS
```

```
*************************************************************************
```

```
         QMOD
         NPN
   IS    100.000000E-18
   BF    100
   NF    1
   BR    1
   NR    1
```

```
******  08/21/  ************************************** 15:12:49 *******
```

DIFFERENTIAL AMPLIFIER, AC ANALYSIS

```
****     SMALL SIGNAL BIAS SOLUTION       TEMPERATURE =  27.000 DEG C
```

```
*************************************************************************
```

```
NODE   VOLTAGE    NODE   VOLTAGE    NODE   VOLTAGE    NODE   VOLTAGE

(   1)    0.0000  (   2)    3.8485  (   3)   -.7955  (   4) -10.0000

(   5)   10.0000  (   6)    3.8485  (   7)    3.8485
```

```
      VOLTAGE SOURCE CURRENTS
      NAME        CURRENT

      VI          -2.278E-05
      VCC         -4.557E-03
      VEE         -4.602E-03

      TOTAL POWER DISSIPATION   9.16E-02  WATTS
```

DIFFERENTIAL AMPLIFIER, AC ANALYSIS

****   AC ANALYSIS   TEMPERATURE =   27.000 DEG C

****

FREQ	VDB(7,0)	VP(7,0)
1.000E+05	4.152E+01	2.046E+00
1.259E+05	4.154E+01	2.568E+00
1.585E+05	4.157E+01	3.217E+00
1.995E+05	4.163E+01	4.020E+00
2.512E+05	4.171E+01	5.002E+00
3.162E+05	4.184E+01	6.178E+00
3.981E+05	4.205E+01	7.543E+00
5.012E+05	4.238E+01	9.036E+00
6.310E+05	4.288E+01	1.047E+01
7.943E+05	4.364E+01	1.143E+01
1.000E+06	4.477E+01	1.095E+01
1.259E+06	4.639E+01	6.945E+00
1.585E+06	4.839E+01	-5.121E+00
1.995E+06	4.952E+01	-3.086E+01
2.512E+06	4.771E+01	-5.982E+01
3.162E+06	4.436E+01	-7.628E+01
3.981E+06	4.115E+01	-8.366E+01
5.012E+06	3.832E+01	-8.699E+01
6.310E+06	3.579E+01	-8.854E+01
7.943E+06	3.345E+01	-8.928E+01
1.000E+07	3.124E+01	-8.965E+01
1.259E+07	2.910E+01	-8.982E+01
1.585E+07	2.702E+01	-8.991E+01
1.995E+07	2.496E+01	-8.996E+01
2.512E+07	2.293E+01	-8.998E+01
3.162E+07	2.091E+01	-8.999E+01
3.981E+07	1.889E+01	-8.999E+01
5.012E+07	1.688E+01	-9.000E+01
6.310E+07	1.488E+01	-9.000E+01
7.943E+07	1.288E+01	-9.000E+01
1.000E+08	1.087E+01	-9.000E+01
1.259E+08	8.872E+00	-9.000E+01
1.585E+08	6.871E+00	-9.000E+01
1.995E+08	4.870E+00	-9.000E+01
2.512E+08	2.870E+00	-9.000E+01
3.162E+08	8.696E-01	-9.000E+01
3.981E+08	-1.131E+00	-9.000E+01
5.012E+08	-3.131E+00	-9.000E+01
6.310E+08	-5.131E+00	-9.000E+01
7.943E+08	-7.131E+00	-9.000E+01
1.000E+09	-9.131E+00	-9.000E+01

```
******  08/21/  ************************************* 15:12:49 *******


DIFFERENTIAL AMPLIFIER, AC ANALYSIS


****      AC ANALYSIS                    TEMPERATURE =   27.000 DEG C


*************************************************************************

 LEGEND:

*: VDB(7,0)
+: VP(7,0)

  FREQ        VDB(7,0)

(*)----------  -2.0000E+01  0.0000E+00  2.0000E+01  4.0000E+01  6.0000E+01
(+)----------  -1.0000E+02 -5.0000E+01  0.0000E+00  5.0000E+01  1.0000E+02

  1.000E+05  4.152E+01 - - - - - - .- - - - - - - .+- - - .* - - - .
  1.259E+05  4.154E+01 .           .            .+         .*        .
  1.585E+05  4.157E+01 .           .            .+         .*        .
  1.995E+05  4.163E+01 .           .            .+         .*        .
  2.512E+05  4.171E+01 .           .            .+         .*        .
  3.162E+05  4.184E+01 .           .            . +        .*        .
  3.981E+05  4.205E+01 .           .            . +        .*        .
  5.012E+05  4.238E+01 .           .            . +        . *       .
  6.310E+05  4.288E+01 .           .            .  +       . *       .
  7.943E+05  4.364E+01 .           .            .  +       . *       .
  1.000E+06  4.477E+01 .           .            .  +       .  *      .
  1.259E+06  4.639E+01 .           .            .  +       .   *     .
  1.585E+06  4.839E+01 .           .          +.          .    *    .
  1.995E+06  4.952E+01 .           .    +       .          .     *   .
  2.512E+06  4.771E+01 .         +  .            .          .    *   .
  3.162E+06  4.436E+01 .      +     .            .          .   *    .
  3.981E+06  4.115E+01 .   +        .            .          .  *     .
  5.012E+06  3.832E+01 . +          .            .        *  .       .
  6.310E+06  3.579E+01 . +          .            .       *  .        .
  7.943E+06  3.345E+01 . +          .            .      *   .        .
  1.000E+07  3.124E+01 . +          .            .    *     .        .
  1.259E+07  2.910E+01 . +          .            .   *      .        .
  1.585E+07  2.702E+01 . +          .            .  *       .        .
  1.995E+07  2.496E+01 . +          .           . *        .        .
  2.512E+07  2.293E+01 . +          .           .*          .        .
  3.162E+07  2.091E+01 . +          .          .*           .        .
  3.981E+07  1.889E+01 . +          .         .*            .        .
  5.012E+07  1.688E+01 . +          .        * .            .        .
  6.310E+07  1.488E+01 . +          .       *  .            .        .
  7.943E+07  1.288E+01 . +          .     *    .            .        .
  1.000E+08  1.087E+01 . +          .    *     .            .        .
  1.259E+08  8.872E+00 . +          .  *       .            .        .
  1.585E+08  6.871E+00 . .+         . *        .            .        .
  1.995E+08  4.870E+00 . +        . *         .            .        .
  2.512E+08  2.870E+00 . +        .*          .            .        .
  3.162E+08  8.696E-01 . +        .*          .            .        .
  3.981E+08 -1.131E+00 . +      *. .           .            .        .
  5.012E+08 -3.131E+00 . +     * . .           .            .        .
  6.310E+08 -5.131E+00 . +    * . .           .            .        .
  7.943E+08 -7.131E+00 . +  *    .           .            .        .
  1.000E+09 -9.131E+00 . +  *     .           .            .        .
                        - - - - - - - - - - - - - - - - - - - - - - -

        JOB CONCLUDED

        TOTAL JOB TIME        6.81
```

A.5   TRANSIENT ANALYSIS

An example of the SPICE input file for a transient simulation is shown below:

```
DIFFERENTIAL AMPLIFIER, TRANSIENT ANALYSIS
VI      1       0       PWL     (0  0   1U  0
+ 1.001U   0.1   5U  0.1   5.001U  0   6U  0)
Q1      2       1       3       QMOD
RC1     5       2       2.7K
RE      3       4       2K
Q2      7       0       3       QMOD
L1      6       7       300U
RC2     5       6       2.7K
C1      7       0       20P
VCC     5       0       DC      10
VEE     0       4       DC      10
.MODEL  QMOD    NPN
.TRAN   0.15U   6U
.PRINT  TRAN    V(1,0)    V(7,0)
.PLOT   TRAN    V(1,0)    V(7,0)
.END
```

In the above SPICE input file, attention should be paid to the following statements:

1. The second statement

    ```
    VI      1       0       PWL     (0  0   1U  0
    + 1.001U       0.1   5U  0.1   5.001U  0    6U  0)
    ```

    specifies that the voltage source V_I is connected between nodes 1 and 0. The waveform of V_I (the voltage at node 1 with respect to 0) is described by a piecewise linear function that joins the coordinate pairs given within the brackets. The actual waveform is shown graphically in Fig. A.2.

2. The control statement,

    ```
    .TRAN   0.15U   6U
    ```

    specifies that a transient simulation is to be carried out. The time step for printing/plotting the output is 0.15 μs, and the final time is 6 μs.

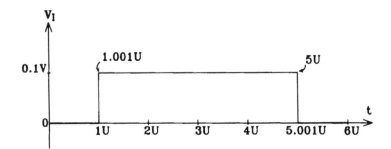

FIGURE A.2   Waveform of V_I for transient simulation.

3.   The control statements,

```
.PRINT   TRAN   V(1,0)     V(7,0)
.PLOT    TRAN   V(1,0)     V(7,0)
```

specify that the waveforms of V(1,0) and V(7,0) are to be printed
and plotted as output.

After the SPICE simulation is run, results such as those given on the next
few pages can be obtained. It should be noted that, because of the large
number of calculations involved, transient simulations take longer to complete than the dc or ac analysis. (Note that if the program is to be run
under the PSpice environment, a .PROBE statement should be added
before the .END statement.)

DIFFERENTIAL AMPLIFIER, TRANSIENT ANALYSIS

****   CIRCUIT DESCRIPTION

****

```
VI 1 0 PWL(0 0   1U 0   1.001U 0.1   5U 0.1   5.001U 0   6U 0)
Q1 2 1 3 QMOD
RC1 5 2 2.7K
RE 3 4 2K
Q2 7 0 3 QMOD
L1 6 7 300U
RC2 5 6 2.7K
C1 7 0 20P
VCC 5 0 DC 10
VEE 0 4 DC 10
.MODEL QMOD NPN
.TRAN 0.15U 6U
.PRINT TRAN V(1,0) V(7,0)
.PLOT TRAN V(1,0) V(7,0)
.END
```

DIFFERENTIAL AMPLIFIER, TRANSIENT ANALYSIS

****   BJT MODEL PARAMETERS

****

```
        QMOD
        NPN
   IS   100.000000E-18
   BF   100
   NF     1
   BR     1
   NR     1
```

DIFFERENTIAL AMPLIFIER, TRANSIENT ANALYSIS

****   INITIAL TRANSIENT SOLUTION   TEMPERATURE =   27.000 DEG C

****

NODE	VOLTAGE	NODE	VOLTAGE	NODE	VOLTAGE	NODE	VOLTAGE
(1)	0.0000	(2)	3.8485	(3)	-.7955	(4)	-10.0000
(5)	10.0000	(6)	3.8485	(7)	3.8485		

```
        VOLTAGE SOURCE CURRENTS
        NAME        CURRENT

        VI          -2.278E-05
        VCC         -4.557E-03
        VEE         -4.602E-03

        TOTAL POWER DISSIPATION   9.16E-02  WATTS
```

```
****** 08/21/ ************************************** 15:12:49 ******


DIFFERENTIAL AMPLIFIER, TRANSIENT ANALYSIS


****      TRANSIENT ANALYSIS              TEMPERATURE =   27.000 DEG C


****************************************************************************
```

```
TIME        V(1,0)      V(7,0)

0.000E+00   0.000E+00   3.848E+00
1.500E-07   0.000E+00   3.848E+00
3.000E-07   0.000E+00   3.848E+00
4.500E-07   0.000E+00   3.848E+00
6.000E-07   0.000E+00   3.848E+00
7.500E-07   0.000E+00   3.848E+00
9.000E-07   0.000E+00   3.848E+00
1.050E-06   1.000E-01   8.810E+00
1.200E-06   1.000E-01   1.347E+01
1.350E-06   1.000E-01   9.176E+00
1.500E-06   1.000E-01   8.725E+00
1.650E-06   1.000E-01   1.006E+01
1.800E-06   1.000E-01   9.979E+00
1.950E-06   1.000E-01   9.562E+00
2.100E-06   1.000E-01   9.653E+00
2.250E-06   1.000E-01   9.774E+00
2.400E-06   1.000E-01   9.743E+00
2.550E-06   1.000E-01   9.692E+00
2.700E-06   1.000E-01   9.705E+00
2.850E-06   1.000E-01   9.726E+00
3.000E-06   1.000E-01   9.721E+00
3.150E-06   1.000E-01   9.712E+00
3.300E-06   1.000E-01   9.714E+00
3.450E-06   1.000E-01   9.718E+00
3.600E-06   1.000E-01   9.717E+00
3.750E-06   1.000E-01   9.715E+00
3.900E-06   1.000E-01   9.716E+00
4.050E-06   1.000E-01   9.716E+00
4.200E-06   1.000E-01   9.716E+00
4.350E-06   1.000E-01   9.716E+00
4.500E-06   1.000E-01   9.716E+00
4.650E-06   1.000E-01   9.716E+00
4.800E-06   1.000E-01   9.716E+00
4.950E-06   1.000E-01   9.716E+00
5.100E-06   0.000E+00   1.413E+00
5.250E-06   0.000E+00   1.279E+00
5.400E-06   0.000E+00   4.999E+00
5.550E-06   0.000E+00   4.535E+00
5.700E-06   0.000E+00   3.289E+00
5.850E-06   0.000E+00   3.528E+00
6.000E-06   0.000E+00   4.074E+00
```

****** 08/21/ *************************************** 15:12:49 ******

DIFFERENTIAL AMPLIFIER, TRANSIENT ANALYSIS

****   TRANSIENT ANALYSIS   TEMPERATURE =   27.000 DEG C

****

```
  LEGEND:

*: V(1,0)
+: V(7,0)

   TIME       V(1,0)
(*)----------  -5.0000E-02   0.0000E+00   5.0000E-02   1.0000E-01   1.5000E-01
(+)----------  -5.0000E+00   0.0000E+00   5.0000E+00   1.0000E+01   1.5000E+01

  0.000E+00  0.000E+00 - - - - - - - -*- - - -+ - - - - - - -.- - - - -
  1.500E-07  0.000E+00 .             *        +  .              .
  3.000E-07  0.000E+00 .             *        +  .              .
  4.500E-07  0.000E+00 .             *        +  .              .
  6.000E-07  0.000E+00 .             *        +  .              .
  7.500E-07  0.000E+00 .             *        +  .              .
  9.000E-07  0.000E+00 .   .         *        +  .              .
  1.050E-06  1.000E-01 .             .           .    +   *     .
  1.200E-06  1.000E-01 .             .           .        *        +  .
  1.350E-06  1.000E-01 .             .           .      +   *     .
  1.500E-06  1.000E-01 .             .           .      +   *     .
  1.650E-06  1.000E-01 .             .           .        X       .
  1.800E-06  1.000E-01 .             .           .        X       .
  1.950E-06  1.000E-01 .             .           .       +*       .
  2.100E-06  1.000E-01 .             .           .       +*       .
  2.250E-06  1.000E-01 .             .           .       +*       .
  2.400E-06  1.000E-01 .             .           .       +*       .
  2.550E-06  1.000E-01 .             .           .       +*       .
  2.700E-06  1.000E-01 .             .           .       +*       .
  2.850E-06  1.000E-01 .             .           .       +*       .
  3.000E-06  1.000E-01 .             .           .       +*       .
  3.150E-06  1.000E-01 .             .           .       +*       .
  3.300E-06  1.000E-01 .             .           .       +*       .
  3.450E-06  1.000E-01 .             .           .       +*       .
  3.600E-06  1.000E-01 .             .           .       +*       .
  3.750E-06  1.000E-01 .             .           .       +*       .
  3.900E-06  1.000E-01 .             .           .       +*       .
  4.050E-06  1.000E-01 .             .           .       +*       .
  4.200E-06  1.000E-01 .             .           .       +*       .
  4.350E-06  1.000E-01 .             .           .       +*       .
  4.500E-06  1.000E-01 .             .           .       +*       .
  4.650E-06  1.000E-01 .             .           .       +*       .
  4.800E-06  1.000E-01 .             .           .       +*       .
  4.950E-06  1.000E-01 .             .           .       +*       .
  5.100E-06  0.000E+00 .        *    +           .              .
  5.250E-06  0.000E+00 .        *     +          .              .
  5.400E-06  0.000E+00 .        *           +    .              .
  5.550E-06  0.000E+00 .        *          +.    .              .
  5.700E-06  0.000E+00 .        *      +       .              .
  5.850E-06  0.000E+00 .        *      +       .              .
  6.000E-06  0.000E+00 .- - - - -*- - - - - + -:- - - - - - -.- - - - -:
```

     JOB CONCLUDED

     TOTAL JOB TIME   12.41

A.6 APPLICATIONS AND LIMITATIONS OF SPICE

SPICE is a very useful numerical analysis tool for the simulation of electronic circuits. It is particularly useful for nonlinear circuits because such circuits are very difficult to analyze manually. SPICE has been used widely in the design of analog circuits and small- to medium-scale digital circuits. However, for very large circuits, such as LSI (large-scale integrated circuits) or VLSI (very-large-scale integrated circuits), SPICE may take a very long time to complete a simulation.

Also, the simulated results from SPICE, because of its numerical nature, do not provide insight into the operation of the circuit. It is the circuit designer who has to be creative in generating new ideas and interpreting the simulated results correctly.

Another problem that is often encountered in the SPICE simulation is the nonconvergence in dc or transient analysis. This is due to the limited ability of the numerical analysis to handle abrupt changes of circuit parameters or to inappropriate circuit models. When such a problem occurs, SPICE will fail to continue with the simulation. Some possible methods for dealing with this problem are discussed in Appendix C.

APPENDIX B

Summary of Commonly Used
SPICE Statements

B.1 INTRODUCTION

The main content of this appendix is a brief summary (in Section B.3) of the important SPICE statements commonly used for the modeling and simulation of electronic circuits. The summary is intended to be a quick reference to remind the reader of the basic formats of the SPICE statements. In order to help the reader appreciate the uses of these statements, many of them are actually quoted from two input file examples. The first example is a behavior model of a bipolar operational amplifier, called Input File 1. The second example is a flyback dc-to-dc power converter, called Input File 2. Whereas the text files are attached to the end of this appendix, the circuits they represent are shown in Figs. B.1 and B.2.

After a brief explanation of the notations and symbols used in the explanatory notes and in the SPICE statements, the actual summary of statements will be given in Section B.3.

Readers who have no previous knowledge of SPICE are advised to read Appendix A before attempting to read the following sections of this appendix.

B.2 NOTATIONS USED

Table B.1 lists the notations used in the explanatory notes for the SPICE statements.

Table B.2 lists the definitions of symbols commonly used in SPICE statements.

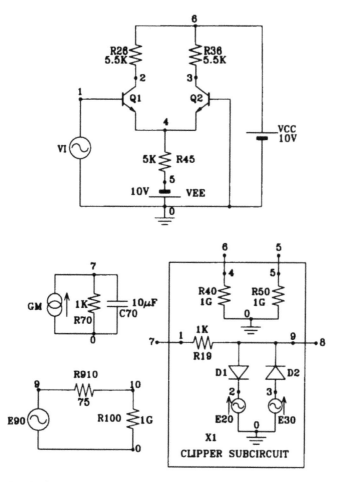

FIGURE B.1   Circuit for Input File 1 (behavior model of operational amplifier).

Since SPICE does not allow the use of subscripts, we shall also not use subscripts in the summary of statements given in Section B.3.

B.3   SUMMARY OF COMMONLY USED SPICE STATEMENTS

In the following summary of SPICE statements, many examples are actually quoted from the two SPICE input files attached to the end of this appendix (Input File 1 and Input File 2). The comments explain the meaning of the

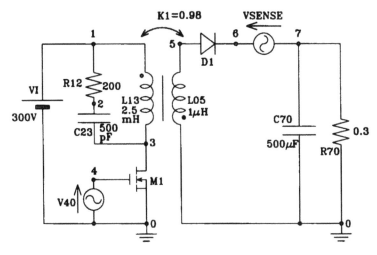

FIGURE B.2   Circuit for Input File 2 (flyback converter).

statements. Note that the numbers assigned to the SPICE statements in the two attached input files are for the purpose of easy reference only. They should not appear in the actual SPICE input file.

B.3.1 Title and End Statements

The first statement of a SPICE input file is always interpreted by SPICE as the title of the file.

Example 1:

```
OP-AMP MODEL
* Quoted from line 1 of Input File 1.
* Meaning:
```

Table B.1   Notations Used in Explanatory Notes for SPICE Statements

Notations used for the explanation of SPICE statements	Meaning
XXXXXXX or YYYYYYY	A string of up to 7 alphanumeric characters
( )	Optional parameters
*	Comment statement

Table B.2   Definitions of Commonly Used SPICE Symbols

Commonly used SPICE symbols	Meaning
V(1)	Voltage at node 1
V(2)	Voltage at node 2
V(3)	Voltage at node 3
. . .	. . .
	(All with respect to ground)
V(1, 2)	Voltage at node 1 with respect to node 2
V(2, 3)	Voltage at node 2 with respect to node 3
V(3, 0)	Voltage at node 3 with respect to node 0
. . .	. . .
VXXXXXXX	Voltage source
EXXXXXXX	Voltage-controlled voltage source
HXXXXXXX	Current-controlled voltage source
I(VXXXXXXX)	Current flowing into voltage source VXXXXXXX
IXXXXXXX	Current source
GXXXXXXX	Voltage-controlled current source
FXXXXXXX	Current-controlled current source
CXXXXXXX	Capacitor
LXXXXXXX	Inductor
RXXXXXXX	Resistor
QXXXXXXX	Bipolar transistor
DXXXXXXX	Diode
JXXXXXXX	Junction FET
MXXXXXXX	MOSFET
F	Femto (10^{-15})
P	Pico (10^{-12})
N	Nano (10^{-9})
U	Micro (10^{-6})
M	Milli (10^{-3})
K	Kilo (10^{3})
MEG (Be sure not to mix M with MEG.)	Mega (10^{6})

Table B.2   Continued

Commonly used SPICE symbols	Meaning
G	Giga (10^9)
T	Tera (10^{12})
*	Comment statement
+	Continuation of last statement
.	Model or command statement

```
* Title of input file is
* OP-AMP MODEL
```

Example 2:

```
FLYBACK CONVERTER
* Quoted from line 1 of Input File 2.
* Meaning:
* Title of input file is
* FLYBACK CONVERTER
```

The last statement of a SPICE input file must be the .END statement.

Example:

```
.END
```

B.3.2   Passive Components

General formats:

```
RXXXXXXX N+ N- VALUE
CXXXXXXX N+ N- VALUE
LXXXXXXX N+ N- VALUE
```

Example 1:

```
R26    2    6    5.5K
* Quoted from line 4 of Input File 1.
```

```
*  Meaning:
*  1.  Resistor R26 is connected between nodes 2
*      and 6.
*  2.  The value of R26 is 5.5 kΩ.
*  Note:  It is not necessary that the resistor
*         connected between nodes 2 and 6 be called
*         R26. This term is used here only for
*         convenience.
```

Example 2:

```
C70    7    0    10U
*  Quoted from line 12 of Input File 1.
*  Meaning:
*  1.  Capacitor C70 is connected between nodes 7
*      and 0.
*  2.  The value of C70 is 10 μF.
*  Note: U is used to mean μ in SPICE.
```

Example 3:

```
L13    1    3    2.5M
*  Quoted from line 5 of Input File 2.
*  Meaning:
*  1.  Inductor L13 is connected between nodes 1
*      and 3.
*  2.  The value of L13 is 2.5 mH.
*  Note:  M is used to mean milli (and MEG to mean
*         mega) in SPICE.
```

B.3.3 Diodes

General format:

```
DXXXXXXX N+ N- MNAME
```

Example:

```
D1    9    2    DMOD
*  Quoted from line 16 of Input File 1.
*  Meaning:
*  1.  Diode D1 is connected between nodes 9 and 2.
```

```
*        (In this example, 9 and 2 are the internal
*        node numbers of the subcircuit CLIPPER.
*        For more information about subcircuits,
*        please see Section B 3.13.)
* 2. The anode of the diode is connected to
*        node 9.
* 3. The cathode of the diode is connected to
*        node 2.
* 4. The model name (MNAME) of the diode is DMOD.
```

B.3.4 Bipolar Transistor

General Format:

```
QXXXXXXX NC NB NE MNAME
```

Example:

```
Q1    2    1    4    QMOD
* Quoted from line 3 of Input File 1.
* Meaning:
* 1. The collector, base, and emitter of bipo-
*        lar transistor Q1 is connected to nodes 2,
*        1, and 4, respectively.
* 2. QMOD (MNAME) is the model name of the bi-
*        polar transistor.
```

B.3.5 MOSFET

General format:

```
MXXXXXXX ND NG NS NB MNAME ⟨W=VALUE⟩ ⟨L=VALUE⟩
```

Example:

```
M1    3    4    0    0    MOD1    W=1
* Quoted from line 8 of Input File 2.
* Meaning:
* 1. The drain, gate, source, and substrate of
*        MOSFET M1 is connected to nodes 3, 4, 0, and
*        0, respectively.
```

```
*  2.  MOD1 is the model name of the transistor.
*  3.  The effective channel width of M1 is 1 m.
```

B.3.6 Independent Voltage and Current Sources

General formats:

```
VXXXXXXX N+ N- ⟨DC VALUE⟩⟨AC VALUE⟩
IXXXXXXX N+ N- ⟨DC VALUE⟩⟨AC VALUE⟩
*  The positive direction of the current source is
*  defined as from N+, through the current source,
*  and then to N-.
```

Example 1:

```
VCC    6    0    DC    10
*  Quoted from line 8 of Input File 1.
*  Meaning:
*  1.  A dc voltage source VCC is connected be-
*      tween nodes 6 and 0.
*  2.  The dc voltage at node 6 (N+) is 10 V with
*      respect to node 0 (N-).
```

Example 2:

```
VEE    0    5    DC    10
*  Quoted from line 9 of Input File 1.
*  Meaning:
*  1.  A dc voltage VEE is connected between nodes
*      0 and 5.
*  2.  The dc voltage at node 0 with respect to
*      node 5 is 10 V.
```

Example 3:

```
VI    1    0    AC    1
*  Quoted from line 2 of Input File 1.
*  Meaning:
*  1.  An ac voltage VI is connected between nodes
*      1 and 0.
*  2.  The ac voltage at node 1 is 1∠0° V, with re-
*      spect to node 0.
```

```
* Note:  This statement is often used to mean one
*        unit of small-signal input voltage in
*        ac analysis.
```

Example 4:

```
I9    18    22    DC    1
* Meaning that I9 is a dc current source flowing
* in the direction from node 18, through current
* source I9, to node 22. The amplitude of I9 is
* 1A.
```

B.3.7 Piecewise Linear Voltage and Current Sources

General formats:

```
VXXXXXXX N+ N- PWL (T1 V1 ⟨T2 V2 T3 V3 T4 V4 . . .⟩)
IXXXXXXX N+ N- PWL (T1 I1 ⟨T2 I2 T3 I3 T4 I4 . . .⟩)
```

Example:

```
V40 4 0 PWL (0 0 10U 0 10.05U 10 20U 10 20.05U 0 30U
+ 0 30.05U 10 40U 10 40.05U 0 50U 0 50.05U 10 60U
+ 10 60.05U 0 70U 0 70.05U 10 80U 10 80.05U 0 90U
+ 0 90.05U 10 100U 10 100.05U 0)
* Quoted from lines 13-16 of Input File 2.
* Meaning that the waveform of V40 is a piecewise
* linear function joining the coordinate pairs
* given within the brackets, as shown in Fig.
* B.3.
```

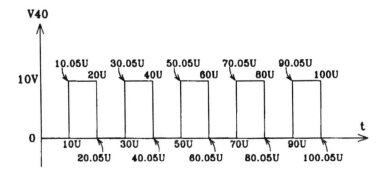

FIGURE B.3 PWL waveform of V40.

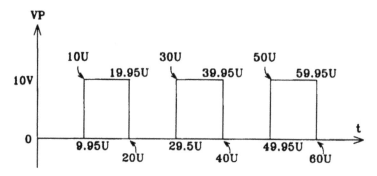

FIGURE B.4 Waveform of VP.

B.3.8 Pulse Voltage and Current Sources

General formats:

```
VXXXXXXX N+ N- PULSE (V1 V2 TD TR TF PW PER)
IXXXXXXX N+ N- PULSE (I1 I2 TD TR TF PW PER)
* Note:  V1 or I1 = Initial value
*        V2 or I2 = Pulse value
*        TD = Delay time
*        TR = Rise time
*        TF = Fall time
*        PW = Pulse width
*        PER = Period
```

Example:

```
VP 5 0 PULSE (0 10 9.95U .05U .05U 9.95U 20U)
* Meaning the voltage waveform as shown in Fig.
* B.4.
```

B.3.9 Voltage-Controlled Voltage and Current Sources with Single Control Variable

General formats for voltage-controlled voltage source:

```
EXXXXXXX N+ N- ⟨POLY(1)⟩ NC+ NC- P1
* Meaning that the voltage at N+ (with respect to
* N-) is equal to the control voltage at NC+ (with
```

```
*  respect to NC-) multiplied by P1.
EXXXXXXX N+ N- (POLY(1)) NC+ NC-
+ (P0 P1 P2 P3 P4 P5 P6 P7 . . .)
* Meaning that if the control voltage at NC+ (with
* respect to NC-) is denoted as A, the voltage at
* N+ (with respect to N-) will be equal to the poly-
* nomial function P0 + P1 A + P2 A² + P3 A³ + P4
* A⁴ + P5 A⁵ + P6 A⁶ + P7 A⁷ + · · ·
```

Example 1:

```
E90    9    0    POLY(1)    8    0    0    1
*      N+   N-                    NC+  NC-  P0   P1
* Quoted from line 24 of Input File 1.
* Meaning:
* 1. E90 is a voltage-controlled voltage
*    source connected between nodes 9 and 0.
* 2. The voltage at node 9 (N+) with respect to
*    node 0 (N-) is controlled by the voltage at
*    node 8 (NC+) with respect to node 0 (NC-).
* 3. E90 = V(8, 0).
```

Example 2:

```
EA1    1    2    POLY(1)    3    4    5    6    7    8    9
*      N+   N-                    NC+  NC-  P0   P1   P2   P3   P4
* Meaning:
* 1. EA1 is a voltage-controlled voltage
*    source connected between nodes 1 and 2.
* 2. The voltage at node 1 (N+) with respect to
*    node 2 (N-) is controlled by the voltage at
*    node 3 (NC+) with respect to node 4 (NC-).
* 3. EA1 is a single-variable polynomial func-
*    tion of V(3, 4):
*    EAI = P0 + P1[V(3, 4)] + P2[V(3, 4)]² +
*    P3[V(3, 4)]³ + P4[V(3, 4)]⁴ = 5 + 6[V(3, 4)] +
*    7[V(3, 4)]² + 8[V(3, 4)]³ + 9[V(3, 4)]⁴
```

General formats for voltage-controlled current source:

```
GXXXXXXX N+ N- (POLY(1)) NC+ NC- P1
GXXXXXXX N+ N- (POLY(1)) NC+ NC-
+ (P0 P1 P2 P3 P4 P5 P6 P7 . . .)
```

Example 1:

```
GM      0   7    POLY(1)     3     2     0      0.1
*       N+  N-                NC+   NC-   P0     P1
* Quoted from line 10 of Input File 1.
* Meaning:
* 1.  GM is a voltage-controlled current source
*     connected between nodes 0 and 7. A current
*     flowing into node 0 (N+), through the cur-
*     rent source GM, and out from node 7 (N-) is
*     considered as positive current.
* 2.  GM is controlled by the voltage at 3 (NC+)
*     with respect to node 2(NC-).
* 3.  GM = 0.1 V(3,2)
```

Example 2:

```
GA1    1   2   POLY(1)    3   4   5   6   7   8   9
*      N+  N-             NC+ NC- P0  P1  P2  P3  P4
* Meaning:
* 1.  GA1  is   a   voltage-controlled   current
*     source connected between nodes 1 and 2.
*     Current flowing into node 1 (N+), through
*     the current source GA1, and out from node 2
*     (N-) is considered as positive current.
* 2.  Current GA1 is controlled by the voltage at
*     node 3 (NC+) with respect to node 4 (NC-).
* 3.  GA1 is a single-variable polynomial func-
*     tion of V(3, 4): GA1 = P0 + P1[V(3, 4)]
*     + P2[V(3, 4)]² + P3[V(3, 4)]³ + P4[V(3, 4)]⁴ = 5
*     + 6[V(3, 4)] + 7[V(3, 4)]² + 8[V(3, 4)]³ + 9[V(3,
*     + 4)]⁴
```

B.3.10 Voltage-Controlled Voltage and Current Sources with Two Control Variables

General format for voltage-controlled voltage source:

```
EXXXXXXX    N+    N-    POLY(2)    NC1+    NC1-
+ NC2+   NC2-   (P0  P1 P2  P3   P4  P5   P6
+ P7   P8 P9 P10 P11 P12 P13   P14 . . .)
* Meaning that if the control voltage at NC1+
* (with respect to NC1-) is denoted as A, and the
```

```
* control voltage at NC2+ (with respect to NC2-)
* is denoted as B, we shall have the voltage at N+
* with respect to N- equal to the sum of all the
* weighted components as listed in the following
* table:
*
* Component        1   A   B   A²  AB  B²  A³  A²B  AB²
*
* ─────────────────────────────────────────────────────
* Multiplied by   P0  P1  P2  P3  P4  P5  P6  P7   P8
*
*
* Component        B³   A⁴   A³B   A²B²   AB³   B⁴  ...
*
* ─────────────────────────────────────────────────────
* Multiplied by   P9  P10  P11  P12  P13  P14  ...
```

Example:

```
EA2 1  2  POLY(2)  3    4    5    6   1   2
+ 3  4  5
* Meaning:
* 1. EA2 is  a  voltage-controlled  voltage
*    source.
* 2. EA2 is controlled by two control voltages,
*    V(3, 4) and V(5, 6).
* 3. EA2 = P0 + P1[V(3, 4)]+ P2[V(5, 6)] + P3[V(3,
*    + 4)]² + P4[V(3, 4)] [V(5, 6)] = 1 + 2[V(3, 4)]
*    + 3[V(5, 6)] + 4[V(3, 4)]² + 5[V(3, 4)][V(5, 6)]
```

General format for voltage-controlled current source:

```
GXXXXXXX   N+   N-   POLY(2)   NC1+   NC1-
+ NC2+   NC2-   (P0  P1  P2  P3  P4  P5  P6
+ P7 . . .)
```

Example:

```
GA2 1  2  POLY(2)  3    4    5    6   1   2
+ 3  4  5
* Meaning:
* 1. GA2 is  a  voltage-controlled  current
*    source.
* 2. GA2 is controlled by two control voltages,
*    V(3, 4) and V(5, 6).
```

```
*  3.  GA2 = PO + P1[V(3, 4)]+ P2[V(5, 6)] + P3[V(3,
*       + 4)]² + P4[V(3, 4)] [V(5, 6)] = 1 + 2[V(3, 4)]
*       + 3[V(5, 6)] + 4[V(3, 4)]² + 5[V(3, 4)][V(5, 6)]
```

B.3.11 Voltage-Controlled Voltage and Current Sources with Three Control Variables

General format for voltage-controlled voltage source:

```
EXXXXXXX    N+     N-      POLY(3)      NC1+
+ NC1-    NC2+    NC2-     NC3+      NC3-
+ (PO   P1    P2    P3    P4    P5    P6    P7    P8
+ P9    P10   P11   P12   P13   P14   P15   P16 . . .)
*  Meaning that if the three control voltages are
*  denoted as A, B, and C, respectively, we shall
*  have the voltage source EXXXXXXX equal to the
*  sum of all the weighted components as listed in
*  the following table:
*
*
*  Component        1    A    B    C    A²   AB   AC   B²   BC   C²
*
*  Multiplied by   PO   P1   P2   P3   P4   P5   P6   P7   P8   P9
*
*
*  Component       A³   A²B   A²C   AB²   ABC   AC²   B³   . . .
*
*  Multiplied by  P10   P11   P12   P13   P14   P15   P16   . . .
```

Example:

```
E300    30    0    POLY(3)     11    0    18    0
+ 22     0    0    1    1    1    0    0    0    0    0    0    0
+ 0    0    0    1
*  Meaning E300=V(11, 0) + V(18, 0) + V(22, 0) +
*  [V(11, 0)][V(18, 0)][V(22, 0)]
*  This statement is very useful for the addi-
*  tion/subtraction    or    multiplication    of
*  three voltages. Otherwise it is not often
*  used.
```

General format for voltage-controlled current source

```
GXXXXXXX    N+      N-      POLY(3)     NC1+
+ NC1-    NC2+    NC2-     NC3+    NC3-
+(P0   P1   P2   P3
+ P4   P5   P6   P7 . . .)
```

Example:

```
GI    0    1    POLY(3)    2    3    4    5    6
+7    8    0    0    0    0    0    0    0    0    0
+0    0    0    0    9
* Meaning that GI=8 + 9[V(2, 3) V(4, 5) V(6, 7)]
```

B.3.12   Current-Controlled Voltage and Current Sources

The formats of current-controlled sources are actually similar to those of the voltage-controlled sources.

General formats for current-controlled voltage source

```
HXXXXXXX N+ N- (POLY(1)) VN1 P1
HXXXXXXX N+ N- POLY(2) VN1 VN2
+(P0 P1 P2 P3 P4 P5 P6 P7 . . .)
```

Example 1:

```
HA    8    7    VSENSE    10
*     N+   N-   VN1       P1
* The is an example with a single control
* variable.
* Meaning:
* 1. HA is a current-controlled voltage source
*      connected between nodes 8 and 7.
* 2. The voltage at node 8 (N+) with respect to
*      node 7 (N-) is controlled by the
*      current in VSENSE.
* 3. HA=10 I(VSENSE), where I(VSENSE) denotes
*      the current in VSENSE. (Current entering
*      into the positive node of VSENSE is taken
*      as positive.)
```

Example 2:

```
HB 9   0    POLY(2) VSENSE1   VSENSE2 1 2 3 4 5
```

```
* This is an example of a current-controlled
* voltage source with two control variables.
* Meaning:
* HB=P0 + P1[I(VSENSE1)] + P2[I(VSENSE2)] +
* P3[I(VSENSE1)]² + P4[(VSENSE1) I(VSENSE2)]=1 +
* 2[I(VSENSE1)] + 3[I(VSENSE2)] + 4[I(VSENSE1)]²
* + 5[I(VSENSE1) I(VSENSE2)]
```

Example 3:

```
HC   5   6   POLY(3)   VSENSE1   VSENSE2
+ VSENSE3   0   0   0   0   0   0   0   0
+ 0   0   0   0   0   0   1
* This is an example of a current-controlled
* voltage source with three control variables.
* Meaning:
* HC=I(VSENSE1) I(VSENSE2) I(VSENSE3)
```

General formats for current-controlled current source

```
FXXXXXXX N+ N- ⟨POLY(1)⟩ VN P1
FXXXXXXX N+ N- POLY(2) VN1 VN2
+ ⟨P0 P1 P2 P3 P4 P5 P6 P7 . . .⟩
```

Example:

```
FA 9 8 POLY(2) VSENSE1 VSENSE2 7 6 5 4 3
* Meaning:
* FA=P0 + P1[I(VSENSE1)] + P2[I(VSENSE2)] +
* P3[I(VSENSE1)]² + P4[(VSENSE1) I(VSENSE2)]=
* 7  + 6[I(VSENSE1)]  + 5[I(VSENSE2)]  +
* 4[I(VSENSE1)]² + 3[I(VSENSE1) I(VSENSE2)]
```

B.3.13 Subcircuits

In the SPICE program, a frequently used circuit may be predefined as a subcircuit, which can then be repeatedly called as components to build up larger circuits.

The general format for the definition of a subcircuit is as follows:

General format of subcircuit definition

```
.SUBCKT  SUBNAM  N1  ⟨N2  N3  N4 . . .⟩
(Insert subcircuit statements here.)
.ENDS ⟨SUBNAM⟩
```

```
* Note:
* 1. SUBNAM=Subcircuit name
* 2. N1, N2, N3, N4 ... refer to the internal nodes
*      that need to be connected to the external
*      circuit.
* 3. It is allowable to have external node numbers
*      overlapping with internal node numbers.
```

Example:

```
.SUBCKT CLIPPER  1  4  5  9
R19   1   9   1K
D1    9   2   DMOD
E20   2   0   POLY(1)  4  0  -1  1
D2    3   9   DMOD
E30   3   0   POLY(1)  5  0  1  1
R40   4   0   1G
R50   5   0   1G
.MODEL  DMOD  D
.ENDS   CLIPPER
* Note:
* 1. This example is quoted from lines 14 to 23
*      of Input File 1.
* 2. The internal nodes 1, 4, 5, and 9 are to be
*      connected to the external circuit.
```

General format of subcircuit call

```
XYYYYYYY  N1  (N2  N3  N4 ...)  SUBNAM
* Meaning that the nodes N1 (N2  N3  N4 ...) (of
* the external circuit) are connected to the
* internal subcircuit nodes in the sequence as
* defined in the .SUBCKT statement of the
* subcircuit definition.
```

Example:

```
X1  7  6  5  8  CLIPPER
* Quoted from line 13 of Input File 1.
* Meaning that nodes 7, 6, 5, and 8 (of the ex-
* ternal circuit) are connected to the in-
* ternal nodes 1, 4, 5, and 9 (the sequence as
* they appear in the subcircuit definition) of
* the subcircuit CLIPPER, respectively.
```

B.3.14 Model Statements

General format:

```
.MODEL   MNAME   TYPE   (PNAME1=PVAL1
+ PNAME2=PVAL2)
* Note: MNAME=Model name
*       TYPE=Model type
*       PNAME=Property name
*       PVAL=Property value
```

Example 1:

```
.MODEL   DMOD   D (VJ=0.6   BV=40   IS=1.0E-14
+ CJO=10P)
* Meaning that DMOD is a diode with the proper-
* ties given within the brackets.
```

Example 2:

```
.MODEL   QMOD   NPN (BF=100   RC=1   CJC=10P
+ TF=1N)
* Meaning that QMOD is an NPN bipolar transistor
* with   the   properties   given   within   the
* brackets.
```

B.3.15 .DC Statement

This statement requests a dc analysis to be carried out in the SPICE program.

General format:

```
.DC   SRCNAM   VSTART   VSTOP   VINCR
* SRCNAM=Voltage (current) source name
* VSTART=Start voltage (current)
* VSTOP=Stop voltage (current)
* VINCR=Incremental voltage (current)
```

Example:

```
.DC   VI   -0.1   0.1   0.005
* Quoted from line 28 of Input File 1.
* Meaning that dc analyses for a VI stepping up
* from -0.1 to +0.1 V in steps of 0.005 V are to
* be performed.
```

B.3.16 .AC Statement

This statement requests an ac analysis to be carried out in the SPICE program.

General formats:

```
.AC   DEC   ND   FSTART   FSTOP
.AC   OCT   NO   FSTART   FSTOP
.AC   LIN   NP   FSTART   FSTOP
* DEC=Decade
* ND=No. of points per decade
* OCT=Octave
* NO=No. of points per octave
* LIN=Linear
* NP=No. of points
* FSTART=Start frequency for ac analysis
* FSTOP=Stop frequency for ac analysis
```

Example:

```
.AC   DEC   5   0.1   10MEG
* Quoted from line 29 of Input File 1.
* Meaning that an ac analysis is to be carried
* out with the following specifications:
* 1.  The sweep of frequency is in logarithmic
*     scale of decades, with 5 points in each
*     decade.
* 2.  The sweep of frequency is from 0.1 Hz to 10
*     MHz.
```

B.3.17 .TRAN Statement

This statement requests a transient simulation to be carried out in the SPICE program.

General format:

```
.TRAN   TSTEP   TSTOP   ⟨TSTART   ⟨TMAX⟩⟩ ⟨UIC⟩
* TSTEP=Time step for printing or plotting out-
* puts
* TSTOP=Stop time for the transient analysis
* TSTART=Start time for printing or plotting out-
* puts
* TMAX=Maximum time step to be used in the tran-
```

```
* sient analysis
* UIC=Use initial condition
```

Example:

```
.TRAN   0.5U   100U   0
* Quoted from line 20 of Input File 2.
* Meaning:
* 1. A transient simulation is to be carried out
*      for 100 μs, the print/plot output is to be in
*      steps of 0.5 μs.
* 2. The start time for printing/plotting the
*      output is 0 μs.
```

B.3.18   .IC Statement

This statement is for setting initial conditions for transient analysis. It has two different implications, depending on whether the UIC option is specified in the .TRAN statement:

1. If UIC is not specified, a biasing point analysis based on the .IC conditions will first be carried out before the transient analysis.
2. If UIC is specified, no biasing point analysis will be carried out before the transient analysis.

General format:

```
.IC   V(NODNUM1)=VAL1   (V(NODNUM2)=VAL2 . . .)
* NODNUM=Node number
* VAL=Value
```

Example:

```
.IC   V(7)=5
* Quoted from line 19 of Input File 2.
* Meaning that the initial voltage at node 7 is
* forced to 5 V in the transient simulation.
```

B.3.19   .SENS Statement

This statement requests a dc sensitivity analysis to be carried out in the SPICE program.

General format:

```
.SENS   OV1   (OV2   OV3   OV4 . . .)
* OV1=Output voltage 1
* OV2=Output voltage 2
* OV3=Output voltage 3
* OV4=Output voltage 4
```

Example:

```
.SENS   V(10)
* Quoted from line 30 of Input File 1.
* Meaning that a sensitivity analysis is to be
* carried out for the output voltage V(10).
```

B.3.20 .PRINT and .PLOT Statements

The .PRINT statement requests the simulated result to be printed as computer output.

The .PLOT statement requests the simulated result to be plotted as computer output.

Note that, in PSpice, the .PRINT and .PLOT statements are not required. Instead, a .PROBE statement should be included so that all waveforms can be traced later.

General format:

```
.PRINT   PRTYPE   OV1   (OV2   OV3   OV4 . . .)
* PRTYPE=Print type (DC, AC, TRAN, etc.)
* OV1, OV2, OV3, OV4 are output voltage 1, output
* voltage 2, output voltage 3, output voltage 4
* . . . .
```

Example 1:

```
.PRINT   DC   V(10)
* Quoted from line 31 of Input File 1.
* Meaning to print the dc voltage at node 10 (as
* a function of VI) as computer output.
```

Example 2:

```
.PLOT   DC   V(10)
* Quoted from line 32 of Input File 1.
* Meaning to plot the dc voltage at node 10 (as a
* function VI) as computer output.
```

Example 3:

```
.PRINT   AC   VDB(10)   VP(10)
.PLOT    AC   VDB(10)   VP(10)
* Quoted from lines 33 and 34 of Input File 1.
* Meaning to print and plot the following ac
* analysis outputs:
* 1. Amplitude of voltage V(10) in dB scale.
* 2. Phase of V(10).
```

Example 4:

```
.PRINT   TRAN   V(3)   V(4)   V(7)   I(VSENSE)
.PLOT    TRAN   V(3)   V(4)   V(7)   I(VSENSE)
* Quoted from lines 21 and 22 of Input File 2.
* Meaning to print and plot the transient wave-
* forms of V(3), V(4), V(7), and I(VSENSE).
```

INPUT FILE 1: BEHAVIOR MODEL OF OPERATIONAL AMPLIFIER

```
1      OP-AMP MODEL
       * INPUT STAGE
2      VI 1 0 AC 1
3      Q1 2 1 4  QMOD
4      R26 2 6 5.5K
5      Q2 3 0 4 QMOD
6      R36 3 6  5.5K
7      R45 4 5  5K
8      VCC 6 0  DC 10
9      VEE 0 5  DC 10
       *
       * FREQ. COMPENSATION
10     GM 0 7 POLY(1) 3 2 0 0.1
11     R70 7 0  1K
12     C70 7 0  10U
       *
       * CLIPPER
13     X1 7 6 5 8  CLIPPER
       *
       * CLIPPER SUBCIRCUIT
       *                   INPUT   VCC   VEE   OUTPUT
14     .SUBCKT CLIPPER     1       4     5     9
15     R19 1 9  1K
16     D1 9 2 DMOD
17     E20 2 0 POLY(1) 4 0  -1 1
18     D2 3 9 DMOD
19     E30 3 0 POLY(1) 5 0   1 1
20     R40 4 0 1G
21     R50 5 0 1G
```

```
22      .MODEL DMOD D
23      .ENDS  CLIPPER
        *
        *OUTPUT CIRCUIT
24      E90 9 0  POLY(1) 8 0  0 1
25      R910 9 10 75
26      R100 10 0 1G
        *
27      .MODEL QMOD NPN
28      .DC VI -0.1 0.1 0.005
29      .AC DEC 5 0.1 10MEG
30      .SENS V(10)
31      .PRINT DC V(10)
32      .PLOT DC V(10)
33      .PRINT AC VDB(10) VP(10)
34      .PLOT AC VDB(10) VP(10)
35      .END
```

INPUT FILE 2: FLYBACK CONVERTER

```
1       FLYBACK CONVERTER
        *
        *INPUT VOLTAGE
2       VI 1 0 DC 300
        *
3       R12 1 2   200
4       C23 2 3   500P
5       L13 1 3   2.5M
6       L05 0 5   1U
7       K1 L13 L05  0.98
8       M1 3 4 0 0  MOD1  W-1
        *
        *OUTPUT STAGE
9       D1 5 6  DMOD
10      VSENSE 6 7
11      C70 7 0  500U
12      R70 7 0  0.3
        *
        *CONTROL VOLTAGE
13      V40 4 0  PWL(0 0   10U 0   10.05U 10   20U 10   20.05U 0
14      + 30U 0 30.05U 10   40U 10  40.05U 0   50U 0   50.05U 10
15      + 60U 10   60.05U 0 70U 0   70.05U 10   80U 10  80.05U 0
16      + 90U 0   90.05U 10   100U 10   100.05U 0)
        *
17      .MODEL MOD1 NMOS
18      .MODEL DMOD D
19      .IC V(7)-5
20      .TRAN 0.5U 100U 0
21      .PRINT TRAN V(3) V(4) V(7) I(VSENSE)
22      .PLOT TRAN V(3) V(4) V(7) I(VSENSE)
23      .END
```

APPENDIX C
Nonconvergence and Related Problems in SPICE

C.1 INTRODUCTION

This appendix is intended to help the reader identify and solve problems commonly encountered in SPICE simulations. The main topics of discussion will include:

1. The nonconvergence problem in dc biasing point (operating point) analysis
2. The "internal time step too small" problem in transient analysis
3. Other related problems

C.2 NONCONVERGENCE IN dc OR TRANSIENT ANALYSIS

When an ac or a transient simulation is performed in SPICE, a dc biasing point analysis is normally carried out before the actual ac or transient analysis is started. For some circuits, SPICE may fail to determine the initial biasing point, giving an error message to say that there is no convergence in the dc analysis. For some other circuits, there is no problem in finding an initial biasing point but, in the course of transient simulation, SPICE may also encounter the nonconvergence problem. When this occurs, the computer will simply stop there, giving an error message to say that the internal time step is too small in transient analysis. (The wording of the message may differ in different versions of SPICE.)

The three main causes of nonconvergence in the dc or transient analysis are:

1. Careless mistakes made in the input file. (A large percentage of "nonconvergence problems" are actually due to careless mistakes!)

2. The limited capability of numerical analysis.
3. Inappropriate circuit models.

The last two problems will be discussed in Subsections C.2.1 and C.2.2.

C.2.1 Limited Capability of Numerical Analysis

The numerical method used in SPICE to deal with nonlinear circuits is the Newton–Raphson algorithm. To someone unfamiliar with this method of analysis, it may be understood as an iterative trial-and-error method. An associated problem with this method is the need for an initial guess of the solution. If the initial guess is very far from the true solution, the Newton–Raphson method may fail to converge to a solution, even after many iterations. Various algorithms have been designed and implemented in SPICE to help solve the nonconvergence problems. However, the complicated combination of nonlinear characteristics of circuit components and improper initial guesses may still cause nonconvergence problems in both dc and transient analyses.

C.2.2 Inappropriate Circuit Models

Nonconvergence in dc or transient analysis may also be caused by inappropriate circuit models. A typical example is a circuit with heavy dc positive feedback but without an amplitude limiter. In theory, the output voltage of such a circuit should be infinitely large. In practice, what appears in the computer output is a nonconvergence message.

Positive feedback may exist implicitly in controlled sources. It may also result from careless mistakes in the input file.

C.3 METHODS OF SOLVING NONCONVERGENCE PROBLEMS IN dc BIASING POINT ANALYSIS

When a nonconvergence message appears while the dc biasing point analysis is running in SPICE, possible methods of solving the problem are:

1. Increase the dc iteration limit (ITL1) using a .OPTIONS statement.

    Example:

    ```
    .OPTIONS ITL1 = 1000
    * This statement requests the dc iteration
    * limit to be set to 1000,
    * The default value of ITL1 is 100.
    ```

2. Use a .NODESET statement to input manually a set of reasonable initial guesses at selected node voltages. This will enable SPICE to start the iteration from a better starting point. These voltages need not be very accurate, but they should be closer guesses than those made by the computer.

   Example:

   ```
   .NODESET V(1)=10 V(3)=7 V(10)=5 V(20)=0
   * This statement requests the Newton-Raph-
   * son iteration to start with an initial
   * guess of:
   * Voltage at node 1 = 10 V, voltage at node
   * 3 = 7 V,
   * voltage at node 10 = 5 V, voltage at node
   * 20 = 0 V.
   ```

   It should be noted that the voltage given in the .NODESET statement do not affect the final results of analysis. They only help SPICE to produce a set of converged initial biasing conditions.

3. Recheck the circuit to ensure that there is no careless mistake.
4. Relax the allowable relative tolerance RELTOL for the analysis.

   Example:

   ```
   .OPTIONS RELTOL=0.01
   * This    statement    relaxes    the    rela-
   * tive-error    tolerance    of    analysis    to
   * 0.01.
   * The default value of RELTOL is 0.001.
   ```

5. Use a .TRAN statement with UIC (use initial conditions) to initiate a transient simulation without any dc biasing point analysis and, then, use the resultant steady-state dc voltages as .IC or .NODESET voltages to start the normal simulation.

   Example:

   ```
   .TRAN  1M    100M  UIC
   * This statement requests SPICE to carry
   * out a transient simulation for 100 ms in
   * steps of 1 ms without calculating the
   * biasing point first.
   ```

It should be noted that when a .TRAN UIC statement is used, no .AC analysis statement should be included in the input file because such a statement implies a request for biasing point analysis.

C.4 METHODS OF SOLVING NONCONVERGENCE OR "INTERNAL TIME STEP TOO SMALL" PROBLEMS IN TRANSIENT ANALYSIS

The transient analysis in SPICE is effectively a series of dc analyses. Therefore, all problems that appear in the dc analysis can appear equally in the transient analysis. Most of the methods used to solve the dc convergence problems, such as increasing the iteration limit and relaxing the relative tolerance, as described in Section C.3, are also applicable to transient analysis.

There is a mechanism in SPICE to control the size of the iternal time step used for transient analysis. If a converged solution cannot be achieved within the preset transient-analysis iteration limit (referred to as ITL4, the default value of which is 10), the step size will be automatically reduced. This process will continue until either a converged solution is obtained within the iteration limit or the time step is considered to be too small. In the latter case, SPICE will stop the simulation and send out the massage "Internal time step too small in transient analysis." (In case of PSpice, the message is "Convergence problem in Transient Analysis.") To solve this time-step problem, the following methods can be tried:

1. Use a .OPTIONS statement to increase the transient-analysis iteration limit ITL4. This will improve the accuracy of the iterated result and thus enable it to meet the tolerance requirement more readily.

   Example:

   ```
   .OPTIONS   ITL4=100
   * This   statement   sets   the   transient-
   * analysis iteration limit ITL4 to 100.
   * The default value of ITL4 is 10.
   ```

2. Use a .OPTIONS statement to request a more accurate integration method in order to improve the accuracy of iteration.

   Example:

   ```
   .OPTIONS   METHOD=GEAR   MAXORD=6
   * GEAR is an alternative method of integra-
   * tion in SPICE.
   * The default method is trapezoidal.
   ```

```
* Note: These options are not available in
* PSpice.
```

3. Recheck the circuit model to ensure that there is no careless mistake in the input file and that it is not an inappropriate model for SPICE simulation as described in Subsection C.2.2.
4. Lengthen the rise and fall times of input signals, or add parasitic capacitances to the circuit or devices. This slows down the rate of change of the node voltages, thus allowing a relatively larger time step to be used in the transient analysis.
5. Relax the relative tolerance RELTOL.

Example:

```
.OPTIONS    RELTOL=0.01
* The default value of RELTOL is 0.001.
```

6. Reduce the total simulation time specified in the .TRAN statement. This will reduce both the total simulation time and the minimum time step used in the simulation.
7. Set a small TMAX in the .TRAN statement. This will also reduce both the maximum and the minimum allowable time steps in the simulation. It will, however, increase the simulation time.

Example:

```
.TRAN    1M     40M      0      1U
*        TSTEP  TSTOP  TSTART  TMAX
* This statement requests the transient
* simulation to be carried out with a max-
* imum step size (TMAX) of 1 μs.
* The actual internal time step used in the
* numerical iteration will thus be much
* smaller than 1 μs.
```

It should be noted that there is no direct method to set the minimum internal time step in SPICE.

8. If necessary (as implied by an error message), use a .OPTIONS statement to set the allowable pivot tolerance PIVTOL to a smaller value.

Example:

```
.OPTIONS    PIVTOL=10E-15
* The default value of PIVTOL is 10E-13.
```

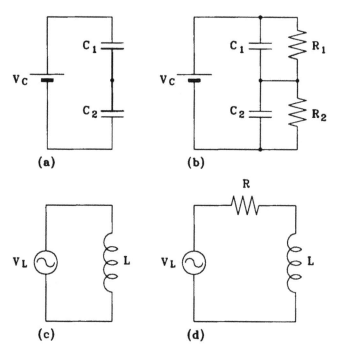

FIGURE C.1   Illegal and legal SPICE circuits.

C.5   OTHER RELATED PROBLEMS

In addition to the precautions mentioned above, attention should also be paid to the following problems:

1. SPICE does not allow any circuit node to have an infinitely large resistance to the ground because this implies that the node can have a floating dc voltage. As an example, the circuit shown in Fig. C.1(a) is illegal. A simple solution to this problem is to add a very large resistance in parallel with the capacitor, as shown in Fig. C.1(b).
2. SPICE does not allow any voltage source to be connected across an inductance without series resistance because this implies that the dc current in the inductance can be of any value. As an example, the circuit shown in Fig. C.1(c) is illegal. One solution to this problem is to add a very small resistance R in series with the inductance, as shown in Fig. C.1(d).
3. SPICE does not allow any node to which there is only one circuit element connected. A method for solving the problem is to add a large dummy resistance (e.g., 1 MEGΩ or 1 GΩ) between the node concerned and the ground.

APPENDIX D
About PSpice*

D.1 INTRODUCTION

PSpice is a commercial version of SPICE, with enhanced graphics and many other useful add-on features. PSpice was originally developed for the IBM PC and its compatibles, but it can now be run on many other machines, including the Macintosh, Sun, and DEC. The software vendor for PSpice, MicroSim Corporation (20 Fairbanks, Irvine, CA 92718, U.S.A.), is prepared to provide class instructors with free evaluation versions of PSpice that can be run on an IBM PC (and its compatibles) or on a Macintosh.

The evaluation version of PSpice is capable of simulating circuits with up to 10 active components and 64 nodes. It can be used to run many of the example programs given in this book. For the full version of PSpice, the

---

*PSpice is a registered trademark of MicroSim Corporation.

maximum number of allowable circuit components is limited only by the memory size of the computer, and this number is approximately given by

Maximum no. of active components

$\approx 400 \times$ [Memory size in megabyte]

D.2 THE MAIN DIFFERENCES BETWEEN SPICE AND PSpice

PSpice was developed based on SPICE2 from the University of California at Berkeley. The simulation algorithm used in PSpice is basically the same as that of SPICE2. The input format of PSpice is also similar to that of SPICE.

However, PSpice has many extra features that are very useful to the circuit designer. The following is a summary of those features that are more closely related to the analysis and design of analog circuits:

1. PSpice has a text editor to help the user prepare input files for simulation.
2. MicroSim provides a Probe software that is effectively an intelligent software oscilloscope to enable the user to display and manipulate the simulation results on a high-resolution color monitor.
3. There is a power down process in PSpice in the dc biasing point analysis to help solve the nonconvergence problem.
4. PSpice allows the use of mathematical expressions to model linear and nonlinear controlled sources. This feature is very useful for the behavior modeling of power converters.
5. The .SAVEBIAS and .LOADBIAS statements in PSpice (for 4.05 or upward versions) can be used to save and load the biasing conditions when necessary for repeated simulations to speed up the simulation process.
6. The implementation of the Jiles–Atherton ferromagnetic equations in PSpice makes possible the simulation of magnetic saturation, hysteresis, remanence, coercivity, and eddy-current losses in inductors and transformers.
7. Voltage-controlled and current-controlled switch models are available in PSpice.
8. A large number of library models of commercial devices are available in PSpice.

It should, however, be noted that the distortion analysis in SPICE is not supported by PSpice.

D.3 OPTIONAL SOFTWARE FOR USE WITH PSpice

PSpice is offered by MicroSim as a component of its Circuit Analysis software package. Other components of the Circuit Analysis package that can be usefully applied to the analysis and design of analog circuits include Parts, Monte-Carlo, Analog Behavioral Modeling, and Digital Simulation. The following is a brief summary of their uses:

1. *Parts*: This is a program to help the user develop SPICE/PSpice models for circuit components based on information from data sheets.

2. *Monte-Carlo*: This is an optional feature of PSpice that enables the user to run multiple runs of dc, ac, or transient analysis based on user-specified component-value tolerance distribution. Monte-Carlo analysis is particularly useful for predicting yields or for verifying circuit performance under the production environment. Worst-case analyses can also be performed to determine the worst-case performance of the circuit.

3. *Analog Behavioral Modeling*: This optional feature of PSpice enables the user to define the frequency and phase characteristics of a linear circuit or the output versus input transfer characteristic of a nonlinear circuit, using a formula or a look-up table. In this way, the behavior of the circuit can be defined without specifying the internal components of the circuit. All PSpice analyses, including dc, ac, and transient, can be carried out for circuits or systems using such behavior models.

4. *Digital Simulation*: This optional feature of PSpice enables the user to carry out digital or mixed analog and digital simulation. The mixed analog and digital simulation is particularly useful for the cycle-by-cycle transient simulation of power converters with digital control circuits.

D.4 RUNNING PSpice

When you have installed PSpice (either the evaluation version or the full version) in your computer, you can start a simulation by executing the PSPICE.BAT program (which, in turn, invokes the PSPICE1.EXE program). All you need to do is to type

```
PSPICE    FNAME.CIR    [RETURN]
```

where FNAME.CIR is the file name of the circuit to be simulated. If you do not have an input circuit file of your own, you may try the examples given in the PSpice package (e.g., EVAL.CIR, EVALOSC.CIR, or EXAMPLE1.CIR). Note that the extension .CIR is used to mean an input circuit file.

When the simulation is completed, the following two output files will be generated:

```
FNAME.OUT
PROBE.DAT
```

The FNAME.OUT is a text file similar to the output file of a SPICE simulation. The PROBE.DAT is a PSpice graphic file, which can be viewed by typing

```
PROBE    [RETURN]
```

and then selecting the options as suggested on the display.

If you want to prepare/edit your own input circuit file, you may execute the PS.EXE program (which contains a file editor) by typing

```
PS       [RETURN]
```

and then selecting the appropriate procedures as suggested on the display.

On-line help manuals are available in the PSpice software package.

Detailed information regarding installation, operation, and technical details about the PSpice package can be obtained from MicroSim's Circuit Analysis User's Guide.

References and Bibliography

1. Y. S. Lee, "Modelling and Analysis of dc to dc Converter Using Y-Parameters," *The Radio and Electronic Engineer*, Vol. 54, No. 3, pp. 129–136, March 1984.
2. Y. S. Lee, "A Systematic and Unified Approach to Modelling Switches in Switch-Mode Power Supplies," *IEEE Transactions on Industrial Electronics*, Vol. IE-32, No. 4, pp. 445–448, November 1985.
3. Y. S. Lee and Y. C. Cheng, "Design of Switching Regulator with Combined FM and On-Off Control," *IEEE Transactions on Aerospace and Electronic Systems*, Vol. AES-22, No. 6, pp. 725–731, November 1986.
4. Y. S. Lee and Y. C. Cheng, "Calculation of Resonance Current in High-Frequency Series-Resonant dc-dc Converters," *IEEE Transactions on Aerospace and Electronic Systems*, Vol. 24, No. 2, pp. 124–132, March 1988.
5. Y. S. Lee and Y. C. Cheng, "Computer-Aided Analysis of Electronic dc-dc Transformers," *IEEE Transactions on Industrial Electronics*, Vol. 35, No. 1, pp. 148–152, February 1988.
6. Y. S. Lee and Y. C. Cheng, "A 580 kHz Switching Regulator Using On-Off Control," *Journal of The Institution of Electronic and Radio Engineers*, Vol. 57, No. 5, pp. 221–226, September/October 1987.
7. Y. S. Lee and Y. C. Cheng, "SPICE Simulation of Switch-Mode Power Supplies," *Proceedings of China 7th Conference on Circuits and Systems*, Shenzhen, China, p. 12–17 to p. 12–20, November 1987.

8. Y. S. Lee and Y. C. Cheng, "High-Frequency and High-Efficiency Switch-Mode Power Supplies," *Proceedings of IEEE Asian Electronics Conference 1987*, Hong Kong, pp. 701–706.

9. F. Barzegar, S. Ćuk, and R. D. Middlebrook, "Using Small Computers to Model and Measure Magnitude and Phase of Regulator Transfer Functions and Loop Gain," *Powercon 8, The Eighth International Solid-State Power Electronics Conference*, Dallas, TX, April 1981.

10. B. D. Bedford and R. G. Holt, *Principles of Inverter Circuits*, John Wiley and Sons, Inc., New York, 1964.

11. V. G. Bello, "Computer-Aided Analysis of Switching Regulators Using SPICE2," IEEE Power Electronics Specialists Conference, *1980 Record*, 80CH1529-7, pp. 3–11.

12. J. J. Biess, L. Y. Inouye, and J. H. Shank, "High-Voltage Series Resonant Inverter Ion Engine Screen Supply," *Proceedings of the 1974 IEEE Power Electronics Specialists Conference*, pp. 97–105, June 1974.

13. R. L. Bonkowski, "A Technique for Increasing Power Transistor Switching Frequency," *IEEE Transactions on Industry Applications*, Vol. IA-22, No. 2, pp. 240–243, March/April 1986.

14. A. R. Brown and R. D. Middlebrook, "Sampled-Data Modelling of Switching Regulators," *Proceedings of IEEE Power Electronics Specialists Conference*, Colorado, pp. 349–369, June 1981.

15. E. E. Buchanan Jr. and E. J. Miller, "Resonant Switching Power Conversion Technique," *PESC'75 Record*, 75-CHO 965-4 AES, pp. 188–193.

16. A. Capel, J. G. Ferrante, and R. Prajoux, "Dynamic Behavior and z Transform Stability Analysis for dc/dc Regulators with a Nonlinear PWM Controlled Loop," IEEE Power Electronics Specialists Conference, *1973 Record*, pp. 149–157 (IEEE Publication CHO 787-2 AES).

17. P. R. K. Chetty, "Current Injected Equivalent Circuit Approach (CIECA) to Modelling of Switching dc-dc Converters in Continuous Inductor Conduction Mode," *IEEE Transactions on Aerospace and Electronic Systems*, Vol. AES-17, pp. 802–808, November 1981.

18. P. R. K. Chetty, "Current Injected Equivalent Circuit Approach (CIECA) to Modelling of Switching dc-dc Converters in Discontinuous Inductor Conduction Mode," *IEEE Transactions on Industrial Electronics*, Vol. IE-29, pp. 230–234, August 1982.

19. P. R. K. Chetty, "Modelling and Analysis of Ćuk Converter Using Current Injected Equivalent Circuit Approach," *IEEE Transactions on Industrial Electronics*, Vol. IE-30, pp. 56–59, February 1983.

20. P. R. K. Chetty, "A New Resonant Mode Amplifier Produces Clean ac Power," *IEEE Transactions on Aerospace and Electronic Systems*, Vol. AES-21, No. 6, pp. 800–803, November 1985.

21. S. Ćuk, "Modelling, Analysis, and Design of Switching Converters," PhD thesis, California Institute of Technology, Pasadena, CA, November 1976.

22. S. Ćuk and R. D. Middlebrook, "A New Optimum Topology Switching dc-to-dc Converter," *Proceedings of IEEE Power Electronics Specialists Conference*, Palo Alto, CA, pp. 160–179, June 1977.

23. S. Ćuk and R. D. Middlebrook, "A General Unified Approach to Modelling Switching dc-to-dc Converters in Discontinuous Conduction Mode," IEEE Power Electronics Specialists Conference, *1977 Record*, pp. 36–57.

24. S. Ćuk, "General Topological Properties of Switching Structures," IEEE Power Electronics Specialists Conference, *1979 Record*, pp. 109–130.

25. W. Ebbinge, "Designing Very High Efficiency Converters with a New High Frequency Resonant GTO Technique," *Proceedings of Powercon 8*, A-1, Power Concepts Inc., pp. 1–7, 1981.

26. D. B. Edwards and T. K. Caughey, "Global Analysis of a Buck Regulator," IEEE Power Electronics Specialists Conference, *1978 Record*, Syracuse, NY, June 1978, pp. 12–16 (IEEE Publication 78CH1337-5 AES).

27. R. W. Erickson, S. Ćuk, and R. D. Middlebrook, "Large-Signal Modelling and Analysis of Switching Regulators," *Proceedings of IEEE Power Electronics Specialists Conference*, MIT, Cambridge, MA, June 1982.

28. S. Y. Feng, T. G. Wilson, and W. A. Sander, "Very High Frequency dc-to-dc Conversion and Regulation in the Low Megahertz Range," IEEE Power Electronics Specialists Conference, *1971 Record*, pp. 58–65.

29. K. Harada and T. Nabeshima, "Large-Signal Transient Response of a Switching Regulator," IEEE Power Electronics Specialists Conference, *1981 Record*, Boulder, CO, pp. 388–394 (IEEE Publication 81CH1652-7).

30. H. Huisman and S. W. H. de Haan, "A dc to 3-Phase Series-Resonant Converter with Low Harmonic Distortion," *IEEE Transactions on Industrial Electronics*, Vol. 32, No. 2, pp. 142–149, May 1985.

31. R. King and T. A. Stuart, "A Normalized Model for the Half-Bridge Series Resonant Converter," *IEEE Transactions on Aerospace and Electronic Systems*, Vol. AES-17, No. 2, pp. 190–198, March 1981.

32. R. J. King and T. A. Stuart, "Modelling the Full-Bridge Series-Resonant Power Converters," *IEEE Transactions on Aerospace and Electronic Systems*, Vol. AES-18, No. 4, pp. 449–459, July 1982.

33. E. E. Landsman, "A Unifying Derivation of Switching Converter Topologies," IEEE Power Electronics Specialists Conference, *1979 Record*, pp. 239–243.

34. F. C. Lee and Y. Yu, "Computer-Aided Analysis and Simulation of Switch dc-dc Converters," *IEEE Transactions on Industry Applications*, Vol. IA-15, No. 5, pp. 511–520, September/October 1979.

35. R. D. Middlebrook and S. Ćuk, "A General Unified Approach to Modelling Switching-Converter Power Stages," IEEE Power Electronics Specialists Conference, *1976 Record*, pp. 18–34.

36. R. D. Middlebrook and S. Ćuk, "Modelling and Analysis Methods for dc-to-dc Switching Converters," IEEE International Semiconductor Power Converter Conference, *1977 Record*, pp. 90–111.

37. R. D. Middlebrook, S. Ćuk, and W. Behen, "A New Battery Charger/Discharger Converter," IEEE Power Electronics Specialists Conference, *1978 Record*, pp. 251–255 (IEEE Publication 78 CH 1337-5 AES).

38. R. D. Middlebrook and S. Ćuk, "Advances in Switched-Mode Power Conversion," Vols. 1, 2, and 3, Teslaco, Pasadena, CA, 1983.

39. E. J. Miller, "Resonant Switching Power Conversion," *PESC 76 Record*, pp. 206–211.

40. D. O. Monteith Jr. and D. Salcedo, "Modeling Feedforward PWM Circuits Using the Non-Linear Function Capabilities of SPICE2," *Proceedings of Powercon 10*, H-4, pp. 1–10, Power Concepts Inc., 1983.

41. E. T. Moore and T. G. Wilson, "Basic Considerations for dc to dc Conversion Network," *IEEE Transactions on Magnetics*, Vol. MAG-2, No. 3, pp. 620–640, September 1966.

42. R. Myers and R. D. Peck, "200 kHz Power FET Technology in New Modular Power Supplies," *Hewlett-Packard Journal*, pp. 3–10, August 1981.

43. W. E. Newell, "Power Electronics—Emerging from Limbo," IEEE Power Electronics Specialists Conference, *1973 Record*, pp. 6–12 (IEEE Publication 73 CHO 787-2 AES).

44. A. I. Pressman, *Switching and Linear Power Supply, Power Converter Design*, Hayden Book Co., New York, 1977.

45. V. T. Ranganathan, P. D. Ziogas, and V. R. Stefanovic, "A Regulated dc-dc Voltage Source Converter Using a High Frequency Link," IEEE Industry Applications Society Conference Record, pp. 917–924, October 1981.

46. R. Redl, B. Molnár, and N. O. Sokal, "Class E Resonant Regulated dc/dc Power Converters: Analysis of Operations, and Experimental Results at 1.5 MHz," *IEEE Transactions on Power Electronics*, Vol. PE-1, No. 2, pp. 111–120, April 1986.

47. R. R. Robson, "Designing a 25-Kilowatt High Frequency Series Resonant dc/dc Converter," *Proceedings of Powercon 11, H1-3*, Power Concepts Inc., 1984, pp. 1–15.

48. F. C. Schwarz, "A Method of Resonant Current Pulse Modulation for Power Converters," *IEEE Transactions on Industrial Electronics and Control Instrumentation*, IECI-17, pp. 209–221, May 1970.

49. F. C. Schwarz, "An Improved Method of Resonant Current Pulse Modulation for Power Converters," *IEEE Transactions on Industrial Electronics and Control Instrumentation*, IECI-23, pp. 133–141, May 1976.

50. F. C. Schwarz and J. B. Klaassens, "A Controllable Secondary Multikilowatt dc Current Source with Constant Maximum Power Factor in Its Three-Phase Supply Line," *IEEE Transactions on Industrial Electronics and Control Instrumentation*, Vol. IECI-23, No. 2, pp. 142–150, May 1976.

51. F. C. Schwarz and J. B. Klaassens, "A 95-Percent Efficient 1-kW dc Converter with an Internal Frequency of 50 kHz," *IEEE Transactions on Industrial Electronics and Control Instrumentation*, Vol. IECI-25, No. 4, pp. 326–333, November 1978.

52. R. Severns and D. Sommers, "Design of High-Efficiency Off-Line Converters above 100 kHz," *Proceedings of Powercon 5*, May 1978, pp. G2-1 to G2-7.

53. R. Severns, "Switch-Mode Converter Topologies—Make Them Work for You!" *Application Bulletin A035*, Intersil, CA, 1980.

54. R. Severns, "The Design of Switchmode Converters above 100 kHz," *Application Bulletin A034*, Intersil, CA, 1980.

55. R. Severns and J. Armijos, "MOSPOWER Applications Handbook," Siliconix Inc., CA, 1985.

56. N. O. Sokal and A. D. Sokal, "Class E–A New Class of High-Efficiency Tuned Single-Ended Switching Power Amplifier," *IEEE Journal of Solid-State Circuits*, Vol. SC-10, pp. 168–176, June 1975.

57. *SPICE Version 2.G User's Guide*, University of California, Berkeley, 1981.

58. R. L. Steigerwald, "High-Frequency Resonant Transistor dc-dc Converters," *IEEE Transactions on Industrial Electronics*, Vol. IE-31, No. 2, pp. 181–191, May 1984.

59. V. Vorpérian and S. Ćuk, "A Complete dc Analysis of the Series Resonant Converter," *Proceedings of the IEEE Power Electronics Specialists Conference*, June 1982, MIT, Cambridge, MA, pp. 85–100.

60. V. Vorpérian and S. Ćuk, "Small-Signal Analysis of Resonant Converters," IEEE Power Electronics Specialists Conference, June 6–9, 1983, Albuquerque, NM, pp. 269–282.

61. G. W. Wester and R. D. Middlebrook, "Low-Frequency Characterization of Switched dc-to-dc Converters," IEEE Power Processing and Electronics Specialists Conference, *1972 Record*, pp. 9–20.

62. A. F. Witulski and R. W. Erickson, "Steady-State Analysis of the Series Resonant Converter," *IEEE Transactions on Aerospace and Electronic Systems*, Vol. AES-21, No. 6, pp. 791–799, November 1985.

63. J. K. A. Everard and A. J. King, "Broadband Power Efficient Class E Amplifier with a Non-Linear CAD Model of the Active MOS Device," *Journal of the Institution of Electronic and Radio Engineers*, Vol. 57, No. 2, pp. 52–58, March/April 1987.

64. R. Oruganti and F. C. Lee, "Resonant Power Processors, Part I–State Plane Analysis," *IEEE Transactions on Industry Applications*, Vol. IA-21, No. 6, pp. 1453–1460, November/December 1985.

65. R. Oruganti and F. C. Lee, "Resonant Power Processors, Part II—Methods of Control," *IEEE Transactions on Industry Applications*, Vol. IA-21, No. 6, pp. 1461–1471, November/December 1985.

66. C. Q. Lee and K. Siri, "Analysis and Design of Series Resonant Converter by State-Plane Diagram," *IEEE Transactions on Aerospace and Electronic Systems*, Vol. AES-22, No. 6, pp. 757–763, November 1986.

67. V. G. Bello, "Using the SPICE2 CAD Package for Easy Simulation of Switching Regulators in Both Continuous and Discontinuous Conduction Modes," *Proceedings of Powercon 8, H-3*, Power Concepts, Inc., 1981, pp. 1–14.

68. R. Redl, B. Molnár, and N. O. Sokal, "Small-Signal Dyanmic Analysis of Regulated Class E dc/dc Converters," *IEEE Transactions on Power Electronics*, Vol. PE-1, No. 2, pp. 121–128, April 1986.

69. N. V. Tilgenkamp, S. W. H. de Haan, and H. Huisman, "A Novel Series-Resonant Converter Topology," *IEEE Transactions on Industrial Electronics*, Vol. IE-34, No. 2, pp. 240–246, May 1987.

70. R. Redl and N. O. Sokal, "A 14-MHz 100-W Class E Resonant Converter: Principles, Design Considerations and Measured Performance," Power Electronics Conference, October 1986, San Jose, CA.

71. D. M. Divan, "Design Considerations for Very High Frequency Resonant Mode dc/dc Converters," *IEEE Transactions on Power Electronics*, Vol. PE-2, No. 1, pp. 45–54, January 1987.

72. R. Redl and N. O. Sokal, "Near-Optimum Dynamic Regulation of

dc-dc Converters Using Feed-Forward of Output Current and Input Voltage with Current-Mode Control," *IEEE Transactions on Power Electronics*, Vol. PE-1, No. 3, pp. 181–192, July 1986.

73. K. H. Liu, R. Oruganti, and F. C. Lee, "Resonant Switches—Topologies and Characteristics," *PESC'85*, pp. 106–116, 1985.

74. K. H. Liu and F. C. Lee, "Zero-Voltage Switching Technique in dc/dc Converters," *PESC'86*, pp. 58–70, 1986.

75. W. A. Tabisz and F. C. Lee, "Zero-Voltage-Switching Multi-Resonant Technique—A Novel Approach to Improve Performance of High-Frequency Quasi-Resonant Converters," *PESC'88*, pp. 9–17, 1988.

76. W. A. Tabisz and F. C. Lee, "DC Analysis and Design of Zero-Voltage-Switched Multi-Resonant Converters," *IEEE PESC'89*, pp. 243–251, 1989.

77. D. T. Ngo, "Generalization of Resonant Switches and Quasi-Resonant dc-dc Converters," *PESC'87*, pp. 395–403, 1987.

78. V. Vorpérian, "Quasi-Square-Wave Converters: Topologies and Analysis," *IEEE Transactions on Power Electronics*, Vol. 3, No. 2, pp. 183–191, April 1988.

79. A. H. Weinberg and L. Ghislanzoni, "A New Zero Voltage and Zero Current Power Switching Technique," *IEEE PESC'89*, pp. 909–919, 1989.

80. R. W. Erickson, A. F. Hernandez, A. F. Witulski, and R. Xu, "A Nonlinear Resonant Switch," *IEEE Transactions on Power Electronics*, Vol. 4, No. 2, pp. 242–252, April 1989.

81. I. Barbi, J. C. Bolacell, D. C. Martins, and F. B. Libano, "Buck Quasi-Resonant Converter Operating at Constant Frequency:Analysis, Design and Experimentation," *IEEE PESC'89*, pp. 873–880, 1989.

82. T. Higashi, K. Fujimoto, T. Ninomiya, and K. Harada, "Improvement of Power Efficiency and Operation Range in Voltage-Mode Resonant Converters," *IEEE PESC'89*, pp. 235–242, 1989.

83. Y. S. Lee, S. C. Wong, and Y. M. Lai, "SPICE Modeling and Simulation of Power Converters," *Proceedings of China 9th Conference on Circuits and Systems*, China, October 1990, pp. 139–145.

84. J. G. Cho and G. H. Cho, "Cyclic Quasi-Resonant Converter: A New Quasi-Resonant dc/dc Converter with Improved Characteristics," *IEE Electronics Letters* (UK), Vol. 26, No. 21, pp. 1821–1822, 11 October 1990.

85. B. S. Jacobson and R. A. DiPerna, "Series Resonant Converter with Clamped Tank Capacitor Voltage," *APEC '90*, Fifth Annual Applied Power Electronics Conference and Exposition, Conference

Proceedings (IEEE Cat. No. 90CH2853-0), Los Angeles, CA, 11–16 March 1990, pp. 137–146.

86. S. Kelkar, "Comparison of Two Circuit-Level Simulation Techniques for Power-Processing Circuits," *APEC '90*, Fifth Annual Applied Power Electronics Conference and Exposition, Conference Proceedings (IEEE Cat. No. 90CH2853-0), Los Angeles, CA, 11–16 March 1990, pp. 605–611.

87. B. Lau, "Computer-Aided Design of a dc-to-dc Switching Converter," *APEC '90*, Fifth Annual Applied Power Electronics Conference and Exposition, Conference Proceedings (IEEE Cat. No. 90CH2853-0), Los Angeles, CA, 11–16 March 1990, pp. 619–628.

88. G. A. Franz, "Multilevel Simulation Tools for Power Converters," *APEC '90*, Fifth Annual Applied Power Electronics Conference and Exposition, Conference Proceedings (IEEE Cat. No. 90CH2853-0), Los Angeles, CA, 11–16 March 1990, pp. 629–633.

89. R. Farrington, M. M. Jovanovic, and F. C. Lee, "Constant-Frequency Zero-Voltage-Switched Multi-Resonant Converters: Analysis, Design, and Experimental Results," *PESC '90 Record*, 21st Annual IEEE Power Electronics Specialists Conference (IEEE Cat. No. 90CH2873-8), San Antonio, TX, 11–14 June 1990, pp. 197–205.

90. W. Tang, W. A. Tabisz, A. Lotfi, F. C. Lee, and V. Vorpérian, "DC Analysis and Design of Forward Multi-Resonant Converter," *PESC '90 Record*, 21st Annual IEEE Power Electronics Specialists Conference (Cat. No. 90CH2873-8), San Antonio, TX, 11–14 June 1990, pp. 862–869.

91. K. Billings, *Switchmode Power Supply Handbook*, McGraw-Hill, New York, 1989.

92. C. Deisch, "Simple Switching Control Method Changes Power Converter into a Current Source," *PESC '78 Record* (IEEE Publication 78CH1337-AES), pp. 300–306.

93. R. Redl and N. O. Sokal, "Current-Mode Control, Five Different Types, Used with the Three Basic Classes of Power Converters: Small-Signal ac and Large-Signal dc Characterization, Stability Requirements, and Implementation of Practical Circuits," *PESC '85 Record* (IEEE Publication 85CH2117-0), pp. 771–785.

94. R. Redl and I. Novak, "Stability Analysis of Constant-Frequency Current-Mode Controlled Power Converters," *Proceedings of the Second International Power Conversion Conference*, 1980, Munich, p. 4B2-1–p. 4B2-17.

95. B. Holland, "Modelling, Analysis and Compensation of the Current-Mode Converter," *Proceedings of Powercon 11*, 1984.

96. S. Hageman, "Behavioral Modeling and PSpice Simulate SMPS Control Loops: Part 1 and Part 2," *PCIM*, April and May 1990.

97. Y. G. Kang, A. K. Upadhyay, and D. L. Stephens, "Analysis and Design of a Half-Bridge Parallel Resonant Converter Operating above Resonance," *IEEE Transactions on Industry Applications*, Vol. 27, No. 2, pp. 386–395, March/April 1991.

98. M. G. Kim and M. J. Youn, "A Discrete Time Domain Modeling and Analysis of Controlled Series Resonant Converter," IEEE *Transactions on Industrial Electronics*, Vol. 38, No. 1, pp. 32–40, February 1991.

99. T. Ninomiya, M. Nakahara, T. Higashi, and K. Harada, "A Unified Analysis of Resonant Converter," *IEEE Transactions on Power Electronics*, Vol. 6, No. 2, pp. 260–270, April 1991.

100. A. F. Witulski, A. F. Hernandez, and R. W. Erickson, "Small Signal Equivalent Circuit Modeling of Resonant Converters," *IEEE Transactions on Power Electronics*, Vol. 6, No. 1, pp. 11–27, January 1991.

101. B. A. Miwa, L. F. Casey, and M. F. Schlecht, "Copper-Based Hybrid Fabrication of a 50 W 5 MHz 40 V–5 V dc/dc Converter," *IEEE Transactions on Power Electronics*, Vol. 6, No. 1, pp. 2–10, January 1991.

102. D. Kimhi and S. Ben-Yaakov, "A SPICE Model for Current Mode PWM Converters Operating under Continuous Inductor Current Conditions," *IEEE Transactions on Power Electronics*, Vol. 6, No. 2, pp. 281–286, April 1991.

103. R. B. Ridley, "A New, Continuous-Time Model for Current-Mode Control," *IEEE Transactions on Power Electronics*, Vol. 6, No. 2, pp. 271–280, April 1991.

104. P. O. Lauritzen and C. L. Ma, "A Simple Diode Model with Reverse Recovery," *IEEE Transactions on Power Electronics*, Vol. 6, No. 2, pp. 188–191, April 1991.

105. R. S. Scott, G. A. Franz, and J. L. Johnson, "An Accurate Model for Power DMOSFET's Including Interelectrode Capacitances," *IEEE Transactions on Power Electronics*, Vol. 6, No. 2, pp. 192–198, April 1991.

106. Unitrode, *Power Supply Design Seminar*, Unitrode Corp., Lexington, MA, 1990.

107. G. Chryssis, *High-Frequency Switching Power Supplies: Theory and Design*, 2nd ed., McGraw-Hill, New York, 1989.

108. K. Kit Sum, *Switch Mode Power Conversion: Basic Theory and Design*, Marcel Dekker, New York, 1984.

109. A. Ioinovici, *Computer-Aided Analysis of Active Circuits*, Marcel Dekker, New York, 1990.

110. A. S. Kislovski, R. Redl, and N. O. Sokal, *Dynamic Analysis and Control-Loop Design of Switching-Mode dc/dc Converters*, Van Nostrand Reinhold, Princeton, NJ, 1991.

111. "Smart Design for Power Conversion," a collection of papers edited by R. M. Martinelli, Intertec Communications Inc., 1988.

112. D. F. Haslam, M. E. Clarke, and J. A. Houldsworth, "Simulating Power MOSFETs with SPICE," Fifth International High Frequency Power Conversion Conference, Santa Clara, CA, May 1990, pp. 296–305.

113. J. H. Chan, A. Vladimirescu, X.-C. Gao, P. Liebmann, and J. Valainis, "Non-Linear Transformer Model for Circuit Simulation," *IEEE Transactions on Computer-Aided Design*, Vol. 10, No. 4, pp. 476–482, April 1991.

114. A. J. Cioffi, "Paralleling Power Converters," *Electrical Manufacturing*, Vol. 5, No. 1, pp. 27–31, January 1991.

115. S. Lorenzo, J. M. Ruiz, F. Aldana, and M. Shaker, "A New Modeling and Simulation CAD Package for Power Converter Design," *IEEE Transactions on Industrial Electronics*, Vol. 37, No. 5, pp. 387–397, October 1990.

116. C. D. Manning, "Off-Line Resonant Power Supplies," *IEE Colloquium on "High Frequency Resonant Power Supplies"* (Digest No. 120), London, pp. 1/1–4, 5 June 1991.

117. J. H. Mulkern, C. P. Henze, and D. S. Lo, "A High-Reliability, Low-Cost, Interleaved Bridge Converter," *IEEE Transactions on Electronic Devices*, Vol. 38, No. 4, pp. 777–783, April 1991.

118. Y. S. Lee, D. K. W. Cheng, and Y. C. Cheng, "Design of a Novel AC Regulator," *IEEE Transactions on Industrial Electronics*, Vol. 38, No. 2, pp. 89–94, April 1991.

119. Y. S. Lee, M. H. L. Chow, and J. S. L. Wong, "SPICE Simulation of Non-Linear Equations and Circuits," *IEE Proceedings–G*, Vol. 138, No. 2, pp. 273–281, April 1991.

120. M. H. L. Chow, Y. S. Lee, and J. S. L. Wong, "Use of SPICE in Analogue Computation and Simulation of Systems Consisting of Circuit Components and Transfer Functions," *IEE Proceedings–G*, Vol. 138, No. 2, pp. 282–288, April 1991.

121. K. W. Ma and Y. S. Lee, "A Novel Uninterruptible dc-dc Converter for UPS Applications," *Conference Record of 1991 IEEE Industry Applications Society Annual Meeting*, Dearborn, Michigan, pp. 1074–1080.

122. K. T. Chau, Y. S. Lee, and A. Ioinovici, "Computer-Aided Model-

ing of Quasi-Resonant Converters in the Presence of Parasitic Losses by Using MISSCO Concept," *IEEE Transactions on Industrial Electronics*, Vol. 38, No. 6, pp. 454–461, December 1991.

123. Y. Amran, F. Huliehel, and S. Ben-Yaakov, "A Unified SPICE Compatible Average Model of PWM Converters," *IEEE Transactions on Power Electronics*, Vol. 6, No. 4, pp. 585–594, October 1991.

124. R. P. Severns and G. E. Bloom, *Modern dc-to-dc Switchmode Power Converter Circuits*, Van Nostrand Reinhold, Princeton, NJ, 1985.

125. MicroSim, "Circuit Analysis User's Guide," MicroSim Corp., Fairbanks, Irvine, CA.

126. B. K. Bose, "Evaluation of Modern Power Semiconductor Devices and Future Trends of Converters," *IEEE Transactions on Industry Applications*, Vol. 28, No. 2, pp. 403–413, March/April 1992.

127. J. Jia, "An Improved Power MOSFET Macro Model for SPICE Simulation," *IEE Colloquium on CAD of Power Electronic Circuits* (Digest No. 084), London, UK, April 1992, pp. 3/1–8.

128. H. L. Thanawala, "CAD of Power Electronic Convertors for High Voltage Transmission Systems," *IEE Colloquium on CAD of Power Electronic Circuits* (Digest No. 084), London, UK, April 1992, pp. 7/1–6.

129. E. B. Patterson, P. G. Holmes and D. Morley, "Electronic Design Automation (EDA) Techniques for the Design of Power Electronic Control Systems," *IEE Proceedings-G*, Vol. 139, No. 2, pp. 191–198, April 1992.

Symbols

A	voltage gain
A_c	voltage gain of converter; or center-pole area of magnetic core
A_e	voltage gain of error amplifier
A_L	loop gain; or inductance factor of magnetic core
A_p	gain factor of pulse-width modulator
A_{pc}	gain of converter and pulse-width modulator together
A_w	winding area of magnetic core
B	magnetic flux density
ΔB	peak-to-peak swing of magnetic flux density
ΔB_m	maximum allowable peak-to-peak swing of magnetic flux density
B_{max}	maximum allowable magnetic flux density
B_w	bandwidth
C, C_1, C_2 C_X, C_Y	capacitances
C_{DS}	drain to source capacitance
C_{GD}	gate to drain capacitance

C_{GS}	gate to source capacitance
C_L, C_{L1}, C_{L2}	output filtering capacitances
D	duty cycle
D_{max}	maximum duty cycle
D_{min}	minimum duty cycle
D_2	duty cycle of flywheel diode
dc	direct current
emf	electromotive force
E	voltage-controlled voltage source
ESR	effective series resistance
f	frequency
F	form factor; or current-controlled current source
f_r	resonant frequency
f_P, f_{P1}, f_{P2}	pole frequencies
f_s	switching frequency
f_T	gain-bandwidth product
f_z, f_{z1}, f_{z2}	zero frequencies
G	$G = \dfrac{T}{2L}$; or gain factor of current-controlled converter; or voltage-controlled current source
H	current-controlled voltage source
$i_1, i_2, i_A,$ $i_B, i_C, i_D,$ $i_{Df}, i_i, i_{i1},$ $i_{i2}, i_L, i_{LA},$ $i_{LB}, i_o, i_{o1},$ $i_{o2}, i_{SW} \cdots$	instantaneous currents as defined in circuit diagrams
I_A, I_B, I_C	averaged value of i_A, i_B, i_C respectively
I_{av}	average current

I_c	current control input
I_{DS}	drain to source current
I_F	forward current
I_i	input current
I_{irms}	rms value of I_i
i_L	instantaneous value of inductor current
I_L	averaged value of inductor current; or initial value of inductor current
i_{LC}	instantaneous current in LC circuit
I_{LC}	peak value of i_{LC}
I_{LD}, I_{LAD}, I_{LBD}	averaged value of discontinuous-mode current in L, L_A, and L_B respectively
I_{LL}, I_{LAL}, I_{LBL}	averaged value of inductive current in L, L_A, and L_B respectively
ΔI_m	maximum allowable peak-to-peak swing of current
I_o, I_{o1}, I_{o2}	output currents
I_{omax}	maximum value of output current I_o
I_{omin}	minimum value of output current I_o
$I_{OA}, I_{OB}, I_{OC} \cdots$	peak amplitudes of resonant current components
I_{pk}	peak value of current
I_R	reverse current
I_s	diode saturation current
J	current density
k	Boltzmann constant ($1.38 \times 10^{-23} \, J/K$)
K	coupling coefficient; or product of K_u and K_p
$K_{C1}, K_{C2}, K_{C3} \cdots$	polynomial coefficients of nonlinear capacitance

K_E	eddy-current loss coefficient
K_H	hysteresis loss coefficient
K_{L1}, K_{L2}, K_{L3} $\cdots$	polynomial coefficients of nonlinear inductance
K_p	transconductance coefficient; or primary area factor of transformer
K_{R1}, K_{R2}, K_{R3} $\cdots$	polynomial coefficients of nonlinear resistance
K_u	winding area utilization factor
L	effective channel length of MOS transistor
L, L_A, L_B, L_F, L_M, L_P, L_S, L_P', L_S', L_1, L_2	inductances
L_{crit}	critical inductance between continuous and discontinuous modes of operation
l_g	length of air gap
M	mutual inductance; or voltage ratio V_o/V_i
m_1, m_2	slopes of current waveforms
m_c	compensation slope
mmf	magnetomotive force
N	turns ratio
N_P	primary turns
N_S	secondary turns
P	power
P_c	power loss per cm^3 of magnetic core
P_d	power dissipation
q	electronic charge (1.602×10^{-19}C)
Q	Q factor

R, R_1, R_2, R_A, R_B, R_C, $R_S \cdots$	resistances
R_L, R_{L1}, R_{L2}	loading resistances
rms	root mean square
R_o	small-signal output resistance
R_{TH}	thermal resistance
s	complex frequency
S	scaling factor
T	switching period; or period of ac mains; or temperature
T_F	forward transit time of bipolar transistor
t_{OFF}	off-time period
t_{ON}, T_{ON}	on-time periods
T_r	oscillation period of resonant current
t_{rr}	reverse recovery time
T_T	transit time of diode
ΔT	rise of temperature
v_A, v_B, v_C, v_D, v_{DS}, v_{GS}, v_p, $v_s \cdots$	instantaneous voltages as defined in circuit diagrams
V_A, V_B, V_C	averaged voltage at A, B, C respectively
V_{ac}	rms value of ac mains voltage
V_{acmin}	minimum value of V_{ac}
V_{acmax}	maximum value of V_{ac}
V_B	back emf
V_c	control voltage; or volume of magnetic core
V_C	voltage across energy-storage capacitor; or control voltage

V_{CO}	dc voltage across capacitor C
V_{DS}	drain to source voltage of MOS transistor
V_{GS}	gate to source voltage of MOS transistor
V_i, V_{in}	input voltages
V_L	averaged voltage across inductor
V_{max}	maximum voltage
V_{min}	minimum voltage
$V_o, V_{o1}, V_{o2},$ $\quad V_{oL1}, V_{oL2}$	output voltages
ΔV_{oc}	change of V_o under closed-loop condition
ΔV_{oo}	change of V_o under open-loop condition
V_{ref}	reference voltage
V_{TO}	threshold voltage of MOS transistor
W	effective channel width of MOS transistor
Z_1, Z_2	impedances
$\alpha, \beta, \phi, \psi, \theta$	phase angles
β	feedback ratio
λ	delay factor of current-controlled converter
Δ, δ	incremental changes
η	efficiency
θ_m	phase margin
μ_0	permeability factor ($4\pi \times 10^{-7}$ H/m)
μ_r	relative permeability
ϕ	magnetic flux
ω	angular frequency

Index

Milton Keynes UK
Ingram Content Group UK Ltd.
UKHW031124141024
449569UK00006B/455